Fuzzy Control and Fuzzy Systems

SECOND, EXTENDED, EDITION

Witold Pedrycz
Department of Electrical and Computer Engineering,
University of Manitoba,
Winnipeg, Canada

RSP
RESEARCH STUDIES PRESS LTD.
Taunton, Somerset, England

JOHN WILEY & SONS INC.
New York · Chichester · Toronto · Brisbane · Singapore

RESEARCH STUDIES PRESS LTD.
24 Belvedere Road, Taunton, Somerset, England TA1 1HD

First published 1989
Second, extended, edition 1993

Marketing and Distribution:

Australia and New Zealand:
Jacaranda Wiley Ltd.
GPO Box 859, Brisbane, Queensland 4001, Australia

Canada:
JOHN WILEY & SONS CANADA LIMITED
22 Worcester Road, Rexdale, Ontario, Canada

Europe, Africa, Middle East and Japan:
JOHN WILEY & SONS LIMITED
Baffins Lane, Chichester, West Sussex, England

North and South America:
JOHN WILEY & SONS INC.
605 Third Avenue, New York, NY 10158, USA

South East Asia:
JOHN WILEY & SONS (SEA) PTE LTD.
37 Jalan Pemimpin 05-04
Block B Union Industrial Building, Singapore 2057

Library of Congress Cataloging-in-Publication Data

Pedrycz, Witold, 1953–
Fuzzy control and fuzzy systems / Witold Pedrycz. — 2nd, extended, ed.
p. cm. — (Electronic & electrical engineering research studies. Control theory and applications series ; 4)
Includes bibliographical references and index.
ISBN 0-86380-131-5 (Research Studies Press). — ISBN 0-471-93475-5 (Wiley)
1. Automatic control. 2. Fuzzy systems. I. Title. II. Series: Electronic & electrical engineering research studies. Control theory and applications studies series ; 4.
TJ213.P363 1992
629.8—dc20 92-41429
CIP

British Library Cataloguing in Publication Data

A catalogue record for this book
is available from the British Library.

ISBN 0 86380 131 5 (Research Studies Press Ltd.)
✓ ISBN 0 471 93475 5 (John Wiley & Sons Inc.)

Typeset by Abracadabra!, Milton Keynes, England
Printed in Great Britain by SRP Ltd., Exeter

To Eve, Adam, and Barbara

Editorial Preface — 1993 Edition

It is always a great pleasure, as a book series editor, to be in a position to write a second Editorial Preface, since it means that the first edition has been well received and that the topic continues to command considerable interest. Some four years after the publication of the first edition, Fuzzy Control and Fuzzy Systems research is still an important area within control engineering: indeed, the topic is perhaps more widely accepted than it was when the then Dr Pedrycz wrote the first edition. His determination to assemble the important concepts within fuzzy control into a cohesive textbook has borne fruit. Researchers were ready for such a reference text and in consequence the first edition surpassed even my expectations. My cautious view expressed to the publishers, Research Studies Press, prior to acceptance of the first manuscript was that the topic was timely but that sales might be restricted owing to the relatively small number of researchers then involved in the topic. I was clearly too cautious but I am relieved that on balance I came down in favour of publication. The first edition sold well and the topic has expanded both in scope and importance within control engineering.

All credit to Professor Pedrycz for his pioneering efforts; and with this, the second significantly extended edition, I feel sure he will have another success. To write a book on a complex topic is difficult enough, but to do so in a language other than one's native tongue requires particular dedication. Professor Pedrycz deserves success.

Professor M.J.H. Sterling

Brunel University,
April 1993.

Preface to the Second Edition

Fuzzy sets constitute collections of objects with gradual transitions between complete membership and complete exclusion in a class. This primordial feature of a fuzzy set plays a key role in emulating an activity of a human being involved in solving ill-defined problems. The knowledge-representation capabilities of fuzzy sets emerging in this way become useful in supporting a new paradigm of computations called soft computing. Soft computing encompasses several faculties of science and engineering, including neural networks, chaotic systems, fuzzy sets, and knowledge-based systems. Soft computing is characterised by approximate rather than categorical reasoning, acceptance of imprecise and incomplete information, and the presentation of outcomes of computations in terms of linguistic variables. Bearing this in mind, fuzzy sets fit ideally into this niche of information processing. The areas of application of fuzzy sets in soft computing or related topics nowadays spread broadly from various control problems (including those arising in controlling home appliances, camcorders, cameras, and industrial installations) to user-friendly decision-making facilities.

This research monograph provides extensive and up-to-date coverage of the methodology and algorithms of fuzzy sets considered mainly in the context of control engineering and system modelling and analysis.

Conceptually, the book can be split into three general parts, each of them consisting of several chapters. This partition reflects the way in which the overall material has been organised.

The first section (Chapters 1 – 3) deals with the fundamentals of fuzzy sets. The aim of this part is to introduce notions of fuzzy sets, study the motivation behind their introduction and develop basic technical issues. Those include a thorough discussion of operations on fuzzy sets, their geometrical interpretations, the functional and relational calculus of fuzzy sets, fuzzy numbers and fuzzy relations (Chapter 1). Chapter 2 concentrates on diverse characterisations of fuzzy sets carried out in terms of measures of fuzziness, introduces concepts of a fuzzy measure and a fuzzy integral, and discusses linguistic variables as formal constructs with a well defined syntax and semantics. The main differences between fuzziness and randomness are highlighted, and several generalised constructs conveying both the aspects of uncertainty such as probabilistic sets and fuzzy random variables are introduced. The essential issues of membership-function elicitation are also studied. Finally, Chapter 3 addresses how fuzzy sets contribute towards fundamental aspects of knowledge representation. The key notion of a cognitive perspective introduced there makes it possible to adjust the specificity of the resulting algorithms and modify them according to the level of problem complexity at hand.

The second section of the book is devoted to fuzzy controllers. First, in Chapter 4, one puts fuzzy controllers in the broader context of expert control. This domain supports a better understanding of the fuzzy controller as a general paradigm of intelligent control. The basic construct is studied, and examples of applications and four modes of utilisation of the controller (depending on the nature of the available information and generated results) are covered, in Chapter 4. Before proceeding with detailed design aspects, we discuss the fuzzy-relation structures and equations that play a primordial role in developing fuzzy controllers and fuzzy models (Chapter 5).

Depending on the character of the input information to the fuzzy controller, the character of the information produced by it and the resulting modes of operation, the controller calls for conceptually different design methods. In general, one can refer to a logic-design level at which the fuzzy controller realises different inference schemes. At this level, one looks carefully at the control rules of the controller, analysing their completeness and consistency and the interaction between them. This also involves different schemes of inference (reasoning) based on Lukasiewicz logic, relational structures and the hierarchies of fuzzy controllers. Regarding fuzzy controllers applied in a numerical mode, where closed-loop control structures start playing an exclusive role, the design tasks should be focused on studies of the dynamic properties of the systems, developing control analogies (such as relay-like structures) and incorporating mechanisms of self-organisation. It is also of fundamental importance here to develop mechanisms of analysis at a higher conceptual level (i.e. at the level of fuzzy sets incorporated in the fuzzy controller). One of these is the use of the linguistic phase plane embracing this level of analysis. Regarding final implementation, it is worth distinguishing between interpreted and compiled versions of the controllers and studying tradeoffs arising there. Chapters 8 and 9 are focused on nonlinear mappings realised as fuzzy neural networks.

Despite the significant diversity of the approaches used in the development of the fuzzy controller, all of them share the same property of trial-and-error methodology of design. This methodology is effective in many cases, mainly owing to the set-theoretic character of the basic pieces of information (linguistic labels) used in the construction of the controller. Nevertheless, the methodology prohibits us from making many detailed and systematic studies, including those of a design nature such as developing fuzzy control under given goals and constraints. A systematic and well structured analysis requires fuzzy models and modelling to be carried out at the conceptual level of fuzzy sets. Then fuzzy controllers can be systematically designed to be fully supported by a detailed closed-loop analysis. To accomplish this task, fuzzy modelling becomes an inevitable component of the design methodology of the fuzzy controller or any other control structure, and it is likely that it will constitute a basic framework for designing fuzzy-control algorithms to pursue the development of this research area.

The last part of the book concentrates on fuzzy models and on system analysis in fuzzy-relational models, and gives a comprehensive coverage of paradigms and algorithms of fuzzy modelling. Chapter 10 forms an introduction to identification problems and carefully defines the sequence of the phases of the identification process, starting from structure selection and ending with model evaluation. Several classes of fuzzy models are then introduced along with detailed identification procedures. The discussion is augmented by numerical examples.

The main problems of system analysis studied in Chapter 11 involve prediction, stability of fuzzy models and a variety of control algorithms.

The book is aimed at several groups of readers interested in both the theory of fuzzy control and fuzzy systems and their applications. The material can be of interest to researchers working on fuzzy-information processing in the areas of control, system modelling, pattern recognition, knowledge-based systems and neural networks. The engineering community and practitioners in many existing or potential areas of the application of fuzzy controllers and fuzzy sets in general will find the material useful in studying the theory of fuzzy sets, in designing fuzzy controllers and in incorporating more advanced topics of the construction of fuzzy models and fuzzy modelling in particular.

The book can also serve as a textbook for one-term graduate courses on fuzzy sets, intelligent control or modern control engineering.

Considering the particular needs and interest of the reader, several tracks are available. They are displayed below. For those who are not familiar with fuzzy sets, Chapters 1, 2 and 3 constitute a solid introduction to the topic. Furthermore, Chapters 5 and 8 can provide suitable material for further studies. Afterwards, depending on more specific areas of interest, two more specialised tracks are worth pursuing.

The reader already familiar with fuzzy sets and interested mainly in fuzzy controllers can follow Chapters 4, 6, 7, 8, 9 and partially 11, which constitute a systematic exposition of their fundamentals, design methodology and algorithms.

Those interested in fuzzy modelling can proceed with Chapters 3, 8, 10 and 11.

The diagram below highlights several main areas of interest (shown vertically) as well as arranges Chapters with respect to the level of advancement of the specialised area.

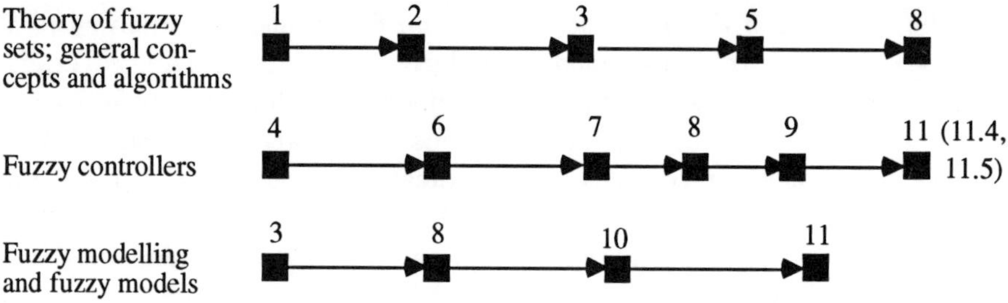

In comparison to the first edition, the book has undergone a series of essential modifications, updates and extensions. Since the first edition, new research trends have become apparent and new interesting results have emerged, particularly in distributed computations completed with the aid of fuzzy neural networks. Being more specific, these pertain to the new Chapters on knowledge representation (Chapter 3), fuzzy neural networks (Chapter 8), and designing fuzzy controllers with the aid of neural networks (Chapter 9). The remaining Chapters have been significantly expanded and new Sections added (e.g. 4.3, 4.5, 5.10, 6.9, 7.3, 11.5). Some introductory chapters have been substantially revised to present a new organisation reflecting the state of the art of the development of fuzzy sets. The book contains original material that has not been published elsewhere, in particular in Chapter 2 and Sections 8.3, 8.5, 9.3, 9.4 and 11.5.

The book is self-contained. In the first Chapters, the reader is acquainted with the basic notions, specific notation and relevant techniques of fuzzy sets and their calculus. These techniques are intensively exploited in the remaining more technical oriented Chapters. The mathematical background, which can be beneficial in studying the material, includes basic aspects of set theory, fundamentals of two-valued logic and linear algebra.

The author gratefully acknowledges financial support from the Natural Sciences and Engineering Research Council of Canada and the Network of Excellence on Microelectronic Devices, Circuits and Systems for Ultra Large Scale Integration MICRONET.

Editorial Preface to the First Edition

The assumption that all control engineering analysis and design can be reduced to the solution of a set of deterministic algebraic and differential equations has progressively been challenged by the emergence of stochastic and heuristic techniques. The recognition that, in terms of the plant model, the measurements and the control action applied to a process, the data are rarely if ever exact and that consequently the degree of uncertainty introduced becomes an important factor in the quality of the system performance is now the driving force behind much control engineering research.

In many situations, the observation that control schemes involving a human input often performed better than full closed loop automatic control gave rise to the suspicion that self adaption and the ability to reject bad measurement data, filter noisy data and to effectively derive control action in the presence of uncertainty were key features which needed to be incorporated in analysis and design techniques. Such a methodology has been developed over the last two decades initially in a highly mathematical form which was not well suited to traditional high level language implementation on computer systems. While equality identities had direct representations in languages such as Fortran, probabilistic relationships were more difficult to handle. However, the emergence of rule based languages with associated knowledge sets has given added impetus to the research in what has become known as fuzzy control and fuzzy systems. Within the Intelligent Knowledge Based Systems approach to computer representation of systems, the subject of Expert Systems has attracted much publicity and the obvious benefits of replicating all the desirable features of human input while retaining all the advantages of closed loop automation control are clearly desirable objectives. In fact such all-embracing concepts are difficult to realise but by providing an environment which is more suited to fuzzy relationships, the expert systems approach is a significant addition to conventional techniques.

Thus the understanding of fuzzy control and fuzzy systems is now an important adjunct to classical techniques. This book therefore provides a sound basis for the study of fuzzy systems beginning with fundamentals and developing the ideas through to comprehensive fuzzy control systems. Professor Pedrycz's presentation of the material leads the reader through the sometimes involved mathematics and illustrates the important concepts with clear examples. This book therefore represents a valuable contribution to the development of fuzzy system methodology by bringing together the concepts in a coherent book form at a time when little comparable material is available.

Professor M.J.H. Sterling

University of Durham,
April 1989.

Contents

CHAPTER 1

Introduction to Fuzzy Sets

1.1 Introductory Remarks

The aim of this Chapter is to provide an introduction to the fundamental notions and concepts of fuzzy sets. We present the idea of fuzzy sets introduced by L.A. Zadeh in his seminal papers published around 1965 [28,29], highlight their properties and explain major reasons for the emergence of this notion. Then we study generic operations on fuzzy sets, carefully expose the motivation behind their introduction, and address some aspects of the semantics of fuzzy sets.

Fuzzy sets have departed from a traditional niche of set theory, two-valued logic, and many-valued logic. Some well established mathematical constructs still apply, but in a different context. New, distinct mathematical structures have emerged as well. The generalisations achieved thereby have opened new conceptual avenues and have allowed us to look at some facets of uncertainty and complexity existing within systems from a new perspective.

1.2 An Idea of Fuzzy Sets: Origin and Basic Notions

To introduce the idea of the fuzzy set, let us remind ourselves of two-valued logic, which forms a cornerstone of any mathematical tool used. A fundamental point arising from this logic is that it imposes a dichotomy of any mathematical model of a given concept. In other words, taking any object, no matter how complex it is, we are forced to assign it to one of two prespecified, rigid and complementary categories (for instance, good-bad, normal-abnormal, odd-even, black-white etc.).

Sometimes this classification makes sense; in others it causes some serious and obvious dilemmas.

For example, in the study of integer numbers, two categories such as odd and even numbers can be defined without hesitation; and any integer number can then be classified within this frame without difficulty.

Nevertheless, in many engineering tasks, we are faced with classes that are ill-defined and do not possess clear and well defined boundaries. Consider, for instance, such categories as *tall* man, *high* speed, *significant* error, *medium* pressure etc. All these convey a useful meaning that is obvious for a certain community. However, the borderline between inclusion and exclusion of a given object for such a class is not evident. Here, it is obvious that two-valued logic, exploited in describing these classes, is not well-suited. The remarks of Russell [20], Black [3] and Popper [19], to name only a few, shed light on a genuine problem that two-valued logic started to confront in order to accommodate the semantics of these terms.

An historical example showing how models based on two-valued logic can give rise to undesired phenomena appeared in a work by Borel [4] (see also [10]). Borel discussed the ancient Greek sophism of a pile of seeds,

> "... one seed does not constitute a pile nor two nor three ... from the other side everybody will agree that 100 million seeds constitute a pile. What therefore is the appropriate limit? Can we say that 325 647 seeds don't constitute a pile but 325 648 do?"

At a glance, it is evident that the 'yes-no' type of answer cannot be a satisfactory solution to the problem. Definitely, each limit (border) point x_0 used in the definition of the predicate pile(x):

$$\text{pile}(x) = \begin{cases} 1 \ , & \text{if } x \in x_0 \\ 0 \ , & \text{if } x \notin x_0 \end{cases}$$

could contribute to an extremely simplified and unrealistic model of this term. The conceptual shortcomings of this Boolean model are there despite fruitless attempts to optimise this limit value.

The key issue of fuzzy sets is that one significantly extends the meaning of a set admitting different grades of belonging, also called membership values. This alleviates the previous problem by embracing all intermediate situations between complete (total) membership and total nonmembership. In contrast, as mentioned above, in set theory one is restricted to the two limit values 0 and 1.

Also, even in mathematics, we can encounter some notions with gradual rather than abrupt boundaries. Examples of such well known terms are: *sparse* matrix, a linear approximation of a function in a *small* neighbourhood of a point x_0, or an *ill-conditioned* matrix. Here, we accept these notions as conveying useful information about the problem to be studied. Obviously, they are not viewed as essential defects of our everyday language but rather as an advantageous feature, indicating our ability to generalise and conceptualise knowledge. Nevertheless, we should stress that these notions are strongly context-dependent and that their definitions are by no means universal. For example, the notion of 'ill-defined matrix' could depend heavily on the type of computer used and on the application one has in mind. Notice that fuzziness, being an inherent ingredient of the above concepts, has a totally different character and a different source of origin in comparison with randomness. This topic is studied later in more detail.

From the above general discussion, the need to construct a convenient, flexible, and easy-to-understand tool for handling fuzziness becomes of primary importance. Hence, the basic idea of fuzzy sets formulated by L.A. Zadeh constitutes an attractive option to follow.

As stated, fuzzy sets modify the basic term of a set to allow for the intermediate grades of membership (belonging). The higher the value of the membership of a certain object x to the notion (fuzzy set) A, say A(x), the stronger the link of x to the category described by A.

For instance, defining a fuzzy notion such as a *large* positive steady-state error expressed in %, we can say that 70% of steady-state error fulfills this category at a grade equal to 0.87, while the level of 0.01% is characterised by a grade of membership equal to zero.

Example 1.1

An interesting interpretation of membership function may be found in sensitivity analysis, where fuzzy sets may contribute to its formation in a natural way. Consider a simple electrical circuit consisting of a voltage source with internal resistance r, which is loaded by a resistor of variable resistance R. As simple computations reveal, for R equal to r we obtain the maximal output power (i.e. the power dissipated on R). However, in practice, one usually tries to accept higher values than the optimal value, R = r. In this situation, it appears that the energy obtained at R for this type of selection is slightly lower than that generated at the optimal point. At the same time, this choice gives rise to considerable energy savings, otherwise dissipated on internal resistance r. Treating the normalised relationship of power obtained against variable R as a membership function (this can be accomplished by a simple normalisation), we can come up with an interesting interpretation.

The value $\mu_{power}(R)$ describes the suitability of resistance R as a good candidate in this circuit to generate the highest level of power. The membership function is characterised by significant changes occurring on the left-hand side of the optimal point R_{opt}, while showing a smooth

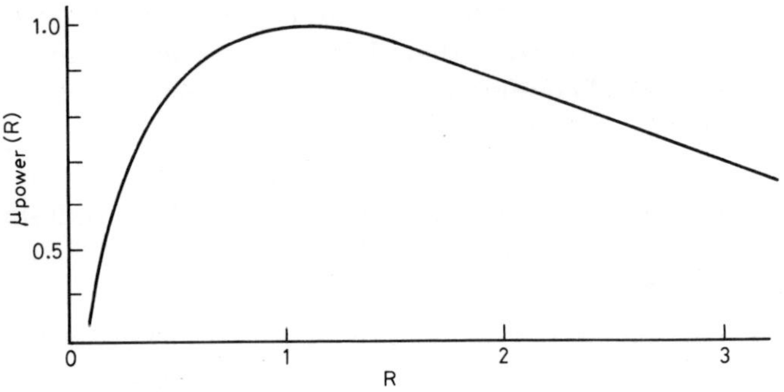

Fig. 1.1 Membership function $\mu_{power}(R)$, r = 1

transition on the opposite side of this point (Fig. 1.1). For instance, the degree of membership at r = 2, $\mu_{power}(2)$ is higher than $\mu_{power}(1/2)$. This highlights an evident preference for r = 2 over the second alternative, r = 1/2, where both of them are viewed as candidates compatible with the concept discussed. Obviously, we are dealing here with an easy and well defined object (electrical circuit). However, if no such model were available, the membership function could be a summary of experimental knowledge gained by anybody experienced enough in the field of basic electrical circuits.

Example 1.2

It is instructive to take a closer look at different formalisms applied to articulate some general concepts used in control or system modelling. Let us assume that a target (goal) of our control algorithm is to maintain a comfortable temperature in a certain room. First, one has to define a meaning of the term *comfortable* temperature and encapsulate it into a certain formal framework. Thus, in all consecutive design steps one can develop a suitable control algorithm to meet this target. This task constitutes an initial design phase for any control algorithm. We can envisage at least three distinct possibilities of representing our knowledge about the goal:

• *pointwise representation of the target.* This is the simplest option. It calls for accepting a single numerical value as the target control value. In a schematic notation, we portray the goal as a single point in the universe of temperatures:

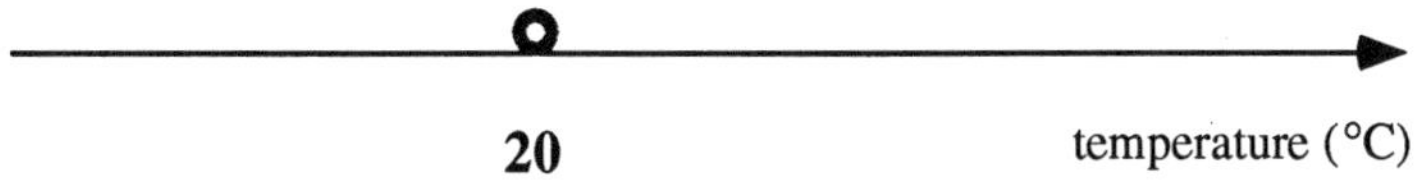

In PROLOG-like programming notation, the target is articulated as a certain attribute of the room that has to be controlled,

room(temperature,20)

This formal notation describes objects as ordered triples, namely

object — attribute — value

The attribute of the room is its temperature, and the value of the attribute is 20°C.

The pointwise control target calls for a sequence of control actions (heating or cooling) that are taken with respect to an error occurring in the system under control. Quite often, the simplest control method in this scenario could require a series of respective on/off control actions.

• *interval-valued control target*. In this case, we admit that the term 'comfortable temperature' embraces all values distributed within a certain numerical interval bounded by a lower and an upper limit. We can express it as follows:

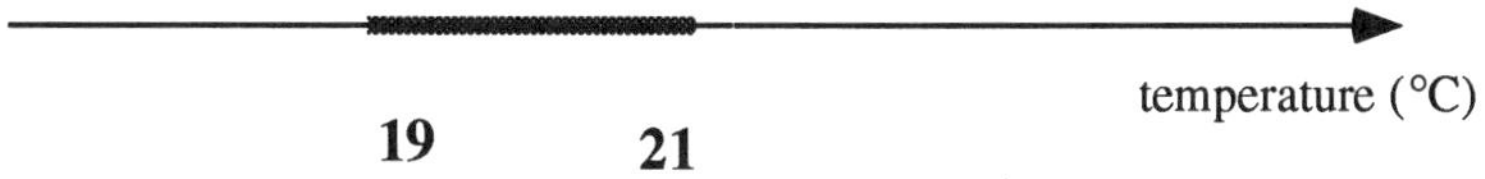

and formally describe it as the following fact in PROLOG:

room(temperature[19,21])

Note that all temperatures lying within this interval are fully accepted. Also, they are not distinguished at all. All the remaining values outside this interval are completely unacceptable. The situations described by them evoke immediate control actions. In comparison to the previous pointwise representation, the frequency of on/off control actions could be significantly reduced. It usually depends upon the width of the interval of the acceptable values of the temperature.

• *fuzzy interval in goal specification*. The previous specification, despite acceptance of a certain level of tolerance build into the definition of the goal represented in this way, still suffers from artificially formulated boundaries. The abrupt transition from 1 (no control action necessary) to 0 (control action required) that is switched by the two boundary values gives rise to some hesitation. The notion discussed is definitely ill-defined in the sense that the level of acceptance of a certain temperature as complying with our perception of *comfortable* temperature should change continuously over slight changes of values. Thus, if 19°C forms an acceptable temperature, values lower than that, say 18.9°C, 18.8°C ... etc., are still acceptable, although to a lesser degree. Therefore, the boundary values are quite artificial.

A better representation is:

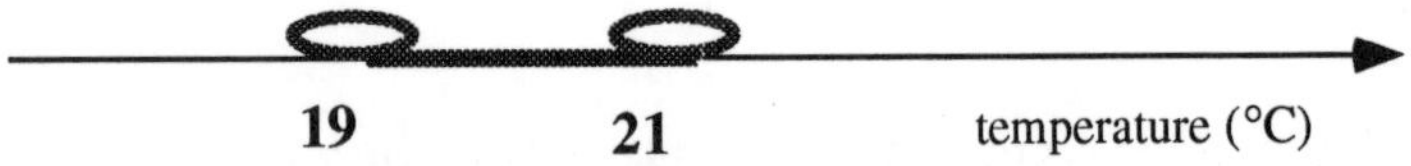

where two small clouds around the limit values underline the hesitation about defining these bounds. In PROLOG notation, we express this goal as

room(temperature, *comfortable*)

assuming that the value of this attribute is a fuzzy set and has to be handled in a special manner.

More formally, a fuzzy set A defined in a universe of discourse **X** is expressed by its membership function

$$A : \mathbf{X} \rightarrow [0, 1]$$

where the degree of membership A(x) expresses the extent to which x fulfills the category described by A.* The condition A(x) = 1 denotes all the elements that are fully compatible with A. The condition A(x) = 0 identifies all elements that definitely do not belong to A.

By the height of a fuzzy set A, hgt(A), we mean a maximal value of its membership function:

$$\mathrm{hgt}(A) = \sup_{x \in \mathbf{X}} A(x)$$

If hgt(A) = 1, then A is called a normal fuzzy set; otherwise, the fuzzy set is named subnormal. This synthetic index simply states that there exist some elements of the universe of discourse fully compatible with the concept characterised by A.

* The original introduction of the notion pertains to a fuzzy set as being *characterised* (but not necessarily *defined*) by a membership function (treated exclusively as a mapping to [0,1]) (see [28], p. 339). This does not confine fuzzy sets to exclusively numerical constructs. Further nonnumerical and symbolically-directed realisations of this fundamental concept are ultimately worth pursuing. Nevertheless, in this book we will focus exclusively on the numerical implementation of the theory.

of membership in A:

$$\mathrm{supp}(A) = \{x \in \mathbf{X} \mid A(x) > 0\}$$

Later, we denote by F(**X**) a family of all fuzzy sets defined in **X**.

Before we proceed with further notions and a formal terminology of fuzzy sets, it is worth while to summarise some basic facts about set theory. In set theory, all objects (sets) are fully described by so-called characteristic functions [16]. The characteristic function of the set A, χ_A, is defined as a two-valued mapping:

$$\chi_A : \mathbf{X} \to \{0,1\}$$

taking its values in the set {0,1} such that

$$\chi_A(x) = \begin{cases} 1 \ , & \text{if } x \in A \\ 0 \ , & \text{otherwise} \end{cases}$$

This definition contributes to the approach equivalent to that in which sets are defined by a straightforward enumeration of their elements.

The basic operations in set theory are defined as follows:

$$A \cup B = \{x \in \mathbf{X} \mid x \in A \text{ or } x \in B\}$$
$$A \cap B = \{x \in \mathbf{X} \mid x \in A \text{ and } x \in B\}$$
$$\overline{A} = \{x \in \mathbf{X} \mid x \notin A\}$$

They could be expressed in an equivalent way by considering characteristic functions of A and B:

$$\chi_{A\cup B}(x) = \max(\chi_A(x), \chi_B(x)) = \chi_A(x) \vee \chi_B(x)$$
$$\chi_{A\cap B}(x) = \min(\chi_A(x), \chi_B(x)) = \chi_A(x) \wedge \chi_B(x)$$
$$\chi_{\overline{A}}(x) = 1 - \chi_A(x)$$
$$x \in \mathbf{X}$$

We refer to them as lattice operations.

In fuzzy sets, the meaning of the fundamental predicate of set theory '$\in$' (element of) is significantly expanded by accepting a partial membership in a set. The basic operations can be defined as before by replacing characteristic functions of sets by membership functions of fuzzy sets. This gives rise to these expressions:

$$(A \cup B)(x) = \max(A(x), B(x))$$

$$(A \cap B)(x) = \min(A(x), B(x))$$

$$\overline{A}(x) = 1 - A(x)$$

$$x \in \mathbf{X}$$

Since the grades of membership lie in the [0,1] interval, not in the two-valued set {0,1}, it is worth recalling the collection of properties essential for set theory and investigate whether they are satisfied for fuzzy sets.

De Morgan law, which is valid for set theory, is preserved for fuzzy sets as well:

$$\overline{(A \cap B)} = \overline{A} \cup \overline{B} \qquad , \qquad \overline{(A \cup B)} = \overline{A} \cap \overline{B}$$

Secondly, distributivity laws are fulfilled as well:

$$A \cap (B \cup C) = (A \cap B) \cup (A \cap C)$$

$$A \cup (B \cap C) = (A \cup B) \cap (A \cup C)$$

Properties of absorption and idempotency hold:

$$(A \cap B) \cup A = A$$

$$(A \cup B) \cap A = A$$

$$A \cup A = A$$

$$A \cap A = A$$

However, an exclusion law is not satisfied, i.e.

$$A \cup \overline{A} \neq \mathbf{X}$$

$$A \cap \overline{A} \neq \varnothing$$

which might be expected, since fuzzy sets do not impose a dichotomy that forms a fundamental feature of set theory. These two relationships express a so-called property of underlap and overlap occurring between a fuzzy set and its complement. They illustrate how well the fuzzy set A and its complement $\overline{A}$ 'cover' the universe of discourse ($A \cup \overline{A}$) or become disjoint ($A \cap \overline{A}$). A maximal departure from the two-valued case is obtained for $A(x) = 1/2$. Then both the union and the intersection produce the same result:

$$\min(1/2,1/2) = \max(1/2,1/2)=1/2$$

and indicate the same level of underlap and overlap.

We thus can state that there is no distinction between the fuzzy set A and its complement.

Fuzzy sets defined in a finite universe of discourse **X** can be represented as lists structuring information about grades of membership at individual points of **X**. For instance, a list composed of two-element sublists:

$$[[x_1,1.0] \quad , \quad [x_2,0.8] \quad , \quad [x_3,0.5] \quad , \quad [x_4,0.0]]$$

denotes a fuzzy set A with these grades of membership:

$$A(x_1) = 1.0 \ , \ A(x_2) = 0.8 \ , \ A(x_3) = 0.5 \ , \ A(x_4) = 0.0$$

Then the logic operations can be defined pointwise (i.e. for each element of the universe of discourse) using the PROLOG-like notation:

$$\text{AND}(a,b) :- \min(a,b)$$

$$\text{OR}(a,b) :- \max(a,b)$$

$$\text{NEG}(a) :- 1.0 - a$$

where 'a' and 'b' are two grades of membership.

Returning to Example 1.2, we can visualise a hierarchy of representation schemes built with the aid of the formal constructs (Fig. 1.2).

• *pointwise target*, which reduces to a single-element set where a characteristic function takes on zero in the entire universe of temperature but a single point

$$\chi_{\text{comfortable temperature}}(x) = \begin{cases} 1 \ , & \text{if } x = 20 \\ 0 \ , & \text{if } x \neq 20 \end{cases}$$

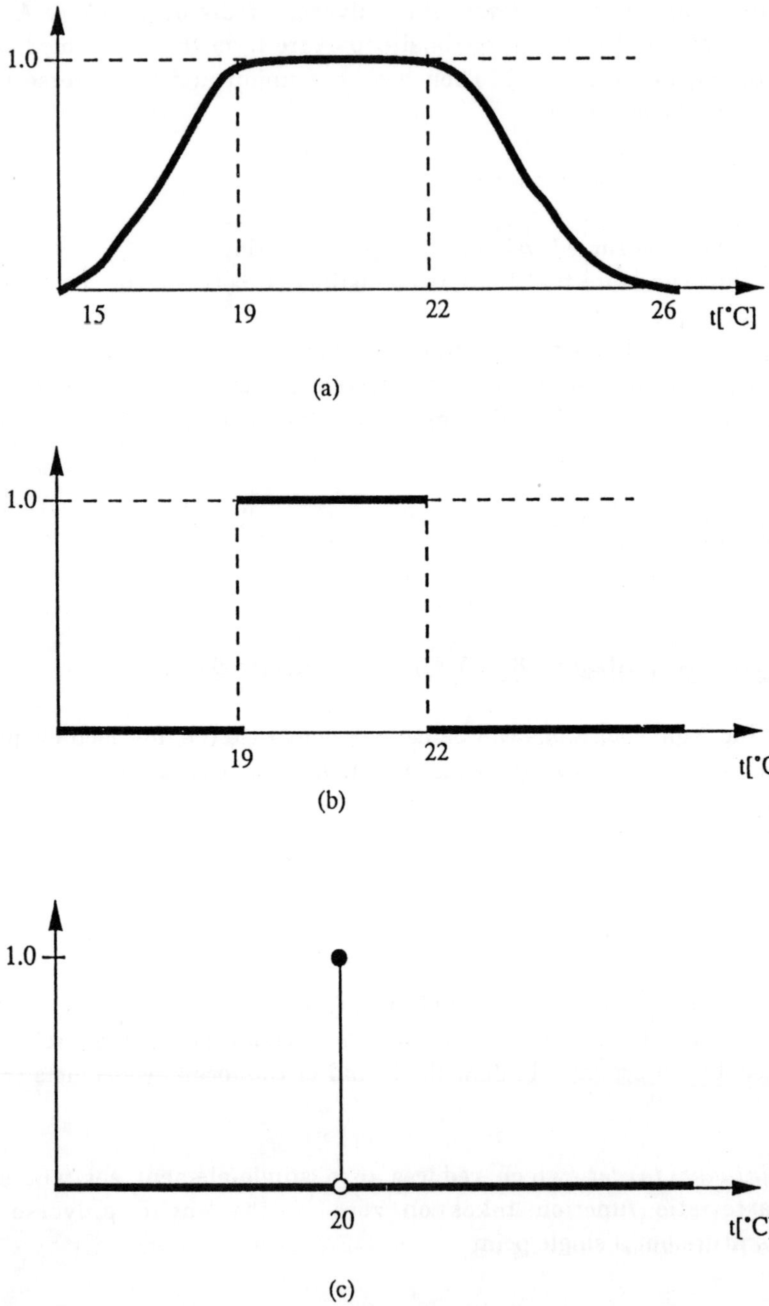

Fig. 1.2 Hierarchy of representation schemes of *comfortable* temperature

• *interval-valued target*,

$$\chi_{\text{comfortable temperature}}(x) = \begin{cases} 1 & , \text{ if } x \in [19,21] \\ 0 & , \text{ if } x \notin [19,21] \end{cases}$$

• *the target specified as a fuzzy set* introduces a gradual transition around the previous boundary points,

$$\textit{comfortable} \text{ temperature}(x) = \begin{cases} f_1(x) & , \text{ if } x \le 19 \\ 1 & , \text{ if } x \in [19,21] \\ f_2(x) & , \text{ if } x \ge 21 \end{cases}$$

where f_1 and f_2 denote smooth functions of x; f_1 defined for all $x < 19$ is an increasing function with the boundary condition $f_1(19) = 1.0$. The second function f_2 illustrates decreasing acceptance of higher temperatures as compatible with the term *comfortable* temperature. Thus, f_2 becomes a decreasing function with $f_2(21) = 1.0$.

By changing the form of functions f_1 and f_2, we can easily model the way in which the notion of *comfortable* temperature is discerned and can be handled numerically.

Two important classes of membership functions called S and Π functions [32] are useful in describing transitions in grades of membership. The S function includes three parameters and is defined piecewise as follows:

$$S(x,\alpha,\beta,\gamma) = \begin{cases} 0 & , \text{ if } x < \alpha \\ 2\left(\dfrac{x-\alpha}{\gamma-\alpha}\right)^2 & , \text{ if } \alpha \le x \le \beta \\ 1-2\left(\dfrac{x-\gamma}{\bar{\gamma}-\alpha}\right)^2 & , \text{ if } \beta \le x \le \gamma \\ 1 & , \text{ if } x > \gamma \end{cases}$$

The parameter $\beta = (\alpha + \gamma)/2$ is called a crossover point at which the membership function is equal to 1/2.

The Π function is build at a superposition of the previous membership function,

$$\Pi(x,\beta,\gamma) = \begin{cases} S(x,\gamma-\beta,\gamma-\beta/2,\gamma) & , \text{ if } x \le \gamma \\ 1-S(x,\gamma,\gamma+\beta/2,\gamma+\beta) & , \text{ if } x > \gamma \end{cases}$$

The parameter β denotes a bandwidth (distance between crossover points of membership function Π). At $x = \gamma$, the membership function attains 1.0.

1.3 Geometrical Interpretation of Fuzzy Sets

Sets and fuzzy sets defined in a finite universe of discourse, card($\mathbf{X}$) = n, have an interesting and transparent geometrical interpretation [14]. It is obvious that any set defined in $\mathbf{X}$ is equivalent to a string of bits (0 or 1) of fixed length 'n'. This in turn can be viewed as a point situated at one of the corners of the unit hypercube $\{0,1\}^n$. For instance, for n = 2, we get 2^2 different strings with the entries

$$\varnothing : (0,0)$$

$$\{x_1\} : (1,0)$$

$$\{x_2\} : (0,1)$$

$$\mathbf{X} = \{x_1, x_2\} : (1,1)$$

They are distributed in four successive corners of the unit square (Fig. 1.3).

Fuzzy sets form finite strings composed of any numbers between 0 and 1 (so-called fits). In contrast to the previous construct, the number of different strings is not finite. All fuzzy sets fill in the entire hypercube, including in particular the corners and edges. For n = 2, we get a similar graphical illustration (Fig. 1.4).

From this geometrical point of view, one observes that fuzzy sets constitute a genuine generalisation of sets, making the interior of the square (or hypercube, in general) 'accessible'. The lattice logical operations (minimum and maximum) produce new points in the hypercube (Fig. 1.5).

The complement (negation) operation applied to A generates a new point situated symmetrically in the cube (Fig. 1.5). The overlap and underlap properties are also well documented. In particular, for $A = [0.5 \quad 0.5]$, A and $\overline{A}$ coincide (Fig. 1.6).

As we will discuss later, the distance between A and $\overline{A}$ can serve as a suitable indicator of fuzziness of this fuzzy set (measure of fuzziness). The lower the distance, the higher the fuzziness. Another phenomenon pertains to approximation of fuzzy sets by a closest set (where the distance is given, for instance, as the Hamming distance). By accepting a threshold level $\lambda = 0.5$, one approximates a fuzzy set A by its Boolean counterpart according to the roundoff formula:

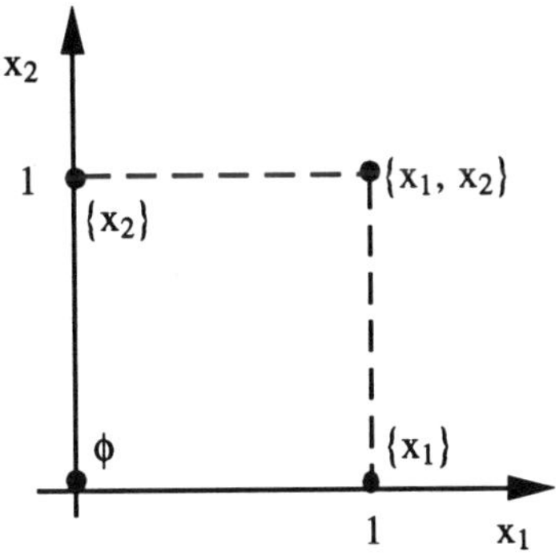

Fig. 1.3 Sets in two-dimensional space as vertices of the unit square

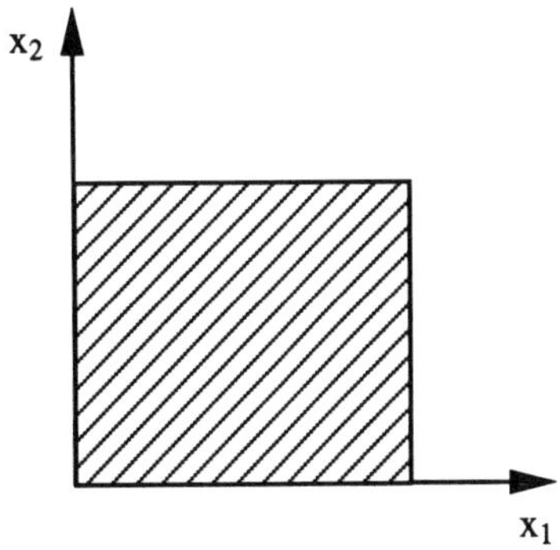

Fig. 1.4 Fuzzy sets cover the entire unit square

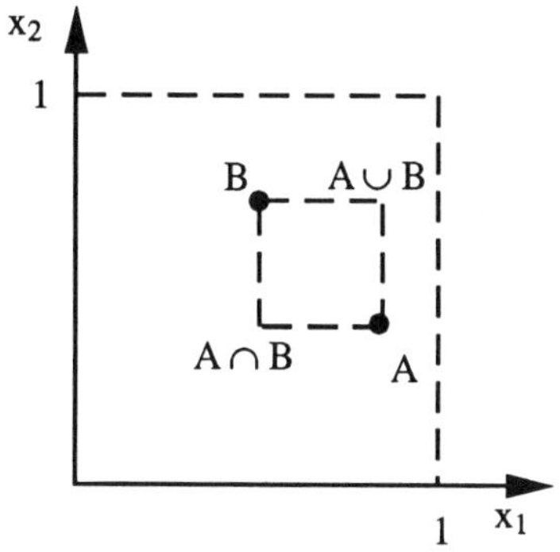

Fig. 1.5 Fuzzy sets and their lattice operations in the unit square

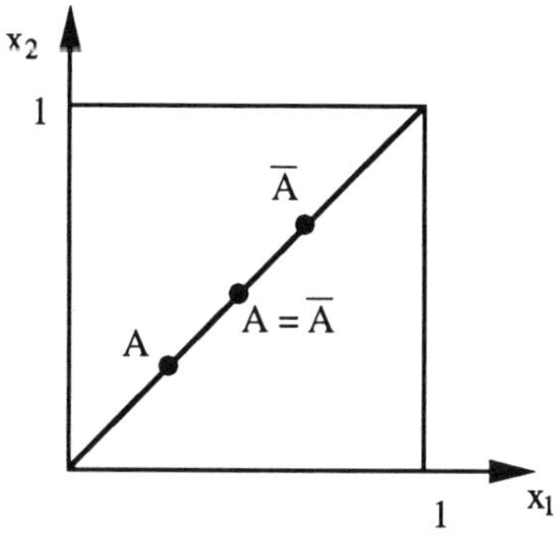

Fig. 1.6 Fuzzy set A and its complement in $[0,1]^2$

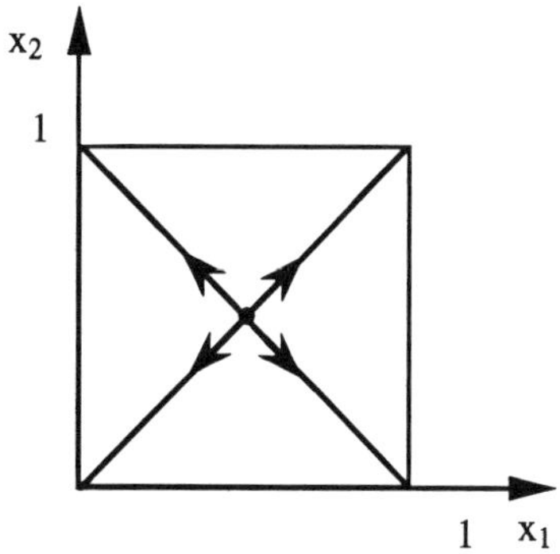

Fig. 1.7 Fuzzy sets and their set approximation

$$\chi_A(x_i) = \begin{cases} 1 \ , & \text{if } A(x_i) > 0.5 \\ 0 \ , & \text{if } A(x_i) < 0.5 \end{cases}$$

i = 1, 2, ..., n. This roundoff approximation recalls a standard procedure applied to real numbers when transforming them into integers. The essence of this process is that any point in the hypercube is 'pushed' to the nearest vertex of the hypercube. Simultaneously, a fuzzy set located in the centre of the hypercube ($A(x_i) = 0.5$ for all i = 1, 2, ..., n) cannot be uniquely approximated by one of the corners of the hypercube. All the corners are viewed as equally acceptable approximations of this fuzzy set (Fig. 1.7).

1.4 Operations on Fuzzy Sets

At an early stage of the development of fuzzy sets, two alternative expressions for union and intersection were proposed [29], namely a product for intersection

$$(A \cap B)(x) = A(x)\, B(x)$$

and a so-called probabilistic sum for a union

$$(A \cup B)(x) = A(x) + B(x) - A(x)\, B(x)$$

In comparison with the lattice (max and min) operations, the degree of membership depends on both the values of membership function A(x) and B(x). In the remainder, we should restrict ourselves to a class of binary operations satisfying a collection of assumptions naturally occurring in set-theoretic operations:

- boundary conditions

$$A \cup \mathbf{X} = \mathbf{X} \quad , \quad A \cap \mathbf{X} = A$$

$$A \cup \varnothing = A \quad , \quad A \cap \varnothing = \varnothing$$

- commutativity

$$A \cap B = B \cap A \quad , \quad A \cup B = B \cup A$$

- associativity

$$(A \cap B) \cap C = A \cap (B \cap C) \quad , \quad (A \cup B) \cup C = A \cup (B \cup C)$$

Observe also that, in interpreting a grade of membership as a grade of truth of any proposition, all the above conditions have a plausible interpretation.

The boundary conditions indicate that the logical connectives for fuzzy sets coincide with those applied in two-valued logic. The property of commutativity reflects a case in which the truth value of a composite expression does not depend on the order of the components used in its formation.

On accepting all the above conditions, a broad class of models for logical connectives (union and intersection) is formed by triangular norms, cf. [15,13,7]. The triangular norms (so-called t- and s-norms) were originally developed in the theory of probabilistic metric spaces.

Definition 1.1

By a t-norm, we mean a function of two arguments

$$t : [0, 1] \times [0, 1] \to [0, 1]$$

such that

(i) it is nondecreasing in each argument

$$\text{for } x \le y, w \le z \quad , \quad x\,t\,w \le y\,t\,z$$

(ii) it is commutative

$$x\,t\,y = y\,t\,x$$

(iii) it is associative

$$(x\,t\,y)\,t\,z = x\,t\,(y\,t\,z)$$

(iv) it satisfies the set of boundary conditions

$$x\,t\,0 = 0 \quad , \quad x\,t\,1 = x$$

$$x, y, z, w \in [0, 1]$$

In a class of t-norms, we distinguish a certain group of t-norms.

Definition 1.2
A t-norm is called Archimedean if, and only if

(i) it is continuous with respect to each argument
(ii) $x\,t\,x < x$ for each $x \in [0,1]$

Definition 1.3
An Archimedean t-norm is called strict if it is strictly increasing in $(0,1) \times (0,1)$.

From these definitions, all the properties of the t-norm can easily be identified with the relevant properties of the operation of the intersection (logic AND).

Possessing t-norms, a union can be modelled by means of s-norms (t-conorms).

Definition 1.4
An s-norm is defined as a function of two arguments

$$s:[0,1]\times[0,1]\to[0,1]$$

satisfying these properties:

(i) s is a nondecreasing function in each argument
(ii) it is commutative
(iii) it is associative
(iv) it satisfies the boundary conditions

$$x\,s\,0 = x \qquad , \qquad x\,s\,1 = 1$$

Looking at this definition, we can see that (i) to (iv) correspond to the properties of the union. An interesting fact is that the s-norm can also be derived from the relationship

$$x\,s\,y = 1-(1-x)\,t\,(1-y)$$

which is nothing but the De Morgan law in set theory.

Some examples of t-norms and s-norms are given below [9,12,25].

t-norms

$$x\ t\ y = \min(x,y) = x \wedge y$$

$$x\ t\ y = 1 - \min\left[1, \left((1-x)^p + (1-y)^p\right)^{1/p}\right] \quad , \quad p \geq 1$$

$$x\ t\ y = \log_w\left[1 + \frac{(w^x - 1)\,(w^y - 1)}{(w-1}\right] \quad , \quad 0 < w < \infty \quad , \quad w \neq 1$$

$$x\ t\ y = x\ y$$

$$x\ t\ y = \frac{x\ y}{\gamma + (1-\gamma)\,(x + y \quad x\ y)} \quad , \quad \gamma \geq 0$$

$$x\ t\ y = \max\left[0, (\lambda + 1)\,(x + y - 1) - \lambda\ x\ y\right] \quad , \quad \lambda \geq -1$$

s-norms

$$x\ s\ y = \min\left(1, (x^p + y^p)^{1/p}\right) \quad , \quad p \geq 1$$

$$x\ s\ y = 1 - \log_w\left[1 + \frac{(w^{1-x} - 1)\,(w^{1-y} - 1)}{(w-1)}\right] \quad , \quad 0 < w < \infty \quad , \quad w \neq 1$$

$$x\ s\ y = x + y - x\ y$$

$$x\ s\ y = \frac{x\ y\,(\gamma - 2) + x + y}{x\ y\,(\gamma - 1) + 1} \quad , \quad \gamma \geq 0$$

$$x\ s\ y = \min\left[1, x + y + \lambda\ x\ y\right] \quad , \quad \lambda \geq -1$$

For each t-norm, we get this inequality:

$$\begin{cases} x & , \text{ if } y = 1 \\ y & , \text{ if } x = 1 \\ 0 & , \text{ otherwise} \end{cases} \leq x\ t\ y \leq x \wedge y$$

Thus, the family of t-norms is bounded: the upper bound is formed by the minimum operator, while the lower one involves a so-called drastic product.*

For s-norms, we get the relationship

$$\max(x, y) \leq x\,s\,y \leq \begin{cases} x &, \text{ if } y = 0 \\ y &, \text{ if } x = 0 \\ 1 &, \text{ otherwise} \end{cases}$$

where the upper bound is known as a drastic sum.

Some of the families of triangular norms include parameters. By choosing their values, one can modify the character of the logical connectives. For instance, from the list of the second family of t- and s-norms, we get:

- if $p = 0$, then $x\,t\,y$ reduces to a drastic product, while $p = \infty$ yields the minimum.

- for $p = 0$, the s-norm becomes a drastic sum; in the limit $p = \infty$, one obtains the maximum operation.

The well known Lukasiewicz connectives originating from the theory of multivalued logic [21] can be derived by setting λ equal to 1 in these t-norms:

$$\lambda = 0 : x\,t\,y = \max(0, x + y - 1)$$

$$\lambda = 0 : x\,t\,y = \min(1, x + y)$$

The lattice types of logical operation frequently found in the literature, $\wedge$ and $\vee$, are continuous but not Archimedean.

The following representation theorem [1], originating from the theory of functional equations, makes it possible to express a certain class of triangular norms with the aid of a so-called additive generator. More

---

* It is worth noting in this context that different logical connectives applied to calculations of a joint probability of the conjunction of two events A, B, say 'A and B', P(A and B), for P(A) and P(B) provided, give rise to these bounds of probability originally formulated by G. Boole, cf. also [11]:

$$\max(0, P(A) + P(B)) \leq P(A \text{ and } B) \leq \min(P(A), P(B))$$

Again, the upper bound is just the maximal t-norm.

precisely, we have:

Proposition 1.1
A t-norm is Archimedean if there exists a decreasing and continuous function f called the additive generator, $f : [0,1] \rightarrow [0,\infty]$, such that

$$x \text{ t } y = \begin{cases} f^{-1}(f(x)+f(y)) & , \text{ if } f(x)+f(y) \in [0, f(0)] \\ 0 & , \text{ otherwise} \end{cases}$$

with $f(1) = 0$.

Proposition 1.2
An s-norm is Archimedean if there exists an increasing and continuous function $g : [0,1] \rightarrow [0,\infty]$ (additive generator), such that

$$x \text{ s } y = \begin{cases} g^{-1}(g(x)+g(y)) & , \text{ if } g(x)+g(y) \in [0, g(1)] \\ 1 & , \text{ otherwise} \end{cases}$$

where $g(0) = 0$. The additive generators f and g are unique up to a positive multiplication constant; hence cf(x), cg(x), $c > 0$ are also additive generators.

Evidently, for the values of arguments restricted to the set {0,1}, all t- and s-norms are equivalent; i.e. they produce the same results.

Having an infinite family of triangular norms enhanced by their parametric versions, one has a broad repertoire of formal models of logical connectives. The choice of a certain AND or OR operator depends on the problem to be solved. Unfortunately, few experiments have been conducted as to which triangular norm is appropriate. In [2], the 'local' character of operations realised on fuzzy sets (or, equivalently, grades of truth in a logical setting) was strongly underlined. Experiments in [24] supported this idea: it appeared that the minimum does not fit an experimental data set well. On the other hand, a product operator produced better results; however, some operators of a compensatory character (i.e. sharing properties of union and intersection) do a good job.

These findings strongly suggest the need for further investigations along this line in order to acquire a deeper knowledge of the semantics of the logic connectives for fuzzy sets.

We can now make some general remarks. First, notice that operators such as minimum and maximum indicate no interaction; i.e. as soon as one of the arguments (a grade of membership function) is greater (smaller, respectively) than the second one, it has no influence on the result.

This phenomenon may be treated as a certain advantage, since no accurate values of the membership functions are required. Thus, the membership function can be estimated roughly, and the errors in the measurements procedure can still be largely ignored. This type of robust performance is obvious in the following example. Assume that the two grades of membership satisfy the condition $a > b$, and define the difference between them as δ; $\delta = a - b$. AND-ing them with the use of the minimum, min(a,b), allows us to tolerate some imprecision in determining the exact value of 'b', say $b \pm \varepsilon$. This type of aggregation tolerates values of ε such that $\varepsilon < \delta$. On the other hand, situations may emerge in which this lack of sensitivity is a disadvantage, since the result obtained does not reflect the character of the data being combined and the way in which they interact and affect one another. Different levels of interaction between two grades of membership could occur. Consider the compound statement:

low control cost **and** *high* tracking accuracy

It consists of two components describing two separate objectives of a certain control policy. Let us assume that these two features (objectives) are satisfied to a certain degree, say 'a' for the control cost and 'b' for the tracking accuracy. The formulation of the problem, with the conjunction (AND), shows interaction between two essentially conflicting control objectives. Thus, the product could be preferred over the minimum as a suitable choice in this application.

On the other hand, in the statement:

new **and** *fast* car

the interaction between the objectives describing the car is not visible and the minimum operator could be a proper choice.

In general, one should be able to measure the level of interaction between the two grades of membership under logical aggregation. Two example measurements of this interaction are listed below. The first one calculates a double integral of the results of a t-norm applied to the arguments and those coming from the use of a minimum, considered to be a reference operator [27]:

$$i(t) = 1 - 3\int_0^1\int_0^1 (x\ t\ y)\, dx\, dy$$

Notice that, the stronger the interactivity of the operation, the higher the reported value of 'i'. One has, for instance,

$$i(\min) = 0 \qquad , \qquad i(\text{product}) = 1/4$$

In the second one, the interactivity index introduced in [17] uses the length of the curves:

$$e(t) = \int_0^1 \left[2(1-c) - L(c)\right] dc$$

where L(c) denotes the length of the curve c = x t y. For t = min, one gets L(c) = 2 (1 − c), and hence e(min) = 0. For another t-norm, e(t) takes positive values. Plots for c = min(x,y) for different values of c viewed as a parameter are given in Fig. 1.8. We refer to them as curves of sensitivity.

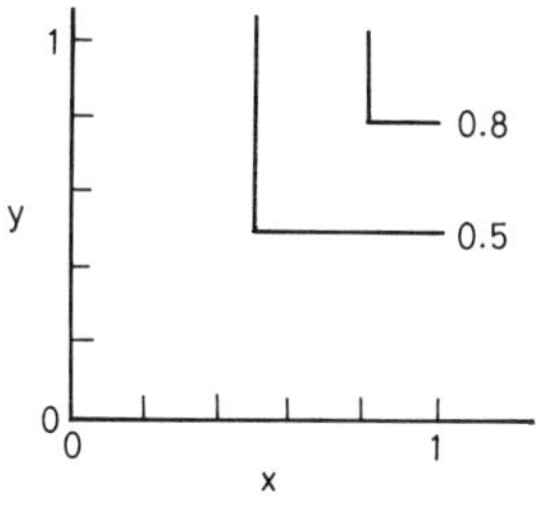

Fig. 1.8 Curves of sensitivity of logic operators

As a byproduct of our discussion of t- and s-norms, it is worth while to pay attention to the notion of inclusion of fuzzy sets. We state that a fuzzy set A is included in a fuzzy set B, $A \subset B$, if this inequality is preserved:

$$A(x) \le B(x) \qquad \text{for all } x \in \mathbf{X}$$

This definition, like the previous ones, defines the property of inclusion pointwise (namely for each x separately) and returns a two-valued predicate. Denote the truth (*T*) a truth value of a given statement *T*.

Then the above condition implies that truth T

$$T(\mathrm{A(x)} \subset \mathrm{B(x)})$$

is either 1 or 0 (the inequality holds or is violated). Because of this two-valued valuation, the definition could sometimes be too restrictive. An evident relaxation relies on the replacement of the Boolean predicate by a grade of inclusion of A in B, say (A ⊂ B)(x) at point x, which is defined as

$$(\mathrm{A} \subset \mathrm{B})(\mathrm{x}) = \mathrm{A(x)}\, \varphi\, \mathrm{B(x)}$$

where the φ-operator denotes a so-called pseudocomplement (or residuation) associated with a given t-norm and defined accordingly:

$$\mathrm{A(x)}\, \varphi\, \mathrm{B(x)} = \sup\left\{\mathrm{c} \in [0, 1] \mid \mathrm{A(x)}\, \mathrm{t}\, \mathrm{c} \le \mathrm{B(x)}\right\}$$

In a logical framework, the operator φ corresponds to the implication → used in many-valued logic. Considering the truth values of two propositions being equal to p and q, respectively, the logical value of the compound statement (implication) p → q is equal to p φ q.

An illustration of union, intersection, and complement is given in Fig. 1.9 with t-norms specified as minimum and product (the corresponding s-norms are maximum and probabilistic sum, respectively). The degree of containment is given in Fig. 1.10.

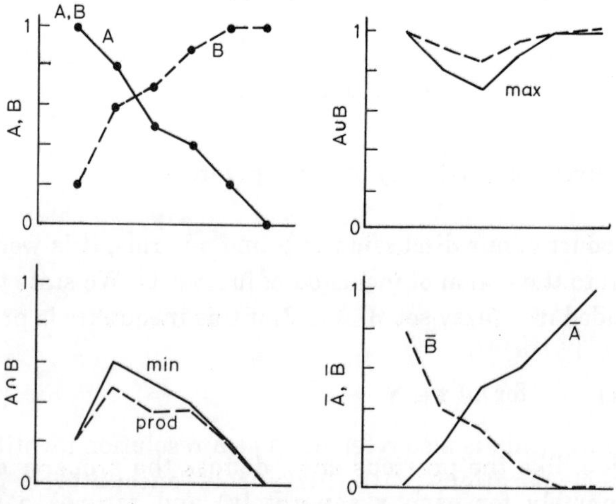

Fig. 1.9 Operations on fuzzy sets

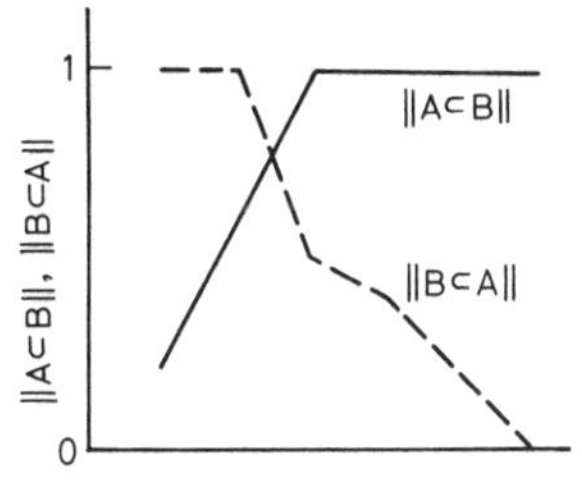

Fig. 1.10 Degree of containment for fuzzy sets

A direct and essential relationship forming a bridge between the notion of a fuzzy set and that of a set is given by the representation theorem [16]. First, we introduce the notion of an α-cut of A:

$$A_\alpha = \{x \in \mathbf{X} \mid A(x) \geq \alpha\} \quad , \quad \alpha \in [0,1]$$

Sometimes, it is convenient to distinguish a strong α-cut, i.e. a collection of elements:

$$\{x \in \mathbf{X} \mid A(x) > \alpha\}$$

The essence of an α-cut operation is that it allows us to convert a fuzzy set into its Boolean counterpart. The elements with grades of membership above or equal to the threshold α are elevated to 1; all the remaining ones are rejected from this α-cut.

The representation theorem states that any fuzzy set A can be represented by a union of its α-cuts, namely:

$$A = \bigcup_{\alpha \in [0,1]} \alpha A_\alpha$$

such that

$$A(x) = \sup_{\alpha \in [0,1]} [\alpha A_\alpha(x)]$$

The above relationship is also referred to as a resolution identity.

Some other operations with fuzzy sets encountered in various applications are briefly summarised below:

- *an operation of normalisation*:

$$\mathrm{Norm(A)(x)} = \frac{\mathrm{A(x)}}{\mathrm{hgt(A)}}$$

(we assume that A is a nonempty fuzzy set, i.e. that at least one element belongs to A with a nonzero degree of membership, hgt(A) = 0). This operation converts a subnormal fuzzy set A into a corresponding normal fuzzy set Norm(A).

- *two operations called concentration and dilation*:

$$\mathrm{CON(A)(x)} = \mathrm{A}^2(\mathrm{x})$$

$$\mathrm{DIL(A)(x)} = \mathrm{A}^{1/2}(\mathrm{x}) \qquad \left(\text{or } 2\mathrm{A(x)} - \mathrm{A}^2(\mathrm{x})\right)$$

yield new fuzzy sets with suppressed or elevated grades of membership. These operations are frequently used in modelling linguistic hedges (such as, for example, *more or less*, *very* etc.). In general, one can introduce a pth power of a fuzzy set, A^p; values of parameter p less than 1 yield dilation; powers greater than 1 cause concentration.

An operation called contrast intensification, INT(A), affects an original fuzzy set more selectively than the two others. It suppresses grades of membership lower than 1/2 and elevates values greater than this threshold:

$$\mathrm{INT(A)(x)} = \begin{cases} 2\mathrm{A}^2(\mathrm{x}) & , \text{ if } \mathrm{A(x)} < 1/2 \\ 1 - 2(1 - \mathrm{A(x)})^2 & , \text{ otherwise} \end{cases}$$

Furthermore, one can add an extra parameter p > 1 that makes the intensification operation more radical:

$$\mathrm{INT(A,p)(x)} = \begin{cases} 2^{\mathrm{p}-1}\mathrm{A}^{\mathrm{p}}(\mathrm{x}) & , \text{ if } \mathrm{A(x)} < 1/2 \\ 1 - 2^{\mathrm{p}-1}(1 - \mathrm{A(x)})^{\mathrm{p}} & , \text{ otherwise} \end{cases}$$

Fuzzification operation has an effect complementary to intensification:

$$\mathrm{FUZZ(A)(x)} = \begin{cases} \sqrt{\mathrm{A(x)}/2} & , \text{ if } \mathrm{A(x)} < 1/2 \\ 1 - \sqrt{(1 - \mathrm{A(x)})/2} & , \text{ otherwise} \end{cases}$$

1.5 Fuzzy-Relation Calculus

1.5.1 Properties and compositions of fuzzy relations

As we have already learned, fuzzy sets constitute a conceptual extension of set theory and so are fuzzy relations. We recall that relations in set theory are treated as elements of the Cartesian products of some spaces. By a fuzzy relation R defined in the Cartesian product $\mathbf{X} \times \mathbf{Y}$, we mean a mapping

$$R: \mathbf{X} \times \mathbf{Y} \to [0, 1]$$

which assigns a grade of membership R(x,y) to each pair (x,y) of the above Cartesian product of the universes. Its interpretation is analogous to that provided for the grades of membership of fuzzy sets. One can think about those membership values as representing strengths of connections (ties) between elements x and y. The closer the value of membership R(x,y) to 1, the stronger the link between x and y viewed in the context of this fuzzy relation.

The relation 'x is similar to y', where x and y are two real numbers, x, $x \in \mathbf{R}$, is an example of a fuzzy relation. Its membership function can be of the form:

$$R(x, y) = \begin{cases} \dfrac{1}{1+(x-y)^4} & , \text{ if } |y-x| \le 5 \\ 0 & , \text{ otherwise} \end{cases}$$

The membership attains 1 for x = y and rapidly disappears for different values of the arguments (Fig. 1.11).

Some other examples of fuzzy relations are those modelling statements like 'x less than y', 'x significantly greater than y' etc.

For a finite universe of discourse, a matrix notation is useful. In this case, fuzzy relation R is treated as a matrix $[R(x_i, y_j)]$ with the entries taking values between 0 and 1. An illustration of the relation defined in $\mathbf{X} = \{x_1, x_2\}$ and $\mathbf{Y} = \{y_1, y_2, y_3\}$ with the entries

$$\begin{bmatrix} 1.0 & 0.7 & 0.0 \\ 0.1 & 0.3 & 0.9 \end{bmatrix}$$

is given in Fig. 1.12, which is displayed as an undirected graph in which the different strengths are expressed symbolically by solid lines connecting the corresponding elements. The extension into multidimensional fuzzy relations is obvious.

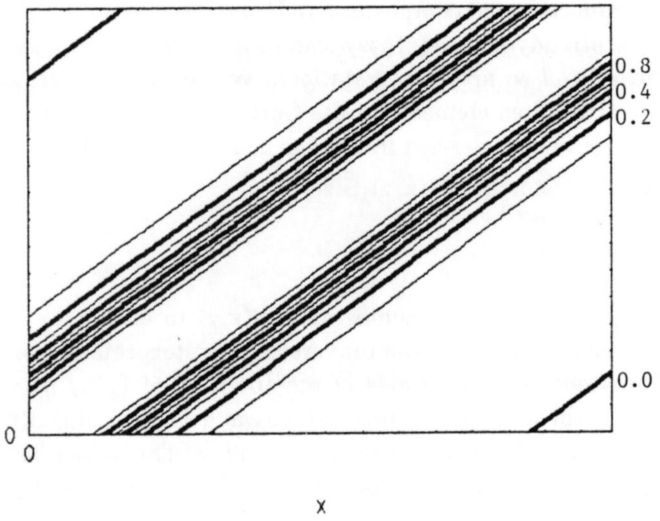

Fig. 1.11 Membership function of the fuzzy relation 'x and y are *similar*'

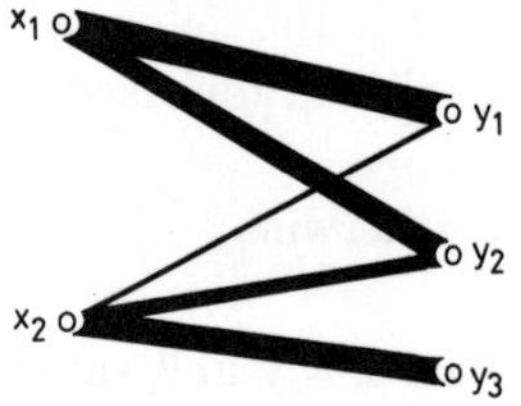

Fig. 1.12 Finite fuzzy relation

In PROLOG notation, we can express fuzzy relations as arrays enumerating all elements of the Cartesian product along with their numerical values. For the above example, one has a list consisting of 2*3 three-element sublists:

$$[[x_1,y_1,1.0],[x_1,x_2,0.7],[x_1,y_3,0.0],[x_2,y_1,0.1],[x_2,y_2,0.3],[x_2,y_3,0.9]]$$

In analogy to fuzzy sets, one can define the support of the fuzzy relation, its α-cuts, and introduce basic operations such as union, intersection, and complement.

From the applicational point of view, it is important to define the operation of the composition of fuzzy relations so that we can process information stored there. The commonest operations are given below:

- sup-min composition of two fuzzy relations $R \in F(\mathbf{X} \times \mathbf{Z})$ and $S \in F(\mathbf{Z} \times \mathbf{Y})$; it is a fuzzy relation $R \bullet S$ defined in $\mathbf{X} \times \mathbf{Y}$ with this membership function:

$$(R \bullet S)(x, y) = \sup_{z \in \mathbf{Z}} \left[\min(R(x, z), S(z, y)) \right] \tag{1.1}$$

- inf-max composition of R and S, $R \otimes S$

$$(R \otimes S)(x, y) = \inf_{z \in \mathbf{Z}} \left[\max(R(x, z), S(z, y)) \right] \tag{1.2}$$

These operations are associated with each other according to the relationship

$$\overline{R \bullet S} = \overline{R} \otimes \overline{S}$$

The sup-min composition has the following properties that result from the relevant operations (max and min) used in this composition:

- distributivity with respect to union

$$(R \cup T) \bullet S = R \bullet S \cup T \bullet S$$

- preservation of the inclusion property, i.e.

$$\text{if } R_1 \subset R_2 \text{, then } R_1 \bullet S \subset R_2 \bullet S$$

$$R_1, R_2, R, T \in F(\mathbf{X} \times \mathbf{Z}) \quad , \quad S \in F(\mathbf{Z} \times \mathbf{Y})$$

However, no distributivity with respect to intersection is reported, and therefore

$$(R \cap T) \bullet S \subset (R \bullet S) \cap (T \bullet S)$$

$$R, T \in F(\mathbf{X} \times \mathbf{Z}) \quad , \quad S \in F(\mathbf{Z} \times \mathbf{Y})$$

A particular case of the compositions is that joining a fuzzy set X and a fuzzy relation R, $X \in F(\mathbf{X})$, $R \in F(\mathbf{X} \times \mathbf{Y})$. This results in a fuzzy set Y in $\mathbf{Y}$:

$$Y = X \bullet R$$

$$Y = X \otimes R$$

A generalisation of the above compositions (1.1) and (1.2) can be obtained by taking t- and s-norms, which leads us to sup-t composition and inf-s composition, respectively:

- sup-t composition

$$(R \blacklozenge S)(x, y) = \sup_{z \in \mathbf{Z}} \left[R(x, z) \, t \, S(z, y) \right]$$

- inf-s composition

$$(R \diamond S)(x, y) = \inf_{z \in \mathbf{Z}} \left[R(x, z) \, s \, S(z, y) \right]$$

The sup-min composition performed for a fuzzy set X and a particular fuzzy relation R is of this form:

$$R(x, y) = \begin{cases} 1 & , \quad \text{if } f(x) = y \\ 0 & , \quad \text{otherwise} \end{cases}$$

(i.e. R is a function). We immediately derive:

$$\begin{aligned} (X \bullet R)(y) &= \sup_{x \in \mathbf{X}} \left[\min(X(x), R(x, y)) \right] = \\ &= \sup_{x \in \mathbf{X}: f(x) = y} \left[X(x) \wedge 1 \right] \quad \sup_{x \in \mathbf{X}: f(x) \neq y} \left[X(x) \wedge 0 \right] = \\ &= \sup_{x \in \mathbf{X}: f(x) = y} X(x) \end{aligned}$$

This formula is nothing but the extension principle (cf. [7,13,16,30,32]) describing the transformation of a fuzzy set X via a given function ‘f’. In other words, a mapping of X via a function f generates a fuzzy set f(X) with a membership function equal to:

$$(f(X))(y) = \sup_{x:f(x)=y} X(x)$$

Yet another generalisation involves both s- and t-norms in the composition operator. We will define s-t and t-s compositions, respectively. The corresponding formulas are expressed accordingly:

- s-t composition operator

$$\underset{z \in Z}{S}[R(x,z)\, t\, S(z,y)]$$

- t-s composition operator

$$\underset{z \subset \mathbf{Z}}{T}[R(x,z)\, s\, S(z,y)]$$

1.5.2 Eigen fuzzy sets generated by a fuzzy relation

A problem that is closely related to any binary relation defined in a finite universe of discourse, $R : \mathbf{X} \times \mathbf{X} \rightarrow [0,1]$, is concerned with a determination of a fuzzy set in $\mathbf{X}$ such that it is invariant under the sup-min composition, namely

$$E \bullet R = E$$

Denote by $\mathbf{E}$ a family of fuzzy sets such that

$$\mathbf{E} = \left\{E \in F(\mathbf{X}) \middle| E \bullet R = E\right\}$$

Each element of $\mathbf{E}$ will be called an eigen fuzzy set. It can be easily shown that $\mathbf{E}$ is nonempty for any fuzzy relation R; for instance, a trivial element of $\mathbf{E}$ is an empty fuzzy set.

In [22,23], the problem of determination of the greatest eigen fuzzy set was studied (in the sense of fuzzy-set inclusion) and some algorithms were proposed. The main idea of one of them was to iterate max-min composition of a fuzzy relation R and a fuzzy set E_1 until the moment when the fuzzy sets resulting in consecutive iteration steps become equal.

The fuzzy set E_1 is defined as

$$E_1(x) = \max_{y \in \mathbf{Y}} R(x, y)$$

while $E_2, E_3, \ldots, E_k, E_{k+1}$ are equal to

$$E_2 = E_1 \bullet R$$

$$E_3 = E_2 \bullet R$$

$$\vdots$$

$$E_{k+1} = E_k \bullet R$$

Since the universe of discourse in which R is defined is a finite one, the algorithm requires a finite number of iterations to terminate.

A significant concept introduced for binary relations is that of a transitive closure.

Definition 1.5
By a transitive closure of a binary relation R, denoted by cl(R), we mean a fuzzy relation constructed as a union of powers of R:

$$cl(R) = R \cup R^2 \cup \ldots \cup \ldots$$

where the kth power of R is defined recursively:

$$R^2 = R \bullet R, \ldots, R^{k+1} = R^k \bullet R$$

1.5.3 Projection of fuzzy relations

An operation that is often used for fuzzy relations relies on their projection on selected subspaces in which they are defined. Consider a fuzzy relation R defined in a Cartesian product of universes of discourse $\mathbf{X}_1 \times \mathbf{X}_2 \times \ldots \times \mathbf{X}_n$. By a projection of R on a subset of spaces **Y**:

$$\mathbf{Y} = \underset{j \in I}{\mathsf{X}} \mathbf{X}_j$$

where $I = \{i_1, i_2, \ldots, i_m\}$, i.e. $I \subset \{1, 2, \ldots, n\}$, we mean an operation that induces another fuzzy relation, denoted by $\mathrm{Proj}_{\mathbf{Y}} R$ and defined in **Y**:

$$G(x_{i_1}, x_{i_2}, \ldots, x_{i_m}) = \mathrm{Proj}_{\mathbf{Y}} R(x_1, x_2, \ldots, x_n)$$
$$= \sup R(x_1, x_2, \ldots, x_n)$$

where supremum is carried over all the elements coming from the subspaces of **X** that do not contribute to the space **Y** on which the projection has to be performed. For instance, for a binary relation defined in $\mathbf{X}_1 \times \mathbf{X}_2$, the above formula implies two projections:

- projection of R on $\mathbf{X}_1$

$$\mathrm{Proj}_{\mathbf{X}_1} R(x_1,x_2) = \sup_{x_2 \in \mathbf{X}_2} R(x_1,x_2)$$

- projection of R on $\mathbf{X}_2$

$$\mathrm{Proj}_{\mathbf{X}_2} R(x_1,x_2) = \sup_{x_1 \in \mathbf{X}_1} R(x_1,x_2)$$

Thus, for example, the resulting projections are fuzzy sets defined in $\mathbf{X}_1$ and $\mathbf{X}_2$, respectively. An illustration of this two-dimensional case is also displayed in Fig. 1.13, where several α-cuts of the fuzzy relation R are marked. The result of the projection can be interpreted as a shadow appearing on a respective coordinate when a source of light has been put behind the object possessing different grades of light absorption.

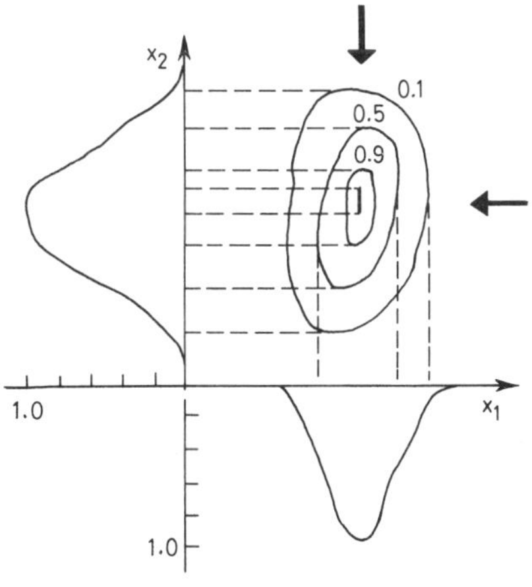

Fig. 1.13 Projection of a fuzzy relation (α-cuts indicated)

One can also interpret projection as a sup-t composition of a fuzzy set X with a constant membership function equal to 1 defined over the relevant space and the fuzzy relation R. This means that the projection on $\mathbf{X}_1$ calls for this transformation:

$$\sup_{x_2 \in \mathbf{X}_2} [X(x_2)\, t\, R(x_1, x_2)] = \sup_{x_2 \in \mathbf{X}_2} [1\, t\, R(x_1, x_2)]$$

The operation of projection is semantically equivalent to the application of an existential quantifier ∀ (there exists one ...). This may sometimes be too weak, causing substantial losses of information; therefore, a generalised formula for projection, called Q-projection, has been proposed and studied in [26].

1.6 Fuzzy Numbers

Fuzzy numbers are a special form of fuzzy sets; they are fuzzy sets defined in the space of real numbers. They possess some additional properties dealt with by the shape of their membership functions. It is instructive to start with a formal definition following considerations extensively discussed in [5,6,7].

Definition 1.6

A fuzzy number A is a fuzzy set defined in $\mathbf{R}$ such that:

(i) A is a normal fuzzy set; i.e. there exists at least one element of $\mathbf{R}$ for which $A(x) = 1$.

(ii) A is convex:

$$\forall_{\lambda \in [0,1]} \forall_{x \in \mathbf{R}} A[\lambda x + (1-\lambda) y] \geq \min(A(x), A(y))$$

(iii) A is upper semicontinuous.

(iv) A has a bounded support.

Recalled the representation theorem, the above definition can be formulated in an analogous way, replacing (ii) and (iii) by two equivalent conditions:

(ii′) All α-cuts of A are convex.

(iii′) All α-cuts are closed intervals of $\mathbf{R}$.

A fuzzy number can be viewed as a suitable model of approximate notions, as, for example, *near* zero, *about* 5 etc. Some shapes of fuzzy numbers are depicted in Fig. 1.14.

The membership function of the fuzzy number visualises a grade of

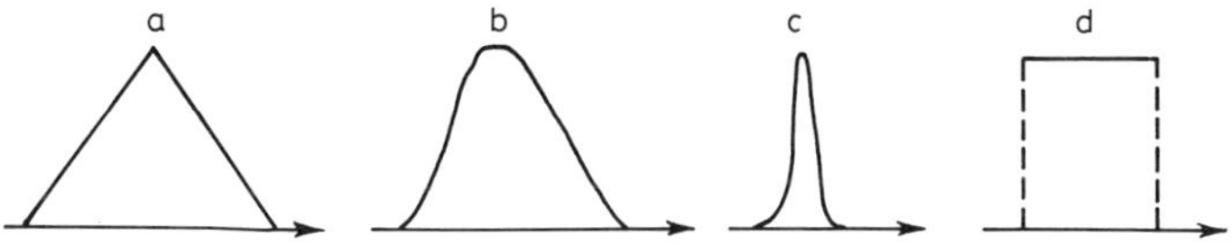

Fig. 1.14 Examples of fuzzy numbers

membership of a given element of a concept (*near*, *about*, ...). Unimodality of A assures us that exactly one region exists, in which the relevant elements have the highest grades of membership. Of course, the shape of a membership function reflects different situations in which a variety of shapes can be observed. For instance, in Fig. 1.14a, we deal with a triangle-like fuzzy number where three characteristic points are interpreted as upper and lower values of the range, and the element of **X** corresponding to the highest grade of membership is called a modal value of the fuzzy number. In Fig. 1.14b, a so-called bell-shaped fuzzy number is given. The fuzzy number displayed in Fig. 1.14c has a narrow membership function centred almost on one point. Finally, in Fig. 1.14d, an interval that is a particular case of a fuzzy number is given.

Having fuzzy numbers at hand, we are interested in some arithmetical operations performed by means of them. For instance, one may like to know the fuzzy number that is a result of, for example, their multiplication, addition or subtraction. In general, we can talk about a function of two given fuzzy numbers.

A routine is defined to derive the resulting fuzzy set using the extension principle, namely

$$C = F(A, B) \tag{1.3}$$

such that

$$C(z) = \sup_{x,y \in \mathbf{R} : z = f(x,y)} [A(x) \, t \, B(y)]$$

where $z \in \mathbf{R}$ and $f : \mathbf{R} \times \mathbf{R} \to \mathbf{R}$ is a function resulting from F such that $F(\{x\},\{y\}) = f(x,y)$. First, we must ensure that C in (1.3) is a fuzzy number (in the sense of the above definition). A sufficient condition that guarantees C to be a fuzzy number has been formulated in [6].

Proposition 1.3

If a t-norm is continuous and f is an upper semicontinuous monotone function, C is a fuzzy number.

We now collect the results of basic algebraic operations performed on fuzzy numbers, such as addition (+), subtraction (–), multiplication (*) and division (/). By means of the extension principle, we get

$$(A+B)(z) = \sup_{x,y:x+y=z} [A(x)\, t\, B(y)]$$

$$(A-B)(z) = \sup_{x,y:x-y=z} [A(x)\, t\, B(y)]$$

$$(A*B)(z) = \sup_{x,y:x*y=z} [A(x)\, t\, B(y)]$$

$$(A/B)(z) = \sup_{x,y:x/y=z} [A(x)\, t\, B(y)]$$

Note that all four operations require the solution of a problem of mathematical programming with constraints resulting from the corresponding operator.

As can be seen from these formulas, the determination of the membership function of a fuzzy number is a rather tedious task, requiring a significant number of computations. On the other hand, it may have been observed in Fig. 1.14 that the shapes of the membership functions are similar in that one modal value and two extreme limits are evident. This enables us to study a parametric representation of the fuzzy numbers. For this purpose, L (left side) and R (right side) fuzzy numbers are introduced. By an L-fuzzy number, we mean a fuzzy number of this membership function:

(i) $L(-x) = L(x)$
(ii) $L(0) = 1$
(iii) L is increasing in $[0,+\infty]$.

The same definition holds for the R-fuzzy number. Some examples of suitable functions of L-fuzzy numbers are:

$$L(x) = \max\left(0,(1-|x|^p\right)$$

$$L(x) = \frac{1}{1+|x|^p}$$

$$p > 0$$

An L-R fuzzy number possesses this membership function:

$$A(x)=\begin{cases} L\left(\dfrac{m-x}{\alpha}\right) & , \quad \text{if } x \le m \\ R\left(\dfrac{x-m}{\beta}\right) & , \quad \text{if } x \ge m \end{cases} \tag{1.4}$$

α, $\beta > 0$ are parameters 'controlling' fuzziness of the fuzzy number. For a degenerated case, $\alpha = \beta = 0$, one gets a genuine real number. Including the above parameters (m, α, β), the fuzzy number defined by (1.4) will be denoted by $A(m,\alpha,\beta)$. For such L-R representation, all basic algebraic operations can be performed using the respective parameters of this representation. For instance, for addition and subtraction of $A(m,\alpha,\beta)$ and $B(n,\delta,\gamma)$, we arrive at the condensed results:

$$A+B=(m+n\,,\alpha+\delta\,,\beta+\gamma)$$

$$A-B=(m-n\,,\alpha+\delta\,,\beta+\gamma)$$

so $A+B$ and $A-B$ are L-R fuzzy numbers.

The relevant formulas for multiplication and division of fuzzy numbers are more complicated; usually, they also return approximate results.

1.7 Conclusions

The purpose of this Chapter was to familiarise the reader with the basic concepts of fuzzy sets. The notion of partial membership of an object in a given category is appealing. Fuzzy sets with their formalisms of graded levels of membership are developed along this line. Departing from the models of two-valued logic connectives, we have reviewed operations on fuzzy sets and pointed out a significant diversity of available definitions. The formal differences between fuzzy sets and sets have also been summarised. Although most of the well known set-theoretic properties are preserved, some of them (owing to the lack of dichotomy in fuzzy sets) do not hold at all (e.g. the underlap and overlap properties). Fuzzy relations and their basic operations, such as projections and composition operations as essential concepts of the calculus of fuzzy relations, have been thoroughly studied. The PROLOG-like notation applied throughout the Chapter indicates the usefulness of logic-based programming as one of the avenues of computer implementation of algorithms using fuzzy sets.

Being acquainted with the basics of fuzzy sets and their calculus, in the next Chapter we proceed to more-advanced notions.

1.8 References

[1] Aczel, J. 1966. 'Lectures on Functional Equations and Their Applications'. Academic Press, New York

[2] Bellman, R.E., & Zadeh, S. 1977. 'Local and fuzzy logics'. *In* 'Modern Uses of Multiple Valued Logic' (Eds. J.M. Dunn & D. Epstein). Reidel, D. Dordrecht, pp. 103–165

[3] Black, M. 1937. 'Vagueness: an exercise in logical analysis'. *Phil. Sci.*, **4**, pp. 427–455

[4] Borel,, E. 1950. 'Probabilité et Certitude'. Press Univ. de France, Paris

[5] Di Nola, A., Pedrycz, W., & Sessa, S. 1986. 'Processing of fuzzy numbers by fuzzy relation equations'. *Kybernetes*, **15**, pp. 43–47

[6] Dubois, D., & Prade, H. 1978. 'Operations on fuzzy numbers'. *Int. J. Syst. Sci.*, **9**, pp. 613–626

[7] Dubois, D., & Prade, H. 1981. 'Fuzzy Sets and Systems: Theory and Applications'. Academic Press, New York

[8] Dubois, D., & Prade, H. 1981. 'Additions of interactive fuzzy numbers'. *IEEE Trans. Autom. Contr.*, **26**, pp. 926–936

[9] Frank, M.J. 1979. 'On the simultaneous associativity of F(x,y) and x + y – F(x,y)'. *Aequationes Math.*, **19**, pp. 194–226

[10] Godal, R.C., & Goodman, T.J. 1980. 'Fuzzy sets and Borel'. *IEEE Trans. Syst., Man & Cybern.*, **10**, p. 637

[11] Hailperin, T. 1976. 'Boole's Logic and Probability'. North Holland, Amsterdam

[12] Hamacher, H. 1978. 'Über logische Verknupfungen unscharfer Aussagen and deren Zugehorige Bewertungsfunktionen'. *In* 'Progress in Cybernetics and Systems Research' (Eds. R. Trappl, G.J. Klir & L. Ricciardi). Hemisphere, Washington, pp. 276–288

[13] Klir, G.J., & Folger, T.A. 1988. 'Fuzzy Sets, Uncertainty, and Information'. Prentice-Hall, Englewood Cliffs, New Jersey

[14] Kosko, B. 1992. 'Neural Networks and Fuzzy Systems: A Dynamical Systems Approach to Machine Intelligence'. Prentice-Hall, Englewood Cliffs, New Jersey

[15] Menger, K. 1942. 'Statistical metric spaces'. *Proc. Nat. Acad. Sci. USA*, **28**, pp. 535–537

[16] Negoita, C.V., & Ralescu, D.A. 1975. 'Applications of Fuzzy Sets to Systems Analysis'. ISR 11, Birkhauser, Basel

[17] Pedrycz, W. 1982. 'Some aspects of fuzzy decision-making'. *Kybernetes*, **11**, pp. 297–301.

[18] Pedrycz, W. 1990. 'Direct and inverse problem in comparison of fuzzy data'. *Fuzzy Sets & Syst.*, **34**, pp. 223–235

[19] Popper, K. 1976. 'Unended Quest'. Fontana Collins, London

[20] Russell, B. 1923. 'Vagueness'. *Aust. J. Phil.*, **1**, pp. 84–92

[21] Rescher, N. 1969. 'Many-Valued Logic'. McGraw-Hill, New York

[22] Sanchez, E. 1977. 'Eigen Fuzzy Sets and Fuzzy Relations'. Memo. UCB/ERL/20, Univ. California

[23] Sanchez, E. 1978. 'Resolution of eigen fuzzy sets equations'. *Fuzzy Sets & Syst.*, **1**, pp. 69–74

[24] Thole, U., Zimmermann, H.J., & Zysno, P. 1979. 'On the suitability of minimum and product operators for the intersection of fuzzy sets'. *Fuzzy Sets & Syst.*, **2**, pp. 167–180

[25] Yager, R.R. 1980. 'On a general class of fuzzy connectives'. *Fuzzy Sets & Syst.*, **4**, pp. 235–242

[26] Yager, R.R. 1985. 'Q-projection of possibility distributions'. *IEEE Trans. Syst., Man & Cybern.*, **6**, pp. 775–777

[27] Yong-Yi, Ch. 1981. 'An approach to fuzzy operators'. *BUSEFAL*, **9**, pp. 113–129

[28] Zadeh, L.A. 1965. 'Fuzzy sets'. *Inf. & Contr.*, **8**, pp. 338–353

[29] Zadeh, L.A. 1965. 'Fuzzy sets and systems'. *Proc. Symp. Syst. Theory*, Polytech. Inst. Brooklyn, pp. 29–37

[30] Zadeh, L.A. 1973. 'Outline of a new approach to the analysis of complex systems and decision processes'. *IEEE Trans. Syst., Man & Cybern.*, **3**, pp. 28–44
[31] Zadeh, L.A. 1976. 'A fuzzy-algorithmic approach to the definition of complex or imprecise concepts'. *Int. J. Man-Mach. Stud.*, **8**, pp. 249–291
[32] Zadeh, L.A. 1979. 'A theory of approximate reasoning'. *In* 'Machine Intelligence' (Eds. J.E. Hayes & L.I. Mikulich). Ellis Horwood/John Wiley, New York, pp. 149–196

CHAPTER 2

Fuzzy Sets — Characterisations, Algorithms and Extensions

2.1 Introduction

In this Chapter, we discuss more advanced issues of fuzzy sets, both conceptual and algorithmic. We are concerned with the global characterisation of fuzzy sets as collections of objects being structured into certain semantic entities. This characterisation includes a variety of indices describing the information content of a fuzzy set, its specificity, and its granularity. Finally, we study more advanced constructions like fuzzy measures and fuzzy integrals.

2.2 Indices of Fuzziness

If a fuzzy set is viewed as a collection of objects belonging to a certain class, we may be interested in expressing the 'fuzziness' of the fuzzy set considered. Attempts to solve this problem were reported relatively early in the development of fuzzy sets. Of course, we need to realise clearly the meaning of the notion we are discussing. Two fundamental concepts, arising from the studies of De Luca & Termini [6] and originating from information theory, are the entropy measure of fuzziness and the energy measure of fuzziness.

The former adapts some general concepts of information theory, while the latter leads us to ideas of determining the cardinality of a fuzzy set.

Let us start from these definitions in their general setting.

Definition 2.1

By the entropy measure of a fuzzy set, A defined in **X**, we mean a number H(A):

$$H(A) = \int_{\mathbf{X}} f(A(x))\,d\nu$$

where ν is any measure defined in **X**, while f is any function $f: [0,1] \rightarrow [0,1]$, such that it is monotonically increasing in [0,1/2] and monotonically decreasing in the [1/2,1] interval.

Evidently, this definition is of great generality. Usually, one is restricted to a function 'f' viewed as a logarithm of base 2:

$$H(A) = \int_{\mathbf{X}} \log_2(A(x))\, dv$$

and then one is faced with the classical notion of entropy, a dominant concept of information theory. A still simpler form of the above function is one that increases (and decreases) in a linear fashion, namely

$$f(x) = \begin{cases} x & , \text{ if } x \in [0, 1/2] \\ 1-x & , \text{ if } x \in [1/2, 1] \end{cases}$$

Of course, a fuzzy set with a membership function equal to 1/2 is characterised by the highest value of entropy among all the fuzzy sets.

By definition, an empty fuzzy set Ø (with a zero membership function) and a set equivalent to the entire universe of discourse have zero entropy.

Definition 2.2

An energy measure of a fuzzy set A, called E(A), is defined as

$$E(A) = \int_{\mathbf{X}} h(A(x))\, dv$$

where v is any measure defined in **X**, and h forms a monotonically increasing function over [0,1].

In the light of this definition, the inclusion $A \subset B$ implies that $E(A) \le E(B)$. In the limit situation formed by a fuzzy set with a membership function of identically 1.0, its energy measure attains a maximum (among all the fuzzy sets defined in the same space).

For $h(v) = v$, the energy measure of fuzziness is expressed as an integral over A(x):

$$E(A) = \int A(x)\, dv$$

and can be interpreted as the cardinality of A.

The two measures of fuzziness are completely distinct, since they evaluate two different global features of fuzzy sets. In particular, the entropy measure of fuzziness expresses the uncertainty residing within the fuzzy set while the energy measure relates to its cardinality.

These two notions can easily be parameterised, introducing an ability to discriminate between different levels of membership. This is easily accomplished by introducing a certain threshold level $\lambda \in [0,1]$. The previous entropy and energy measures of fuzziness are modified through the parameterised versions of the functions 'f' and 'g', thus:

$$H(A,\lambda) = \int f_\lambda(A(x))\, dv$$

$$E(A,\lambda) = \int h_\lambda(A(x))\, dv$$

where

$$f_\lambda(u) = \begin{cases} f(u) & , \text{ if } u \in (\lambda, 1-\lambda) \\ 0 & , \text{ otherwise} \end{cases}$$

and

$$h_\lambda(u) = \begin{cases} h(u) & , \text{ if } u \in (\lambda, 1) \\ 0 & , \text{ otherwise} \end{cases}$$

The role of the threshold λ is to 'filter out' some insignificant degrees of membership that would otherwise have an undesirable impact on the values of these indices of fuzziness. For the entropy measure of fuzziness, these are the values of membership close to 0 and 1. For the energy measure of fuzziness, the lowest grades of membership (below λ) are negligible. For $\lambda = 0$, $H(A,\lambda)$ and $H(A)$ as well as $E(A,\lambda)$ and $E(A)$ coincide.

A different way of measuring the fuzziness of a fuzzy set is by calculation of the absolute difference between a fuzzy set and its complement, and integrating over the universe of discourse [30]. The more 'fuzzy' the fuzzy set, the lower the value of this index:

$$\delta(A) = \int_{\mathbf{X}} \left| A(x) - \overline{A}(x) \right| dx$$

For $A(x) = 1/2$, the index attains zero. The zero difference $A(x) - \overline{A}(x)$ can be interpreted as a lack of distinction between A and its complement; cf. a geometrical interpretation of this fact in Chapter 1. In this sense, any set is characterised by $\delta(A)$ equal to zero.

Another index of a fuzzy set is its granularity.

Definition 2.3

The granularity of a fuzzy set A defined in **X**, G(A), is:

$$G(A) = 1 - \int_0^1 \frac{\text{meas}(A_\alpha)\,\psi(\alpha)\,d\alpha}{\text{meas}(\mathbf{X})}$$

where $\psi : [0,1] \rightarrow [0,1]$ is a functional defined over the grades of membership while meas(.) denotes a measure of an α-cut of A. We assume that the integral does exist and that space **X** is finite. The essence of this operation is that it selectively aggregates successive α-cuts of A. The role of functional ψ is to enhance or eliminate the impact of the different ranges of the grades of membership. For example, ψ could be selected as one among the functionals of energy or entropy measures of fuzziness. This index is used to express the granularity of a fuzzy set: the coarser the fuzzy set, the lower its granularity.

For a degenerated fuzzy set A defined in **L**, where $\mathbf{L} \subset \mathbf{R}$, meas(**L**) $< \infty$, we get

$$A(x) = \delta(x - x_0) = \begin{cases} 1 & , \quad \text{if } x = x_0 \\ 0 & , \quad \text{otherwise} \end{cases}$$

The granularity of A is computed as:

$$\int_0^1 \text{meas}(A_\alpha)\,\psi(\alpha)\,d\alpha = \text{meas}(A_1)\,\psi(1) + \text{meas}(A_0)\,\psi(0)$$

Assuming that $\psi(1) = 1$ and $\psi(0) = 0$ (which implies the energy type of functional), the above integral vanishes and G(A) = 1.

For the fuzzy set viewed as an interval,

$$A(x) = \begin{cases} 1 & , \quad \text{if } x \in [x_1, x_2] \\ 0 & , \quad \text{otherwise} \end{cases}$$

where $\Delta = |x_2 - x_1|$, we obtain:

$$\int_0^1 \text{meas}(A_\alpha)\,\psi(\alpha)\,d\alpha = \int_0^1 \Delta\,\delta(\alpha - 1)\,\psi(\alpha)\,d\alpha + \int_0^1 \text{meas}(\mathbf{L} \setminus \Delta)\,\delta(\alpha)\,\psi(\alpha)\,d\alpha$$

From the previous boundary conditions, one has

$$G(A) = 1 - \frac{\Delta}{\text{meas}(\mathbf{L})}$$

Proceeding with the same functional ψ, we compute the granularity of a triangular fuzzy number with the same support Δ. We define

$$A(x) = \begin{cases} 1 - \dfrac{x}{\Delta} & , \text{ if } x \in [0, \Delta] \\ 0 & , \text{ otherwise} \end{cases}$$

and derive

$$\int_0^1 \text{meas}(A_\alpha)\,\psi(\alpha)\,d\alpha = \int_0^1 \Delta(1-\alpha)\,\psi(\alpha)\,d\alpha = \Delta \int_0^1 (1-\alpha)\,\psi(\alpha)\,d\alpha$$

Since

$$\int_0^1 (1-\alpha)\,\psi(\alpha)\,d\alpha < 1$$

The granularity is greater than for the interval of the length Δ.

We recall that a specificity measure Sp(.), defined in [31], describes a similar property of fuzzy sets but uses only their cardinality:

$$Sp(A) = \int_0^1 \frac{\alpha_{max}}{\text{card}(A_\alpha)}$$

In this definition, we make no distinction between different values of the membership. In the above definition, α_{max} stands for the maximal grade of membership of A.

For the two normal fuzzy sets A and B, these properties are preserved:

$$0 \le Sp(A) \le 1$$

$$Sp(A) = 1 \quad , \quad \text{if A is a singleton in } \mathbf{X} \quad , \quad A = \{x_0\}$$

$$\text{if } A \subset B \quad , \quad \text{then } Sp(A) \ge Sp(B)$$

Thus, the more precise the fuzzy set is, the higher the value attained for its specificity index. The maximum specificity corresponds to a pointwise (x_0) assessment of the value of the variable.

As revealed in [31], the specificity of a fuzzy set when the membership function is ordered in a decreasing sequence is calculated from this formula:

$$Sp(A) = \sum_{i=1}^{n} \frac{A(x_i) - A(x_{i+1})}{i}$$

where card(**X**) = n and we accept the convention that $A(x_{n+1}) = 0$.

2.3 Comparison of Fuzzy Quantities

Comparison of two fuzzy sets forms an important task in many problems where, in general, a matching task occurs. To present the main ways in which this comparison is performed, let us consider two fuzzy sets A and B defined in the same universe of discourse **X**.

A first and self-evident approach is to compare fuzzy sets by taking any distance between the membership functions of A and B. In a general setting, one can take the Minkowski distance:

$$d(A,B) = \left(\int |A(x) - B(x)|^p \, dx \right)^{1/p} \quad , \qquad p \geq 1$$

(assuming the above integral does exist). Very often, instead of such a generalised version, one can use:

- Euclidean distance, p = 2

$$d(A,B) = \left(\int (A(x) - B(x))^2 \, dx \right)^{1/2}$$

- Hamming distance (city-block distance), p = 1

$$d(A,B) = \int |A(x) - B(x)| \, dx$$

- or a so-called normalised distance, $d_n(A,B)$

$$d_n(A,B) = \frac{d(A,B)}{card(\mathbf{X})}$$

if it has any sense. This form of comparison of fuzzy sets has a plausible interpretation in the field of mathematical analysis rather than coming from a set-theoretic approach.

The comparison is completed on the basis of membership functions treated as a specific kind of mapping. This, obviously, neglects specificity and its set-oriented character.

A second approach relies on the comparison of fuzzy sets in a certain logic-oriented way. First, we recall that in set theory the equality of two sets can be studied by verifying that the two conditions:

- A included in B
- B included in A

are simultaneously satisfied. If some differences are observable, we say that these two sets are unequal. Returning to fuzzy sets, this fact is expressed in logical notation in this way:

$$A = B \quad , \quad \text{if } A \subset B \text{ and } B \subset A$$

Modelling inclusion $\subset$ by an implication operator (φ operator) and an 'and' conjunction by a corresponding t-norm, we can define the grade at which A and B are equal to each other at point x, $x \in \mathbf{X}$, as:

$$(A{=}B)(x) = [A(x)\,\varphi\,B(x)]\,t\,[B(x)\,\varphi\,A(x)] \tag{2.1}$$

This equality index (2.1) is quite asymmetrical. To alleviate this demerit, we modify its generic form by adding a new condition binding the complements of A and B:

$$\overline{A} \subset \overline{B} \qquad \text{and} \qquad \overline{B} \subset \overline{A} \tag{2.2}$$

Averaging (2.1) and (2.2), one gets:

$$(A{=}B)(x) = \tfrac{1}{2}\left\{[A(x)\varphi B(x)]t[B(x)\varphi A(x)] + \left[\overline{A}(x)\varphi\overline{B}(x)\right]t\left[\overline{B}(x)\varphi\overline{A}(x)\right]\right\} \tag{2.3}$$

In the remainder, we refer to the equality index as defined by (2.3).

It is instructive to take a look at the equality index for two selected t-norms such as the product and the Lukasiewicz AND connective.

For the product operator, we derive:

$$A(x)\ \varphi\ B(x) = \begin{cases} 1 & ,\ \text{if } A(x) \le B(x) \\ \dfrac{B(x)}{A(x)} & ,\ \text{if } A(x) > B(x) \end{cases}$$

or equivalently:

$$A(x)\ \varphi\ B(x) = \min\left(1, \frac{B(x)}{A(x)}\right)$$

and then:

$$(A{=}B)(x) = \begin{cases} 0.5\left[\dfrac{A(x)}{B(x)} + \dfrac{1-B(x)}{1-A(x)}\right] & ,\ \text{if } A(x) < B(x) \\ 1 & ,\ \text{if } A(x) = B(x) \\ 0.5\left[\dfrac{B(x)}{A(x)} + \dfrac{1-A(x)}{1-B(x)}\right] & ,\ \text{if } A(x) > B(x) \end{cases}$$

The Lukasiewicz AND operation yields a piecewise linear form of the equality index:

$$A(x)\ \varphi\ B(x) = \begin{cases} 1-A(x)+B(x) & ,\ \text{if } A(x) > B(x) \\ 1 & ,\ \text{if } A(x) = B(x) \\ 1-B(x)+A(x) & ,\ \text{if } A(x) < B(x) \end{cases}$$

Their two-dimensional contours in the $(A(x)-B(x))$ two-dimensional space clearly indicate the diversified nature of the regions of equality that depend heavily on the form of the t-norm used in their construction (Figs. 2.1(a) and (b)).

The equality index has a local character; i.e. it depends on the current point in **X**. To take a look at the global property of set equality, we can aggregate the values (A=B)(x) over all the elements of **X**. These three aggregation methods are of interest:

- an optimistic form of aggregation

$$(A{=}B)_{opt} = \sup_{x \in \mathbf{X}} \left[(A{=}B)(x)\right]$$

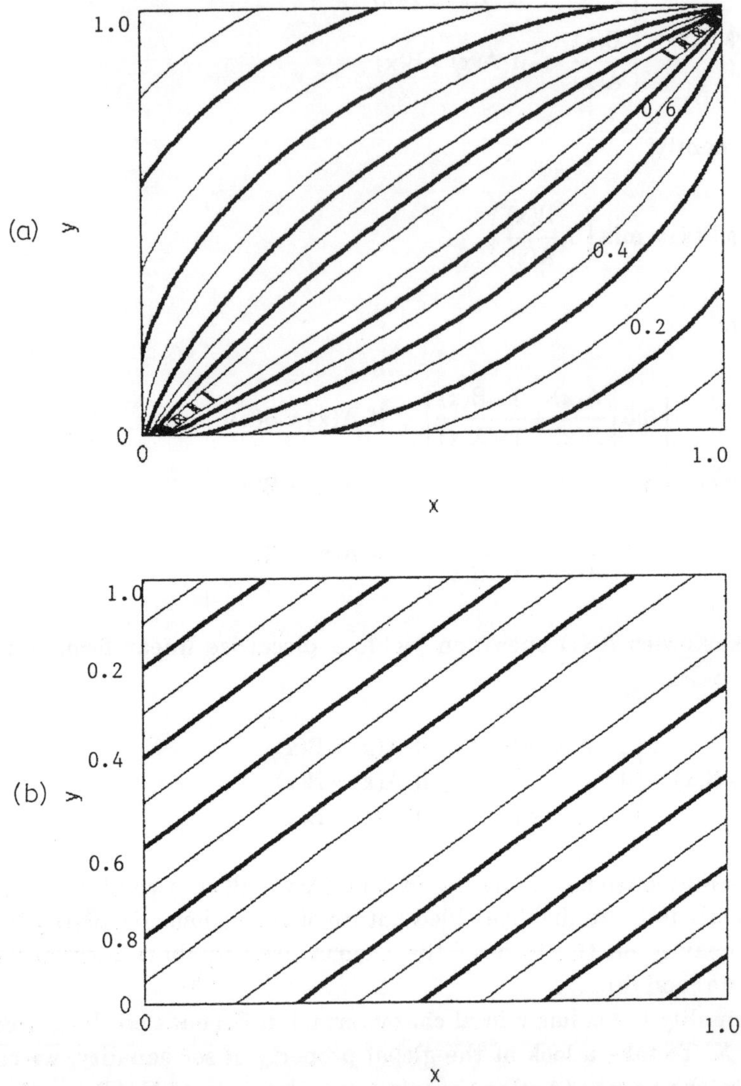

Fig. 2.1 Equality index in A(x) – B(x) coordinates: (a) product; (b) Lukasiewicz AND conjunction

- a pessimistic form of aggregation

$$(A{=}B)_{\text{pess}} = \inf_{x \in \mathbf{X}} \left[(A{=}B)(x)\right]$$

- an averaged form

$$(A{=}B)_{av} = \int_{\mathbf{X}} \frac{(A{=}B)(x)\,dx}{\mathrm{card}(\mathbf{X})}$$

The obvious relationship holds:

$$(A{=}B)_{pess} \leq (A{=}B)_{av} \leq (A{=}B)_{opt}$$

A pointwise definition of the equality of two fuzzy sets A and B (2.3) is visualised in Fig. 2.2.

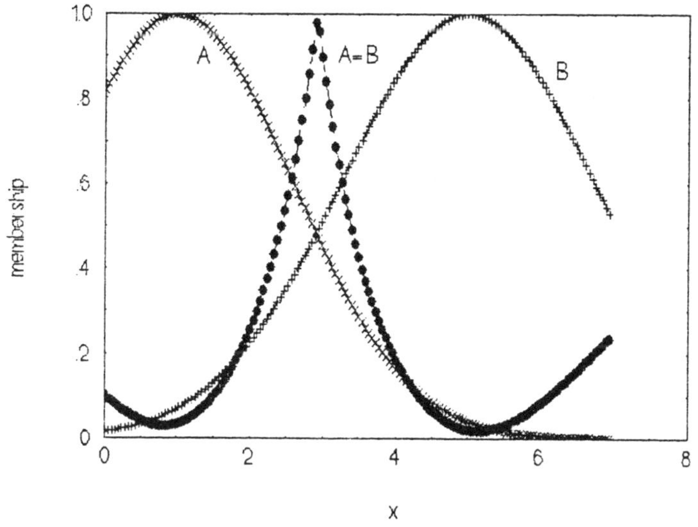

Fig. 2.2 Equality index A(x) = B(x); t-norm specified as product

An interesting problem emerges: For a given membership function A(x) and a level of equality γ, determine the values of membership B(x) such that A(x) and B(x) are equal to a degree not lower than γ. This can be formalised:

$$(A{=}B)(x) \geq \gamma$$

One can verify that, for each value of γ, the set of solutions

$$\Gamma(\gamma) = \left\{B(x) \mid (A{=}B)(x) \geq \gamma\right\}$$

is nonempty and is given in a form of a closed subinterval of the unit interval. One can also observe:

- $\Gamma(\gamma)$ is a nondecreasing function of γ; $\gamma_1 < \gamma_2$ implies that $\Gamma(\gamma_1) \geq \Gamma(\gamma_2)$
- for $\gamma = 0$, $\Gamma(0) = [0,1]$ when, for $\gamma = 1$, $\Gamma(1)$ reduces to a single point, $\Gamma(1) = \{A(x)\}$.

Notice that, for a strict inequality, $(A{=}B)(x) > \gamma$, the set of solutions $\Gamma(\gamma)$ may be an empty one.

Another stream of methods of matching two fuzzy sets is based on two measures of the fit of fuzzy sets, namely possibility and necessity. Recall these definitions [7]:

The possibility measure of A with respect to B, $\Pi(A \mid B)$, is defined as

$$\Pi(A \mid B) = \sup_{x \in \mathbf{X}} \left[\min\left(A(x), B(x)\right)\right]$$

The necessity (certainty) measure of A with respect to B, $N(A \mid B)$, is defined as

$$N(A \mid B) = \inf_{x \in \mathbf{X}} \left[\max\left(A(x), 1 - B(x)\right)\right]$$

To interpret both of these, notice that the possibility measure of A with respect to B reflects the extent to which A and B coincide or overlap, while the necessity expresses a grade to which B is contained in A. The first measure is symmetrical with respect to the arguments

$$\Pi(A \mid B) = \Pi(B \mid A)$$

and for the second, with a reverse order of the arguments, we get a relationship

$$N(A \mid B) = 1 - \Pi(\overline{A} \mid B)$$

These original definitions, which can be found in papers by Zadeh or Dubois & Prade, can be extended a little by replacing minimum and maximum by t- and s-norm, respectively, thus yielding sup-t and inf-s compositions of two fuzzy sets:

$$\Pi(A|B) = \sup_{x \in \mathbf{X}} [A(x)\, t\, B(x)]$$

$$N(A|B) = \inf_{x \in \mathbf{X}} [A(x)\, s\, (1 - B(x))]$$

Since t-norm and s-norm are distributive with respect to maximum and minimum, one has

$$\Pi(A \cup C|B) = \max\{\Pi(A|B), \Pi(C|B)\}$$

and

$$N(A \cap C|B) = \min\{N(A|B), N(C|B)\}$$

An illustration of possibility and necessity measures is given in Fig. 2.3.

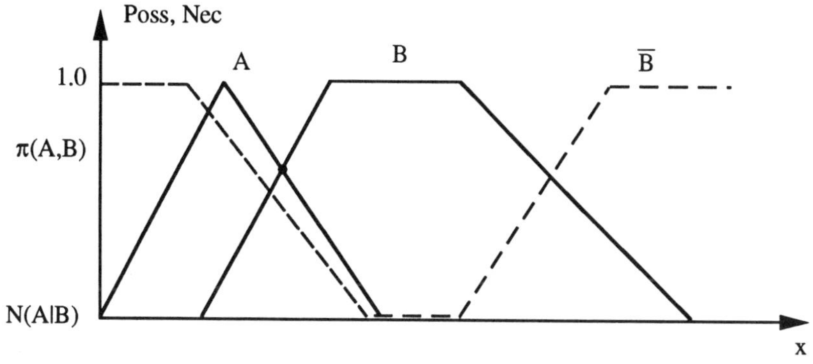

Fig. 2.3 Possibility and necessity measures of fuzzy sets A and B

For a degenerated fuzzy set, i.e. a single point x_0 in **X**, the two measures are related in a straightforward manner:

$$\Pi(A|B) = \sup_{x \in \mathbf{X}} [A(x)\, t\, B(x)] = \max\left\{1\, t\, B(x_0),\ \sup_{x \neq x_0} [0\, t\, B(x)]\right\} = B(x_0)$$

and

$$N(A|B)=\inf_{x\in \mathbf{X}}[A(x)\,s\,(1-B(x))]=\min\left\{1\,s\,(1-B(x_0)),\inf_{x\neq x_0}[0\,s\,(1-B(x))]\right\}=$$

$$=\min\left\{1,\inf_{x\neq x_0}(1-B(x))\right\}=\inf_{x\neq x_0}(1-B(x))=1-B(x_0)$$

Assuming that B has a continuous membership function, one gets:

$$N(A|B)=1-\Pi(A|B)$$

or

$$\Pi(A|B)+N(A|B)=1$$

For a datum A viewed as a set with support $X=[x_0,x_{00}]$, the possibility and necessity measures become:

$$\Pi(A|B)=\sup_{x\in[x_0,x_{00}]}B(x)$$

and

$$N(A|B)=\inf_{x\in[x_0,x_{00}]}[1-B(x)]$$

For $A\subset A'$, we derive $\Pi(A|B)\le\Pi(A'|B)$. The same inclusion has an opposite effect on the necessity measure, $N(A|B)\ge N(A'|B)$ (Figs. 2.4(a) and (b)).

Reference [7] discusses a fuzzy degree of truth, called also a compatibility of A and B denoted by Cp(A,B). It is defined as a fuzzy set in [0,1], where its membership function is equal to:

$$Cp(A,B)(v)=\begin{cases}\sup\limits_{y:B(x)=v}A(y)\\ 0\ ,\ \text{if } B^{-1}(v)=\varnothing\end{cases}\qquad(2.4)$$

$v\in[0,1]$. Notice that, for A equal to B, the compatibility is equal to the identity function:

$$Cp(A,B)(v)=v\quad,\quad\text{for all } v\in[0,1]$$

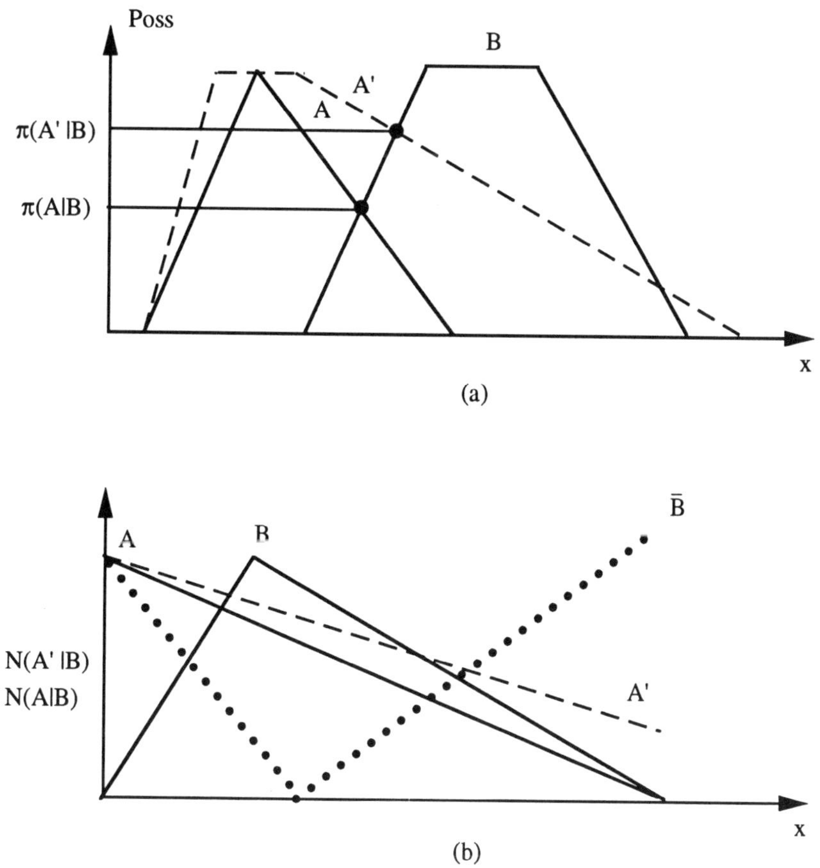

Fig. 2.4(a),(b) Possibility and necessity measures for A, A′ and B

Making use of (2.4), the measures of possibility and necessity are rewritten as:

$$\Pi(A \mid B) = \sup_{v \in [0,1]} \left[v \; t \; Cp(A,B)(v) \right]$$

$$N(A \mid B) = \inf_{v \in [0,1]} \left[v \; s \; (1 - Cp(A,B)(v)) \right]$$

2.4 Fuzzy Measures and Fuzzy Integrals

The notion of a fuzzy measure was introduced and studied by Sugeno [26,27]. Some of the mathematical background stems from the notion of capacity studied by Choquet [3], whose theory was established independently of the development of fuzzy sets.

We briefly outline the main ideas below.

Let **X** be a universe of discourse. By **B**, we denote a family of subsets of **X**, forming a σ-field of **X**.

By a fuzzy measure, we mean a mapping $g : \mathbf{B} \rightarrow [0,1]$, satisfying these conditions:

(i) $g(\emptyset) = 0$, $g(\mathbf{X}) = 1$ (boundary conditions)
(ii) for each A, B ∈ **B**, where $A \subset B$, $g(A) \le g(B)$
(iii) for a monotonic sequence of A_n, $A_n \in \mathbf{B}$, such that

$$\lim_{n\to\infty} A_n \in \mathbf{B}$$

we get

$$\lim_{n\to\infty} g(A_n) = g\left(\lim_{n\to\infty} A_n\right)$$

The first two of these conditions indicate the most characteristic features of the fuzzy measure. The second one especially differs significantly from what one finds in a definition of a probability measure; an additivity property of the probability measure is replaced by the evidently less restrictive property of monotonicity. Condition (iii) applies only to infinite space **X**, and it can be read as a kind of continuity requirement.

As an analogy in probability theory, we can introduce a fuzzy space viewed as a triple (**X**, **B**, g). For comparison, remember that a probability measure P defined in a σ-field of **X**, $P : \mathbf{B} \rightarrow [0,1]$, is characterised by these features:

(i) $$\forall_{A\in\mathbf{B}} P(A) \in [0, 1]$$

and $P(\mathbf{X}) = 1$ (boundary conditions)

(ii) the probability of mutually disjoint events ($A_i \cap A_j = \emptyset$ for $i \neq j$) is read as a sum of the probabilities of individual events, $P(A_i)$:

$$P\left(\bigcup_{i=1}^{\infty} A_i\right) = \sum_{i=1}^{\infty} P(A_i)$$

Note that the additivity property is significantly relaxed in the definition of the fuzzy measure.

As will be seen later in applications, the monotonicity property is definitely much more appealing.

Some illustrative examples of fuzzy measures as used in [26] are:

(1) a fuzzy measure formed on the basis of a characteristic function of a set A for a fixed element $x_0 \in \mathbf{X}$ is described by:

$$g(x_0, A) = \chi_A(x_0)$$

(2) a fuzzy measure constructed by means of any function $h : \mathbf{X} \to [0,1]$ via the formula:

$$g(A) = \sup_{x \in A} h(x)$$

Note, however, that the definition of a fuzzy measure in the form given above does not give us any hint of the way in which a fuzzy measure can be obtained for a given set A of **X**. To make it possible to construct such a fuzzy measure so as to allow us to compute its value for any set, we restrict ourselves to a certain family of fuzzy measures.

By a λ-fuzzy measure, g_λ, defined in **B**, we mean a fuzzy measure for which, for each A and B such that $A \cap B = \emptyset$ and $A \cup B \in \mathbf{B}$, we get

$$g_\lambda(A \cup B) = g_\lambda(A) + g_\lambda(B) + \lambda g_\lambda(A) g_\lambda(B)$$

where $\lambda \in (-1,\infty)$. This fuzzy measure is called an λ-additive one. Furthermore, the fuzzy measure g_λ in **B** is numerically λ-additive if, for each sequence of say $\{A_n\}$ pairwise disjoint sets in **B** such that their union belongs to **B**, one has:

$$g_\lambda\left(\bigcup A_n\right) = \frac{1}{\lambda}\left[\prod_{i=1}^{\infty} (1 + \lambda g_\lambda(A_i)) - 1\right]$$

Note that, for $\lambda = 0$, g_λ converts into a probability measure:

$$g_\lambda(A \cup B) = g_\lambda(A) + g_\lambda(B)$$

which indicates a significant level of generality achieved for the fuzzy measure.

Some interesting properties of λ-fuzzy measures are listed below:

- if $\lambda \geq 0$, then $g_\lambda(A \cup B) \geq g_\lambda(A) + g_\lambda(B)$

- for a difference of two sets A and B, its fuzzy measure is:

$$g_\lambda(A/B) = \frac{g_\lambda(A) - g_\lambda(B)}{1 + \lambda\, g_\lambda(B)} \quad , \quad B \subset A$$

From this, one deduces that a complement of A is expressed as:

$$g_\lambda(\overline{A}) = \frac{1 - g_\lambda(A)}{1 + g_\lambda(A)}$$

The λ-fuzzy measure can be conveniently applied in a finite space **X**, say $\mathbf{X} = \{x_1, x_2, \ldots, x_n\}$. Assume values of the fuzzy measure: each single point x_i equal to $g_\lambda(\{x_i\})$. Then, for each $F \subset \mathbf{X}$, its fuzzy measure is:

$$g_\lambda(F) = \frac{1}{\lambda}\left[\prod_{x_i \in F} \left(1 + \lambda\, g_\lambda(\{x_i\})\right) - 1\right] \tag{2.5}$$

In particular, for a set F consisting of only two elements x_1 and x_2, the above relationship reduces to:

$$g_\lambda(\{x_1, x_2\}) = g_\lambda(\{x_1\}) + g_\lambda(\{x_2\}) + \lambda\, g_\lambda(\{x_1\})\, g_\lambda(\{x_2\})$$

On the basis of the boundary condition, $g_\lambda(\mathbf{X}) = 1$, parameter λ can be derived by solving an algebraic equation of the first order:

$$1 = g_\lambda(\{x_1\}) + g_\lambda(\{x_2\}) + \lambda\, g_\lambda(\{x_1\})\, g_\lambda(\{x_2\})$$

i.e.

$$\lambda = \frac{1 - g_\lambda(\{x_1\}) - g_\lambda(\{x_2\})}{g_\lambda(\{x_1\})\, g_\lambda(\{x_2\})}$$

For a finite space **X**, the value of λ results from an algebraic equation that appears when equating (2.5) to 1. As proved in [26], it has a unique solution that can be obtained, for example, by the Newton-Raphson iterative scheme. The value of this parameter for the k+1th iteration is:

$$\lambda(k+1)=\lambda(k)-\frac{F(\lambda(k))-1}{F'(\lambda(k))}$$

with

$$F(\lambda(k))=\frac{1}{\lambda}\left[\prod_{k=1}^{n}\left(1+\lambda(k)\,g_{\lambda}(\{x_1\})-1\right)\right]$$

and

$$F'(\lambda(k))=\frac{1}{\lambda}\left[\left(1+\lambda(k)\,F(\lambda(k))\right)\sum_{i=1}^{n}\frac{g_{\lambda}(\{x_1\})}{1+\lambda(k)\,g_{\lambda}(\{x_1\})}-F(\lambda(k))\right]$$

The notion of a fuzzy measure having already been introduced, a fuzzy integral of any function $h : \mathbf{X} \to [0,1]$ measurable in the above sense is defined accordingly.

By a fuzzy integral of h with respect to a fuzzy measure $g_{\lambda}(.)$ over a set A:

$$\int_A h \bullet g_{\lambda}(.)$$

we mean a nonnegative number equal to:

$$\int_A h \bullet g_{\lambda}(.)=\sup_{\alpha\in[0,1]}\left[\alpha \wedge g_{\lambda}(A \cap F_{\alpha})\right]$$

with F_{α} being an α-cut of h, $F_{\alpha} = \{x \in \mathbf{X} \mid h(x) \geq \alpha\}$. As studied in [26], the fuzzy integral is a nonlinear functional possessing some properties of an ordinal integral; the most specific ones are listed below:

$$\int_A a \bullet g_{\lambda}(.)=a$$

$$\int_A (a \wedge h) \bullet g_\lambda(.) = a \wedge \int_A h \bullet g_\lambda(.)$$

$$\int_A (h_1 \cup h_2) \geq \int_A h_1 \bullet g_\lambda(.) \wedge \int_A h_2 \bullet g_\lambda(.)$$

$$\int_A (h_1 \cap h_2) \geq \int_A h_1 \bullet g_\lambda(.) \wedge \int_A h_2 \bullet g_\lambda(.)$$

Computations of the fuzzy integral for a finite **X** can be performed after a preliminary arrangement of the values of the membership function h in a nonincreasing order, say

$$h(x_1) \geq h(x_2) \geq \ldots \geq h(x_n)$$

Thus, in what follows we assume that the elements of X are ordered (and renumbered) to comply with this sequence. Then the value of the fuzzy integral results from the max-min composition:

$$\int_A h \bullet g_\lambda(.) = \max_{1 \leq i \leq n} [h(x_1) \wedge g_\lambda(X_i)]$$

where X_i contains 'i' elements of **X**, $X_i = \{x_1, x_2, \ldots, x_i\}$.

Note that, for each X_i, the values taken by the function h are included between $h(x_1)$ and $h(x_i)$; obviously, $h(x_1) \geq h(x_i)$.

Note also that $g_\lambda(X_i)$ forms a nondecreasing sequence of numbers; so in fact the value of the fuzzy integral is taken as the height of the intersection $h(x_i) \wedge g_\lambda(X_i)$. The construction of the fuzzy integral is very efficient. It reduces computations to 'n', as opposed to 2^n combinations required by the original definition.

The notion of the fuzzy integral can be generalised by admitting t-norms in place of the minimum, producing this definition:

$$\int h \bullet g_\lambda(.) = \max_{1 \leq i \leq n} [h(x_i)\, t\, g_\lambda(X_i)]$$

2.5 Linguistic Variables and Linguistic Approximation

The term 'linguistic variable', coined by L.A. Zadeh [35], denotes a variable defined in a given universe of discourse and taking on some linguistic values like *small*, *medium*, *large* etc. These values are modelled by fuzzy sets. Each linguistic variable involves a finite, usually small, collection of generic linguistic terms (sometimes called primary terms). Syntactic and semantic rules describe how to manipulate linguistic variables. By using syntactic rules, one builds well formed sentences and nonprimary terms. The semantic rules specify a way in which the meaning (membership functions) of a term can be obtained.

More formally, we consider a linguistic variable of pressure defined over a range of pressure values of interest in the problem under discussion. We recognise generic terms such as *small*, *medium*, *large* and hedges (modifiers) such as *slightly*, *very*, *more or less* etc. Well formed sentences (wfs) involve these terms combined in accordance with some syntax rules. For example, a compound term *very high* consisting of the hedge *very* and the generic linguistic value *high* is a wfs.

The semantics describes how the membership function of this wfs is computed. Specifically, we defined the semantics of the three basic logic operators (AND, OR, NOT) in Chapter 1. Quite frequently, the semantics of the hedges is defined by accepting selected values of powers of the membership function of the primary term. Selected cases are summarised as follows:

$$\textit{very}\ A = A^2 \quad \textit{plus}\ A = A^{1.25}, \quad \textit{more or less}\ A = A^{1/2} \quad \textit{minus}\ A = A^{3/4}$$

or in general

$$h(A) = A^p \quad , \quad (h(A))(x) = A^p(x)$$

where h denotes a hedge. The hedges with $p > 1$ imply a concentration type of operation (plus, very etc.). For $p < 1$, the corresponding hedges 'dilute' the fuzzy set on which they operate (more or less, minus etc.). The semantics of hedges is extensively studied in [34,18].

The above model of hedges can be generalised by accepting some translation along the universe of discourse, i.e.:

$$(h(A))(x) = A^p(x - \tau)$$

where τ shifts the original membership function. The need for this shift is discussed in [22].

The notion of linguistic approximation is used to express a process of matching (approximating) a given fuzzy set by a collection of primitive (basic) terms available for this linguistic variable. Assuming that we have at our disposal a collection of fuzzy sets (linguistic terms) A_1, A_2, ...,

A_c and a group of hedges $h_1, h_2, \ldots, h_p$, the procedure of approximation of a fuzzy set B expresses it in terms of A_i s and h_j s. The straightforward approach consists of two steps: first, one approximates B by one of the A_i s. The selection is completed following the values obtained of the equality index computed for B and A_i. B is approximated to by A_{i0}, where:

$$\max_{i=1,2,\ldots,n} [av(A_i = B)] = av(A_{i0} = B)$$

To improve this approximation, we now select one among the hedges and apply it in what follows to A_{i0}:

$$\max_{j=1,2,\ldots,p} \left[av\left(h_j(A_{i0}) = B\right)\right] = av\left(h_{j0}(A_{i0}) = B\right)$$

As a result of this approximation, fuzzy set B is represented as one among the properly modified generic fuzzy sets.

2.6 Extensions of Fuzzy Sets

Returning to the basic definition of the fuzzy set, we recall that it was defined as a mapping from the universe of discourse to the unit interval. Soon, it appeared that extensions of fuzzy sets could be of interest as well. The need to study them arises from the theoretical point of view, and their study is also of value for the application of fuzzy sets. Thus, there is an interest in developing more advanced algebraic structures and in having mechanisms that avoid infinite precision in the determination of the values of membership functions. Here we recall some of these extensions.

One of those belonging to the second group is the so-called Φ-fuzzy set or interval-valued fuzzy set, introduced in [25]. Now, in comparison with the original fuzzy set, the grades of membership function form a subinterval of [0,1]. This has to model the uncertainty associated with determination of a precise value of the membership function. Therefore, instead of assigning a single value of the grade of membership to the concept under consideration, one is allowed to specify only a certain interval whose bounds represent the lower and upper estimates of the strength of membership of the concept.

More formally, a Φ-fuzzy set (interval-valued fuzzy set), A_Φ, in **X** is defined as a couple of mappings A_- and A_+, respectively, $A_\Phi = (A_-, A_+)$, such that:

$$A_-, A_+ : \mathbf{X} \to [0,1]$$

and

$$A_-(x) \le A_+(x) \quad , \quad \text{for all } x \in \mathbf{X}$$

For interval-valued fuzzy sets, the basic operations are defined using their upper and lower bounds separately. Thus, following the form of the previous presentation, we have (see [25]) for Φ-fuzzy sets $A_\Phi = (A_-, A_+)$ and $B_\Phi = (B_-, B_+)$:

- union

$$A_\Phi \cup B_\Phi = (A_- \cup B_-, A_+ \cup B_+)$$

- intersection

$$A_\Phi \cap B_\Phi = (A_- \cap B_-, A_+ \cap B_+)$$

- complement

$$\overline{A}_\Phi = \left(\overline{A}_+, \overline{A}_-\right)$$

However, an inclusion of Φ-fuzzy sets is defined as follows. A_Φ is included in B_Φ if, for each element of the universe of discourse, one gets

$$B_-(x) \le A_-(x) \qquad \text{and} \qquad A_+(x) \le B_+(x)$$

An interpretation of the intervals of this kind of Φ-fuzzy set can be carried out in the light of the methods of membership-function determination (estimation) (see Section 2.8.1).

As has been argued in studies of fuzzy sets, there is an apparent inconsistency in using a precise membership function to represent a vague concept. To overcome this paradoxical situation, some generalised constructions of so-called m-fuzzy sets have appeared [23]. These extensions are defined recursively as follows:

- a type-1 fuzzy set is a fuzzy set
- a type-m fuzzy set is a fuzzy set whose membership values are type $m-1$ fuzzy sets.

The value of m is not restricted to a certain specified limit, but in fact type-2 fuzzy sets look like a reasonable construction. This extension of the fuzzy set reflects a case to which it is difficult to assign a simple number expressing the grade of membership, and this can be modelled by a fuzzy set defined in the unit interval. An illustration of a type-2 fuzzy set defined in a finite **X** is given in Figure 2.5.

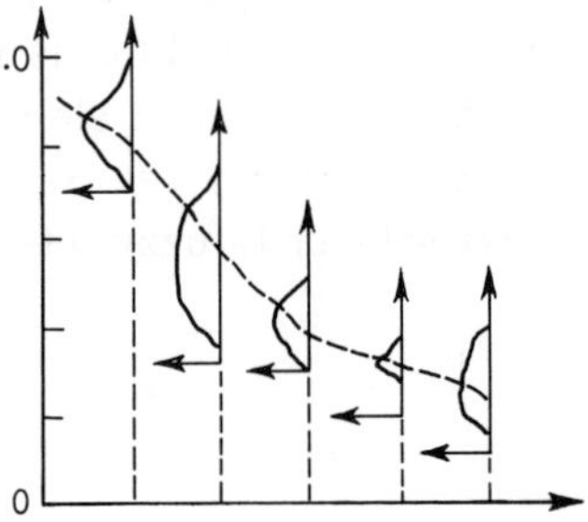

Fig. 2.5 Fuzzy set of type 2

Unfortunately, m-type fuzzy sets suffer from a significant shortcoming because working with them leads to an infinite-regress paradox. The imprecision is put on a second level; at the second level, it is observed that the imprecision is still present and a third level is much required. This procedure could continue *ad infinitum*. From a practical point of view, this approach may be time-consuming. Also, there is no assumption or experience that this imprecision can be reduced by moving from a lower to an upper level; that is to say, the membership function at the upper level is more condensed (less fuzzy) in the sense of the measures of fuzziness discussed in Section 2.2.

An example of generalisation stemming from the first stream of investigations relies on replacing the unit interval by a much more general algebraic structure such as a lattice. This was mentioned as early as in Zadeh's paper [32], and afterwards studied extensively by Goguen [12]. For a detailed explanation, we recall the definition of a lattice [2,10]. To work with it, it is instructive to start with a partially ordered set **X**. We say **X** is a partially ordered set if it is equipped with the ordering relation $\leq$ satisfying the properties:

(i) $x \leq x$
(ii) if $x \leq y$ and $y \leq x$, then $x = y$
(iii) if $x \leq y$ and $y \leq z$, then $x \leq z$ for all $x, y, z \in \mathbf{X}$.

Moreover, we say a partially ordered set **X** contains the greatest and the least elements, say I and o, if these properties are preserved:

$$\forall_{x \in \mathbf{X}} \; x \leq \mathrm{I} \qquad , \qquad \mathrm{o} \leq x$$

Then a partially ordered set containing both I and o is called a lattice. We denote it by L. Note that the unit interval is a simple example of a lattice with I specified as 1, and o treated as 0. Fig. 2.6 presents two examples of partially ordered sets in which the second forms a lattice. The first, seen in Fig. 2.6(a), has o, but there is no greatest element. In Fig. 2.6(b), 'a' is viewed as I while 'h' forms o. The fuzzy sets taking their values in any lattice are called lattice fuzzy sets, or L-fuzzy sets for short.

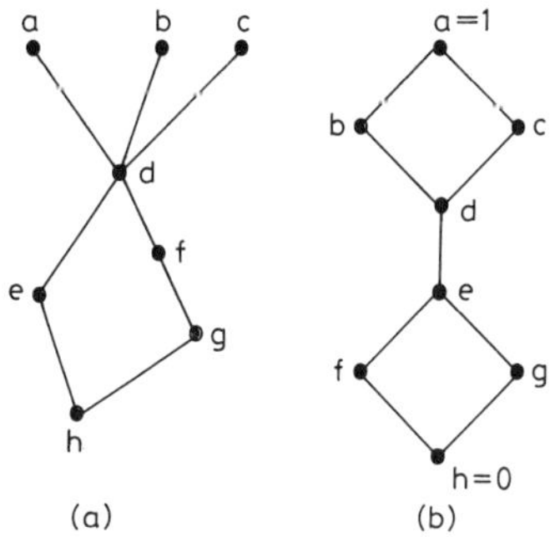

Fig. 2.6 Examples of partially ordered sets

To consider the basic operations on L-fuzzy sets, it is useful to recall the notion of the least upper bound (l.u.b.) and the greatest lower bound (g.l.b.) of the elements of a lattice L (see [2]). By the g.l.b. of $x, y \in L$, we mean an element $z \in L$ satisfying the property:

$$z \leq x \quad \text{and} \quad z \leq y \quad \text{and} \quad \{w \in L \,|\, z \neq w, z \leq w, w \leq z, w \leq y\} = \varnothing$$

The meaning of the l.u.b. is similarly defined. A remarkable fact is that a lattice may not be distributive over the g.l.b. and the l.u.b.

Also, the union and the intersection of A and B, being L-fuzzy sets, are defined as:

$$(A \cup B)(x) = \text{l.u.b.}\,(A(x), B(x))$$

$$(A \cap B)(x) = \text{g.l.b.}\,(A(x), B(x))$$

$$x \in L$$

For more theoretical details, the reader is referred to [12].

The question of the direct application of the links of L-fuzzy sets arises. There are some exceptional results, one of which is covered by [9].

2.7 Extensions of Fuzzy Sets Used to Cope with Fuzziness and Randomness

2.7.1 Different ways of coping with uncertainty — fuzziness and randomness

When referring to fuzziness and randomness, one can sometimes meet statements in which fuzziness is viewed as a certain form of equivalence of randomness. Without moving into the mathematical formulas that are different for fuzzy sets and probability (namely set theory and logics on one hand, and measure theory on the other), we can refer, for instance, to [33], where one finds this explanation:

> 'Randomness has to do with uncertainty concerning membership or nonmembership of an object in a nonfuzzy set, while fuzziness has to do with classes in which there may be grades of membership intermediate between full membership and nonmembership.'

Yet another argument may help to make a philosophical distinction between probability and fuzziness, especially between a grade of membership and a value of probability. Take a finite universe of discourse **X** and write

$A(x) = a$ – a value of the membership function of A for a certain element of X is equal to 'a'

$P\{x \in A\} = a$ – a probability of x belonging to A is equal to 'a'

Let us perform an experiment in which we pick up this element x and observe the outcome. The results are summarised in the Table below:

	Before experiment	After experiment
Fuzziness	$A(x) = a$	$A(x) = a$
Randomness	$P\{x \in A\} = a$	1 , if $x \in A$ 0 , otherwise

On observation of x, the *a priori* probability $P\{x \in A\} = a$ becomes *a posteriori*, i.e. equal either to 1 if $x \in A$ or to 0 otherwise. At the same time, A(x), being a measure of the extent to which x belongs to A, remains the same on observation of x.

In general, we can conclude that randomness deals with models of statistical inexactness due to the occurrence of random events, while fuzziness concerns situations of modelling of inexactness due to perception processes of the human being. However, there are some complex situations in which both factors should be taken into account, since they cannot be conveniently split and discussed separately. This has led to further attempts to develop tools where both the factors of uncertainty are studied. We show below the most important constructions, which also have a significant influence on the methodology of model building. They also play an inspiratory role in membership-function determination. We discuss probabilistic sets and fuzzy random variables, which in fact represent two diverse approaches to the handling of uncertainty: a horizontal one and a vertical one.

2.7.2 Probabilistic sets

The idea behind this concept is tied up with efforts to overcome difficulties in determining a precise value for grades of membership forming the membership function. Instead of treating a fuzzy set as a mapping of $\mathbf{X}$ into the [0,1] interval, the probabilistic set is characterised by a mapping of $\mathbf{X} \times \Omega$ into [0,1], where Ω plays the same role as in probability calculus. Hence, the value of the membership function depends on the element of $\mathbf{X}$ as well as on the element of Ω. To underline this, we put down $A(x,\omega)$ (instead of the notation A(x), the standard for fuzzy sets). The space Ω can be treated as having the meaning of the sample space; so the grades of membership $A(x,\omega)$ can vary from one experiment to another. This allows us to treat $A(x,\omega)$ as a so-called random field (or a random process if x is treated as time parameter), taking its values in the unit interval. This then enables us to use the whole well developed apparatus of random variables. To be more formal, we recall some fundamentals of probabilistic sets as introduced by Hirota, using his own terminology [15,16].

It is instructive to start with a probabilistic space $(\Omega,\mathbf{B},P)$ that is used in any consideration of probability theory and introduce a Borel field based on the [0,1] interval $(\Omega_c,\mathbf{B}_c)$, calling it a characteristic space. This suffices to introduce the notion of a probabilistic set in a formal way.

A probabilistic set A is defined by a mapping

$$A : \mathbf{X} \times \Omega \rightarrow \Omega_c$$

with A(x,.) viewed as a $(\mathbf{B},\mathbf{B}_c)$ measurable mapping for each $x \in \mathbf{X}$. The above mapping is also called a defining function of the probabilistic set A. One can define the same operations for probabilistic sets as for fuzzy sets.

An interesting view of probabilistic sets can be obtained by averaging over the space Ω. We also get:

- the mean-value membership function **E**(A) of the probabilistic set A:

$$\mathbf{E}(A)(x) = \int_{\Omega} A(x,\omega)\,dP(\omega) = m_1(A)(x)$$

- the vagueness function **V**(A) of the probabilistic set A:

$$\mathbf{V}(A)(x) = \int_{\Omega} (A(x,\omega) - m_1(x))^2\, dP(\omega)$$

- the standard deviation:

$$\sigma(A)(x) = (\mathbf{V}(A)(x))^{1/2}$$

In general, the nth monitor of A is given by:

$$m_n(A)(x) = \int_{\Omega} (A(x,\omega))^n\, dP(\omega) \quad , \quad n = 1, 2, \ldots$$

A simple example of a probabilistic set may arise when one is obtaining a membership function of a vague concept, say *about* 3. We may perform an experiment asking a group of people about this idea (some detailed hints concerning this technique may be found in Section 2.8.1). Of course, it is hardly to be expected that all the membership functions will have exactly the same value at all the elements of the universe of discourse. They have a similar shape, but some differences may be reported. An illustration of this idea is given in Fig. 2.7(a); here Ω consists of three elements ω_1, ω_2, ω_3. The functions shown are treated as realisations of the defining function describing this vague concept. Then, performing computations, we get the membership function and the vagueness function visualised in Figs. 2.7(b) and 2.7(c). It should be noted that the vagueness function representing the scattering of the opinions of those under test attains its highest value on the left- and right-hand sides of the element equal to 3. This is natural, since the highest ambiguity concerning the description of the concept 'about 3' is located at some value close to 3, whereas it completely vanishes at this point (since we are sure that the number 3 perfectly fits the vague idea discussed above).

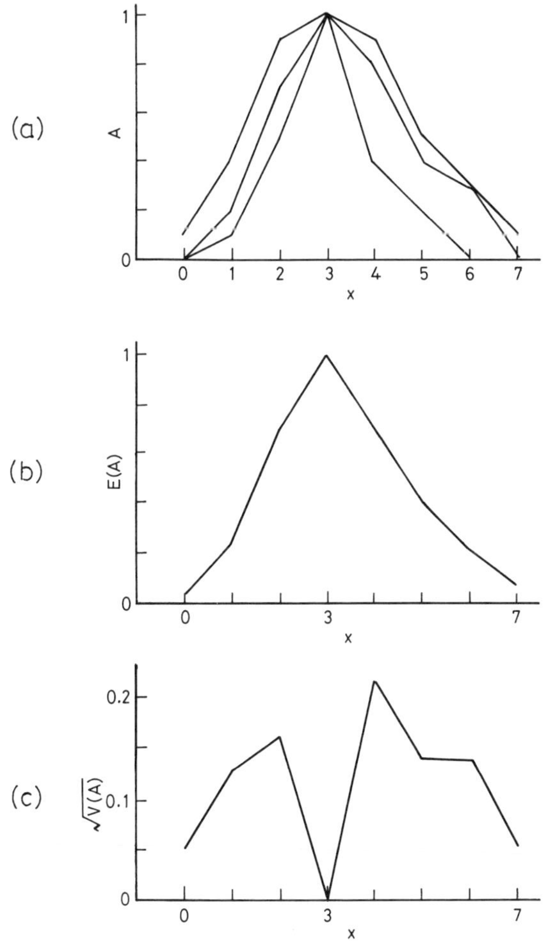

Fig. 2.7 Probabilistic set (a), membership function (b) and vagueness function (c)

Some useful properties of this approach should be mentioned here. First, the formalism of probabilistic sets is well developed (see [15], where all the mechanisms known for fuzzy sets (e.g. operations, mappings, fuzziness measures) are available). Thus, it does not cause any trouble while working in this area. Secondly, probabilistic sets form a remedy for the paradox of infinite precision noted in Section 2.6 when considering fuzzy sets of type 'm'. As was proved in [15], almost all the relevant information is conveyed by lower monitors and, as the order of the monitors increases, their values tend to zero. This gives us a useful result, focusing our attention mainly to the membership function and the vagueness function. It is shown in Section 2.8.1 that the estimation procedures for membership-function determination can easily be equipped with an analogue of the vagueness function.

It should be remembered, however, that, using this concept of probabilistic sets, some approaches may be dangerously misleading. The defining function $A(x,\omega)$, being a random variable, can also be expressed by its probability density function (or a distribution function). However, such a description, if not used in an appropriate way, is completely useless and misleading. For a finite **X** consisting of 'n' elements, the relevant distribution function conveying the entire information about the defining function should take the form:

$$F(x_1, x_2, \ldots, x_n, w_1, w_2, \ldots, w_n) =$$

$$= P\left\{\omega \mid A(x_1, \omega_1) \le w_1, A(x_2, \omega_2) \le w_2, \ldots, A(x_n, \omega_n) \le w_n\right\}$$

Sacrifice of the complexity of this notation and further computations by an assumption of the independence of each random variable $A(x_i, \omega_i)$ for different x_i has no meaning, destroying the idea of the membership function. Simply, it imposes no dependence between the elements of the space **X**. Clearly, no smooth transition from one value of the membership function to another, say x_i and x_{i+1}, is assured.

2.7.3 Fuzzy random variables

Another approach to combining fuzziness and randomness was established by Kwakernaak [19,20]. In this approach, the so-called fuzzy random variables are discussed. They are viewed as an extension of any random variable where the set of its values is viewed not as real numbers but as fuzzy numbers. More formally, we have:

A fuzzy random variable X, defined in a probability space $(\Omega, \mathbf{B}, P)$, is a mapping X:

$$X : \Omega \to S$$

where S is a space of all membership functions of fuzzy numbers.

Additionally, a set of mappings X is restricted in such a way as to fulfil certain measurable conditions usually required in the constructions of probability calculus. To have a feasible picture, we refer to a simple example. A group of people are asked about a possible temperature for the weekend. A few fuzzy responses are reported: e.g. *warm*, *very warm*, *no opinion* etc. Thus we can say that to each response (modelled by a fuzzy set) a certain probability is assigned. The results of the poll are summarised as follows:

$$p_1, X_1$$

$$p_2, X_2$$

$$\vdots$$

$$p_N, X_N$$

with $\sum p_i = 1$. Then, for given membership functions of $X_1, X_2, \ldots, X_N$, an extended list of operations can be worked out [19]. They refer to calculations of the expected value of the fuzzy random variable, fuzzy probabilities, fuzzy conditional expectation etc. The algorithms developed in [20] rely strongly on the representation theorem: the computations are performed taking α-cuts of the fuzzy sets. We reconsider one of the algorithms from [20] that enables us to calculate the fuzzy expected value of the fuzzy random variable. As before, we take a discrete fuzzy random variable given by a set of pairs (p_i, X_i) (assume all X_i are normal fuzzy sets). The procedure of computation of EX proceeds accordingly:

(i) Take any value $\alpha \in [0,1]$.

(ii) For each i = 1, 2, ..., N, determine the numbers

$$x_i{}^* = \inf\{x \in \mathbf{R} \mid X_i(x) \geq \alpha\}$$

$$x_i{}^{**} = \sup\{x \in \mathbf{R} \mid X_i(x) \geq \alpha\}$$

(iii) Determine the level set $EX_\alpha = \{x \in \mathbf{R} \mid EX(x) \geq \alpha\}$ as

$$EX_\alpha = \left[\sum_{i=1}^{N} p_i x_i{}^*, \sum_{i=1}^{N} p_i x_i{}^{**}\right]$$

(iv) Repeat (i) to (iii) for different values of α.

(v) Finally, determine the α-cut forming the support of EX by taking α equal to zero.

A certain assumption of this method is that, when a fraction of those under test give the same answer (namely, *warm*), we assume their response can be modelled by the same membership function. Of course, such an experiment does not allow us to estimate this membership function.

Some further studies concerning this approach can be found in [20], where it has been used in a certain class of decision-making processes.

2.8 Selected Methods of Determination of Membership Functions

2.8.1 General comments on the problem of membership determination

We have already mentioned that the problem of obtaining the values of the membership function, or at least their estimation, is of interest in the further application of fuzzy-set techniques. Following a stream of fuzzy sets as formulas describing vague notions, it is difficult to see how this problem could have a straightforward solution. First, fuzzy sets model a subjective category; therefore, their membership functions can be evaluated in a subjective fashion. We should also bear in mind that notions or categories modelled by fuzzy sets have a local character; that is to say, the meaning of a certain category relies on the context (situation) in which its application is planned. For instance, when talking about the concept *large steady-state error* in a certain community (e.g. in the control of a certain industrial process) and after establishing the relevant membership function, it is not possible to play with the same membership (calibrating) function in a completely different community; usually, at least some scaling will be required. From the measurement-theory point of view, it is not clear which type of scale should be used for the estimation of the membership function. Thole & Zimmermann [28], for instance, used an absolute scale; in Saaty's approach [24], a ratio scale is suggested; while Goguen [13] argues that no stronger scale than an ordinal one may be obtained. Leaving these questions open, and not moving into deep investigations of a psycholinguistic character, which may be found in [14], [29] and [37], we focus our attention on the discussion of some methods that may be used in engineering practice when the membership function has to be estimated. Later, we restrict ourselves in the main to discrete cases: i.e. we consider finite universes of discourse usually of significant practical interest.

We discuss two essential classes of method:

Horizontal approach. The underlying idea of this method is to gather information about grades of membership of selected elements of a universe of discourse in which a fuzzy set is to be defined. The process of elicitation of these membership functions can be concisely stated as follows: Consider a group of researchers (experts) involved in the same area of investigation. They are asked to answer a question having the format:

Can x_0 be viewed as compatible with the concept of the fuzzy set A?

where x_0 is a fixed element of the universe of discourse. The answer is 'yes' or 'no'. Then, counting the fraction of positive ('yes') responses $n(x_0)$ found in the total number of responses, we get the value of the membership function of this element of the universe of discourse x_0:

$$\mu_A(x_0) = \frac{n(x_0)}{N}$$

with N being the total number of responses that dealt with x_0.

Thus, the method is based on a straightforward counting of responses, and in its spirit reminds us of the procedure in the example cited by Borel (cf. Section 1.2). The evident advantage of this method lies in its simplicity. The experiment can be easily designed and new points of the universe of discourse added, if required. Furthermore, one can determine the standard deviation of the estimates obtained. This results from elementary statistical analysis by noting that the yes-no responses constitute realisations of a certain random variable of binomial distribution. The standard deviation of $A(x_0)$, st-dev(A)(x_0), is:

$$\text{st-dev}(A)(x_0) = \left[A(x_0)\frac{1-A(x_0)}{N}\right]^{1/2}$$

The derived values of the standard deviation of A can be utilised in a simple acceptance criterion regarding the grade of membership obtained:

- accept $A(x_0)$ as a sound estimate of the grade of membership if the value of the standard deviation (or the ratio of this deviation and the value of $A(x_0)$) does not exceed a threshold λ.

Usually, the result of this estimation is acceptable if the standard deviation takes on low values in comparison to $A(x_0)$ itself. The horizontal method is mentioned in [1] and [11]; however, no method of expressing its precision has been studied.

Vertical approach. The concept exploited in this method is to fix a certain level of membership α and ask a group of experts to identify a collection of elements in **X** satisfying the concept conveyed by A to a degree not less than α. Thus, a fuzzy set A is constructed by identifying its α-cuts. Again, the choice of levels of α is quite arbitrary and depends on the precision of membership-function determination one wants to achieve.

Comparing these two approaches, one finds that they are conceptually simple. The factor of uncertainty reflected by the fuzzy boundaries of A is distributed either vertically (grades of membership) or horizontally (limit points of α-cuts). The values of α or different elements of the universe of discourse should be selected randomly to avoid any potential systematic bias provided by the testees (experts). One should stress that both the estimation procedures reveal membership functions with single numerical values only in the ideal case. Usually, they give rise to ranges of grades of membership. Bearing this in mind, one can accept some of the previous generalisations of fuzzy sets as experimentally justifiable. Once the pointwise grades of membership have been determined, we can proceed with an obvious parametric fitting of standard membership functions (S or Π).

A quite evident shortcoming of the two methods resides in the 'local' nature of the experiments performed. This means that each grade of membership (at x_0 or at a certain level α) is estimated independently of the other findings of the experiment and the result does not reflect 'continuity' in the transition from full membership to absolute exclusion.

The pairwise comparison method tries to cope with these weaknesses.

2.8.2 Estimation of the membership function in a ratio scale — Saaty's pairwise comparison method

The membership function expressed in a ratio scale can be conveniently estimated by the pairwise comparison proposed by Saaty [24]. Let μ_i denote the degree to which the ith element of the universe of discourse fulfills the fuzzy notion A. Take, now, the ratios μ_i/μ_j for all i, j = 1, 2, ..., n and arrange them in the form of a square matrix $[\mu_i/\mu_j]$. Multiplying it by a vector $\mu = [\mu_1, \mu_2, \ldots, \mu_n]$, we get a system of homogeneous equations:

$$[\mu_i / \mu_j]\mu = n\mu$$

i.e.

$$([\mu_i / \mu_j] - nI)\mu = 0$$

where I denotes an $(n \times n)$ matrix.

From basic algebra, we notice that 'n' forms the largest eigenvalue of the above eigenvalue problem and μ is the corresponding eigenvector. This eigenvector will be accepted as a membership function of A.

Clearly, the entries of the above matrix are not known and have to be determined experimentally. The relevant experiment involves a series of pairwise comparisons of all the objects pertaining to the description of A. Let us first establish a scale; usually it consists of 7 ± 2 levels, say, for instance, 1, 2, ..., 7. Take a pair (i,j) of the objects and evaluate a preference of the ith one as satisfying the concept A, treating the second jth object as a point of reference. The more preference given to the first object (with respect to the second one), the higher the numerical level of preference associated with it. In the case of seven levels of quantisation, the strongest preference is reflected by associating the value 7 to this comparison. Then, automatically, the preference given to the jth object (in the same unchanged pairwise context) is quantified as 1/7. For less visible preference levels, the assigned values become lower. For the same object being evaluated with respect to itself (namely i = j), the degree of preference is neutral and is taken as 1.

All the results of the above pairwise comparison process are arranged in a matrix form. Let us denote this matrix by B. Observe that:

- the total number of comparisons required to fill all the entries of B is equal to $n^*(n-1)/2$.
- owing to the inevitable estimation errors, the transitivity property that holds for the previous matrix $[\mu_i/\mu_j]$ is not preserved any more. Usually one can expect that the approximate relationship

$$b_{ij} \approx \frac{\mu_i}{\mu_j}$$

holds, and therefore

$$b_{ik}\, b_{kj} \neq b_{ij}$$

Interestingly enough, this lack of transitivity, perceived as a sort of inconsistency in the available experimental data, can be well captured by the maximal eigenvalue of B, which in these situations becomes larger than 'n'. The more distant this eigenvalue from the dimension of the matrix, the more significant are the transitivity inconsistencies in the collected data.

There are some numerical schemes useful for the calculation of the membership function on the basis of the matrix A. One of them [4] replaces the eigenvalue problem by that involving the minimisation of a sum of squares:

$$\min_{\mu} \sum_{i=1}^{n} \sum_{j=1}^{n} \left(b_{ij}\,\mu_j - \mu_i\right)^2$$

subject to constraint:

$$\sum_{i=1}^{n} \mu_i = 1$$

Final renormalisation of μ (leading to the maximal value of μ equal to 1) produces a normal fuzzy set.

For more extensions of this method, as well as a thorough and interesting discussion of the foundations of the pairwise comparison method, the reader is referred to [21].

2.8.3 Membership-function estimation with the aid of probabilistic characteristics

The issue of the estimation of membership functions can also be addressed by constructing some bijective mappings between discrete probabilities and grades of membership (probability-possibility transformations). We consider the development of membership functions, making use of the probability-distribution function stemming from available experimental data.

Probability-possibility transformation

A bijective mapping introduced in [8] transforms a probability function into a corresponding membership function and vice versa. It shows how a probabilistic type of information can be effectively utilised in determining membership functions. For the discrete probabilities $p_1, p_2, \ldots, p_n, \sum p_i = 1$, arranged in descending order $p_1 \geq p_2 \geq \ldots \geq p_n$, the values of the membership function $\mu_1, \mu_2, \ldots, \mu_n$ are computed by the formula:

$$\mu_i = 1 - \sum_{j=1}^{i-1} \left(p_j - p_i\right) \quad , \quad i > 1$$

Setting $\mu_1 = 1$, or noting the normalisation condition, one also has:

$$\mu_i = i\,p_i + \sum_{j=i+1}^{n} p_j \qquad \text{(with the convention } \sum_{j=n+1}^{n} p_j = 0)$$

Noting assuming any arrangement of p_i s, one has:

$$\mu_i = \sum_{j=1}^{n} \min(p_i\,, p_j)$$

By inspection, the membership function and the corresponding probability function have the same shape. Therefore, $\mu_i = \mu_j$ implies $p_i = p_j$ and, from $\mu_i \geq \mu_j$, we know that $p_i \geq p_j$. Treating the above formula as an equation with respect to p_i for μ_i known, we get:

$$p_i = \sum_{j=1}^{n} \frac{1}{j}(\mu_j - \mu_{j+1}) \quad , \quad i = 1, 2, \ldots, n \text{ and } \mu_{n+1} = 0$$

In [17], a surjective possibility-probability transformation is proposed in this format:

$$p_i = k\,\mu_i^{\alpha}$$

where $\alpha > 0$. With additional convexity conditions, one can determine parameter 'k', and then

$$p_i = \frac{\mu_i^{\alpha}}{\sum_{j=1}^{n} \mu_j^{\alpha}}$$

In [5], a continuous case is investigated. Given a probability-density function (p.d.f.) defined in **R**, the corresponding membership function is determined in this optimisation problem:

$$\min_{A} \int A^2(x)\,dx$$

subject to these constraints:

(i) $E(A(x)\,|\,x$ distributed according to the p.d.f.$) \geq c$, where $E(\,.\,)$ stands for the expected value of the membership function of A while 'c' denotes a confidence level put close to 1; and
(ii) $0 \leq A(x) \leq 1$

The integral in the above formulation, which is minimised, visualises the fact that the membership function A obtained should be sufficiently 'sharp' (i.e. selective) (note that this integral is just an energy measure of fuzziness). Regarding the constraints: the second one is obvious, while the first one states that the elements that are most likely (in the sense of probability) should possess high values of membership — hence 'c' should be kept close to 1.

The optimal membership function resulting from this optimisation problem reads:

$$A(x) = \begin{cases} \lambda\, p(x) & , \text{ if } \lambda\, p(x) < 1 \\ 1 & , \text{ otherwise} \end{cases}$$

with λ resulting from this equation:

$$\lambda \int_{\lambda p(x)<1} p^2(x)\,dx + \int_{\lambda p(x)\geq 1} p(x)\,dx - c = 0$$

An interesting formalism of 'projectable' random sets (cf. [36]) leads to another constructive method of membership-function determination. When describing vague notions, it is evident that the most difficult situation is to express a grade of membership for intermediate (borderline) elements. The elements that definitely belong to the concept or are absolutely excluded from it introduce no difficulties. We consider regions between two such fuzzy notions (for example, modelling a concept of *high* and *medium*) where uncertainty in the classification of the objects in these categories is essential. Suppose it can be characterised by the probability-density function p(x). This function can be estimated, for instance, by using a histogram of 'don't know' responses distributed along the universe of discourse. It is rational to assume that p(x) takes on nonzero values in a certain closed interval of this universe. Note that $p(x) = 0$ for $x \in [a_1,a_2]$ and zero otherwise. This interval covers the region where a significant hesitation arises about assigning elements to A or B. Then, the membership functions of A and B, say *medium* and *high*, are constructed accordingly (Fig. 2.8):

$$A(x) = \begin{cases} 1 & , \text{ if } x < a_1 \\ \int_x^{a_2} p(u)\,du & , \text{ if } a_1 \leq x \leq a_2 \\ 0 & , \text{ otherwise} \end{cases}$$

and

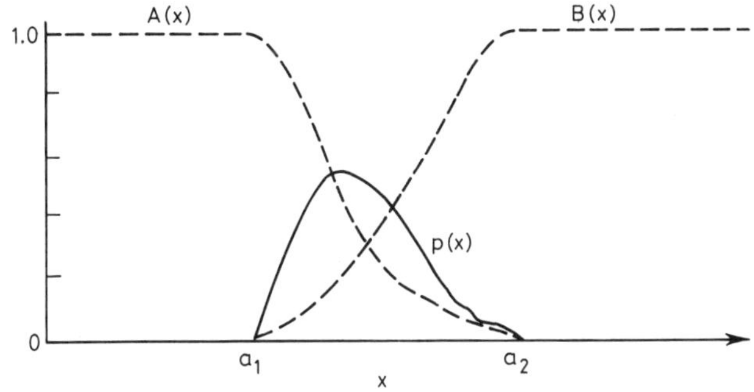

Fig. 2.8 Membership-function construction with the aid of probability-density function

$$B(x) = \begin{cases} 0 & , \text{ if } x < a_1 \\ \int_{a_1}^{x} p(u)\,du & , \text{ if } a_1 \le x \le a_2 \\ 1 & , \text{ otherwise} \end{cases}$$

Some relations of the problem of membership-function estimation to the psychometric scaling techniques discussed by Thurstone [29] are examined in [37].

The algorithm proposed below builds membership functions in order to capture and represent available numerical data. We assume that they are described by a probability-density function (p.d.f.) p(x). To clarify the basic idea, we consider two fuzzy sets A and B. Furthermore, we assume that their maximal values of membership have already been located and our interest is to optimise their spreads. Fig. 2.9 illustrates both the membership functions of A and B and the underlying p.d.f. We maintain an overlap between A and B to underline that the categories represented by A and B have some conceptual elements in common (Chapter 6 will prove that this property becomes necessary for completeness of the knowledge representation expressed by fuzzy sets). We require that this overlap be maintained at a certain level, say ξ. The point of intersection is denoted by x_0, $A(x_0) = B(x_0)$. Each of the membership functions has some parameters that help us to control the distribution of the grades of membership. The two expected values of the grades of membership established with respect to the probability of the data p(x) are:

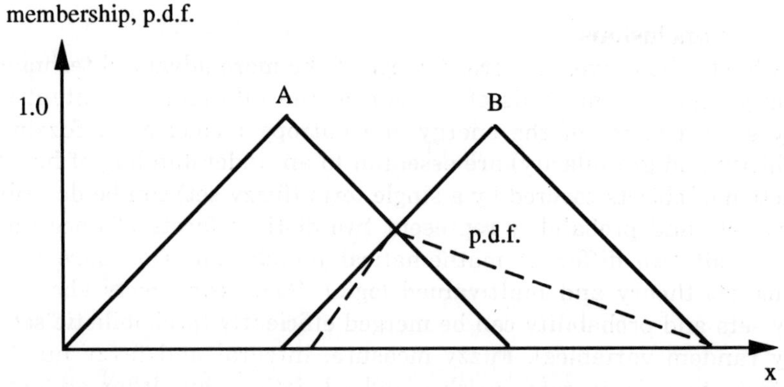

Fig. 2.9 Membership functions A and B with the underlying probability-density function

$$\mu = \int_{-\infty}^{x_0} A(x)\,p(x)\,dx \quad , \quad \nu = \int_{x_0}^{\infty} B(x)\,p(x)\,dx$$

The probability

$$\int_{-\infty}^{A^{-1}(\mu)} p(x)\,dx + \int_{B^{-1}(\nu)}^{\infty} p(x)\,dx$$

characterises the overall explanatory capabilities of the fuzzy sets A and B provided for the data (Fig. 2.9) (we assume that A^{-1} and B^{-1} do exist).

The residual probability is:

$$r = \int_{\rho_1}^{\rho_2} p(x)\,dx$$

where $\rho_1 = \min(B^{-1}(\nu), A^{-1}(\nu))$ and $\rho_2 = \max(B^{-1}(\nu), A^{-1}(\nu))$. This integral characterises all the data that have not been properly captured (explained) by either A or B. If the probability of the data in this region is significant, we should update A and/or B by changing their parameters so that the value of r can be minimised. If the values of r cannot be made negligible (the overlap of A and B should be kept unchanged), we should add an additional fuzzy set C between A and B and repeat the entire procedure for A and C as well as C and B respectively.

2.9 Conclusions

The Chapter has covered a broad range of the more advanced techniques of fuzzy sets. The methods of measuring the information content of a fuzzy set (in terms of the energy and entropy measures of fuzziness, specificity and granularity) are essential to an understanding of how the collection of objects covered by a single term (fuzzy set) can be described. Fuzzy sets and probability represent two distinct facets of uncertainty and exploit two different mathematical foundations (measure theory versus set theory and multivalued logic). Some constructs show that fuzzy sets and probability can be merged efficiently (probabilistic sets or fuzzy random variables). Fuzzy measure, integral and fuzzy numbers illustrate how fuzzy sets enrich standard notions by either adding an extra dimension to them (as happens with fuzzy numbers) or enhancing them in a conceptual way (as occurs with fuzzy measures).

2.10 References

[1] Bharathi Devi, B., & Sarma, V.V.S. 1985. 'Estimation of fuzzy memberships from histograms'. *Inf. Sci.*, **35**, pp. 43–59

[2] Birkhoff, G. 1948. 'Lattice Theory'. American Math. Soc. Colloq. Publ. (25), New York

[3] Choquet, G. 1953. 'Theory of capacities'. *Ann. Inst. Fourier, Univ. Grenoble*, **5**, pp. 131–296

[4] Chu, A.T.W., Kalaba, R.E., & Spingarn, K. 1979. 'A comparison of two methods for determining the weights of belonging to fuzzy sets'. *J. Optimiz. Theory & Appl.*, **4**, pp. 531–539

[5] Civanlar, M.R., & Trussel, H.J. 1986. 'Constructing membership functions using statistical data'. *Fuzzy Sets & Syst.*, **18**, pp. 1–13

[6] De Luca, A., & Termini, S. 1972. 'A definition of a non-probabilistic entropy in the setting of fuzzy sets theory'. *Inf. & Contr.*, **20**, pp. 301–312

[7] Dubois, D., & Prade, H. 1981. 'Fuzzy Sets and Systems: Theory and Applications'. Academic Press, New York

[8] Dubois, D., & Prade, H. 1983. 'Unfair coins and necessity measures: toward a possibilistic interpretation of histograms'. *Fuzzy Sets & Syst.*, **10**, pp. 15–20

[9] Edmonds, E.A. 1980. 'Lattice fuzzy logics'. *Int. J. Man-Mach. Stud.*, **13**, pp. 455–465

[10] Fuchs, L. 1963. 'Partially Ordered Algebraic Systems'. Pergamon Press, Oxford

[11] Godal, R.C., & Goodman, T.J. 1980. 'Fuzzy sets and Borel'. *IEEE Trans. Syst., Man & Cybern.*, **10**, p. 637

[12] Goguen, J.A. 1967. 'L-fuzzy sets'. *J. Math. Anal. & Appl.*, **18**, pp. 145–174

[13] Goguen, J.A. 1969. 'The logic of inexact concepts'. *Synthèse*, **19**, pp. 325–375

[14] Hersh, H.M., & Caramazza, A. 1976. 'A fuzzy-set approach to modifiers and vagueness in natural languages'. *J. Exp. Psychol. Gen.*, **105**, pp. 254–276

[15] Hirota, K. 1979. 'Extended fuzzy expression of probabilistic sets'. *In* 'Advances in Fuzzy Set Theory and Applications' (Eds. M.M. Gupta, R.K. Ragade & R.R. Yager). North Holland, Amsterdam, pp. 201–214

[16] Hirota, K. 1981. 'Concepts of probabilistic sets'. *Fuzzy Sets & Syst.*, **5**, pp. 31–46

[17] Klir, G.J. 1990. 'A principle of uncertainty and information invariance'. *Int. J. Gen. Syst.*, **17**, pp. 249–275

[18] Lakoff, G. 1973. 'Hedges: a study in meaning criteria and the logic of fuzzy concepts'. *J. Phil. Logic*, **2**, pp. 458–508

[19] Kwakernaak, H. 1978. 'Fuzzy random variables: I — Definitions and theorems'. *Inf. Sci.*, **15**, pp. 1–29.

[20] Kwakernaak, H. 1979. 'Fuzzy random variables: II — Algorithms and examples in the discrete case'. *Inf. Sci.*, **17**, pp. 253–278

[21] Lootsma, F.A., Boonekamp, P.G.M., Cooke, R.M., & van Oostroom, F. 1989. 'Choice of a long-term strategy for the national electricity supply via scenario analysis and multi-criteria analysis'. Rep. 89–64, Delft University of Technology, Delft, The Netherlands

[22] Martin-Clouaire, R. 1987. 'Semantics and computation of the generalized modus ponens: Efficient deduction in fuzzy logic'. *In* 'Uncertainty in Knowledge-Based Systems' (Eds. B. Bouchon & R.R. Yager). Springer-Verlag, Berlin, pp. 123–136

[23] Mizumoto, M., & Tanaka, K. 1976. 'Some properties of fuzzy sets of type 2'. *Inf. & Contr.*, **31**, pp. 312–340

[24] Saaty, T.L. 1980. 'The Analytic Hierarchy Processes'. McGraw-Hill, New York

[25] Sambuc, R. 1975. 'Fonctions d'Φ-flous. Application à l'aide au diagnostic en pathologie thyroidienne'. PhD Thesis, Marseille

[26] Sugeno, M. 1972. 'Theory of fuzzy integral and its applications'. PhD Thesis, Tokyo Institute of Technology, Tokyo

[27] Sugeno, M. 1977. 'Fuzzy measures and fuzzy integrals: a survey'. *In* 'Fuzzy Automata and Decision Processes' (Eds. M.M. Gupta, G.N. Saridis & B.R. Gaines). North Holland, Amsterdam, pp. 89–102

[28] Thole, U., Zimmermann, H.J., & Zysno, P. 1979. 'On the suitability of minimum and product operators for the intersection of fuzzy sets'. *Fuzzy Sets & Syst.*, **2**, pp. 167–180

[29] Thurstone, L.L. 1927. 'A law of comparative judgement'. *Psychol. Rev.*, **34**, pp. 273–286

[30] Yager, R.R. 1979. 'On the measure of fuzziness and negation. Part I — Membership in the unit interval'. Rep. RRY 79–016, School of Business Administration, New Rochelle, New York

[31] Yager, R.R. 1981. 'Measurement properties of fuzzy sets and possibility distributions'. Proc. 3rd Int. Seminar on Fuzzy Set Theory (Ed. E.P. Klement). Linz, Austria, pp. 211–222

[32] Zadeh, L.A. 1965. 'Fuzzy sets and systems'. Proc. Symp. on System Theory, Polytech. Inst. Brooklyn, pp. 29–37

[33] Zadeh, L.A. 1968. Probability measures of fuzzy events'. *J. Math. Anal. & Appl.*, **23**, pp. 421–427

[34] Zadeh, L.A. 1972. 'A fuzzy-set-theoretic interpretation of linguistic hedges'. *J. Cybern.*, **2**, pp. 4–34

[35] Zadeh, L.A. 1975. 'The concept of a linguistic variable and its application to approximate reasoning'. *Inf. Sci.*, **8**, pp. 199–249

[36] Pei-Zhuang, W., & Sanchez, E. 1982. 'Treating a fuzzy subset as a projectable random subset'. *In* 'Fuzzy Information and Decision Processes' (Eds. M.M. Gupta & E. Sanchez). North Holland, Amsterdam, pp. 213–219

[37] Zwick, R. 1987. 'A note on random sets and the Thurstonian scaling methods'. *Fuzzy Sets & Syst.*, **21**, pp. 351–356

CHAPTER 3
Fuzzy Sets in the Development of the Cognitive Perspective

3.1 Introductory Remarks

In this Chapter, we introduce and examine the notion of the cognitive perspective realised by fuzzy sets. This notion, as explained later, plays a key rôle in the application of the methodology of fuzzy sets to problems of system modelling, control, and pattern recognition, to name only a few.

We recall that the general methodological intent is to emulate a human-like way of dealing with a variety of control and recognition problems. When a human being is solving a certain complex problem, he first tries to structure the knowledge about it in terms of some general concepts and afterwards to reveal essential relationships between them. This sort of top-down approach allows him to convert these quite general and imprecise relationships into more detailed operational algorithms. Our perspective of the problem does not allow us to discriminate (differentiate) between single numbers (numerical quantities), but leads us to aggregate some objects (e.g. numerical values) into more general categories like sets or fuzzy sets. Psychological findings show that a human being is capable of memorising and actively utilising up to 7 ± 2 items (concepts) no matter how complex they are. In fact, the structure of these concepts is important for their efficient memorisation, appropriate recall, and further utilisation. This Chapter is structured as follows: First, we introduce and study the cognitive perspective (also called the frame of cognition) and its main features (with particular emphasis on its specificity and robustness). Then we consider the problem of translating (coding) any type of information within this frame of cognition.

3.2 Frame of Cognition

We recall (Chapter 1) that fuzzy sets used in describing linguistic labels act as elastic constraints over a given universe of discourse and thus identify some of its regions as compatible to the highest degree with these constraints. Sometimes, the linguistic labels are referred to as information granules [5]. The information granules defined for a certain variable constitute a frame of cognition of this variable [3,4]. More formally, the family of fuzzy sets

$$\{A_1, A_2, \ldots, A_c\} \tag{3.1}$$

constitutes a frame of cognition *A* if these two properties are fulfilled:

• *A* 'covers' the universe **X**; i.e. each element of this universe is assigned to at least one granule with a nonzero degree of membership. Thus:

$$\forall_{x} \exists_{i} A_i(x) > \varepsilon$$

where $\varepsilon > 0$ denotes the level of 'coverage' of **X**. As shown later, this property ensures that any piece of information defined in **X** is then properly represented in terms of the generic linguistic labels (A_i s). The property of 'coverage' is illustrated in Fig. 3.1.

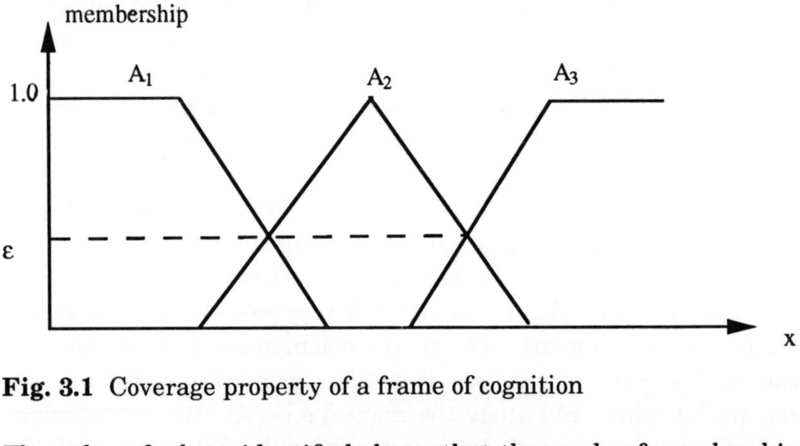

Fig. 3.1 Coverage property of a frame of cognition

The value of ε here identified shows that the grade of membership of any point is not lower than ε across the entire universe of discourse.

• the elements of *A* are unimodal fuzzy sets. In stating this, we identify several regions of **X** (one for each A_i) that are highly compatible with the labels (i.e. with significantly high grades of membership in A_i). The regions defined in this way possess a clear semantic meaning. Sometimes, the frame of cognition is referred to as a fuzzy partition. A fuzzy partition usually satisfies an additional property stating that all the grades of membership of A_i s sum up to 1 at any point of **X**:

$$A_1(x) + A_2(x) + \ldots + A_c(x) = 1 \qquad \text{for each } x \in \mathbf{X}$$

This constraint is automatically satisfied by the so-called partition matrices generated by the clustering algorithms. There is no guarantee that the same calibration condition will hold for the above frame. Despite their technical differences, we use the terms 'fuzzy partition' and 'frame of cognition' interchangeably.

The so-called Boolean partition induced by the fuzzy partition (or the frame of cognition) is constructed as a collection of sets with boundaries formed by the points of intersection of successive A_i s (Fig. 3.2).

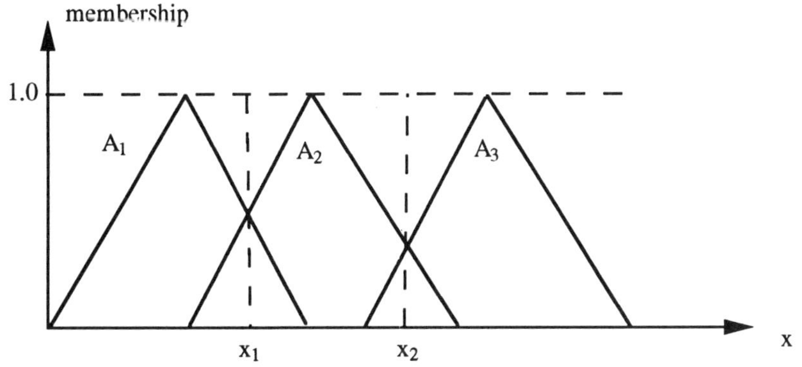

Fig. 3.2 Fuzzy partition and its induced Boolean partition

3.3 Properties of the Frame of Cognition

Considering the linguistic labels standing in (3.1), there are three essential features characterising the frame with respect both to its elements and to the relationships between them:

Specificity

Generally speaking, the frame of cognition A' is more specific than A if all the elements of A' are more specific (e.g. in terms of their specificity measure) than the elements of A. Then, the number of elements of A' is greater than the number of labels in A.

For instance, the frame:

$$A = \{Negative, Zero, Positive\}$$

is less specific than the frame A' containing seven items (fuzzy sets):

$A' = \{$*Negative Large, Negative Medium, Negative Small,*

Zero,

Positive Small, Positive Medium , Positive Large$\}$

The variable defined by A' takes on more levels of this linguistic quantisation. The partition A' is less general than the previous one. The information granularity of A' is finer than that of A (or the information in A is coarser than in the components of A' [5]). As can be seen (Fig. 3.3), the frame with the higher level of specificity contains more fuzzy sets.

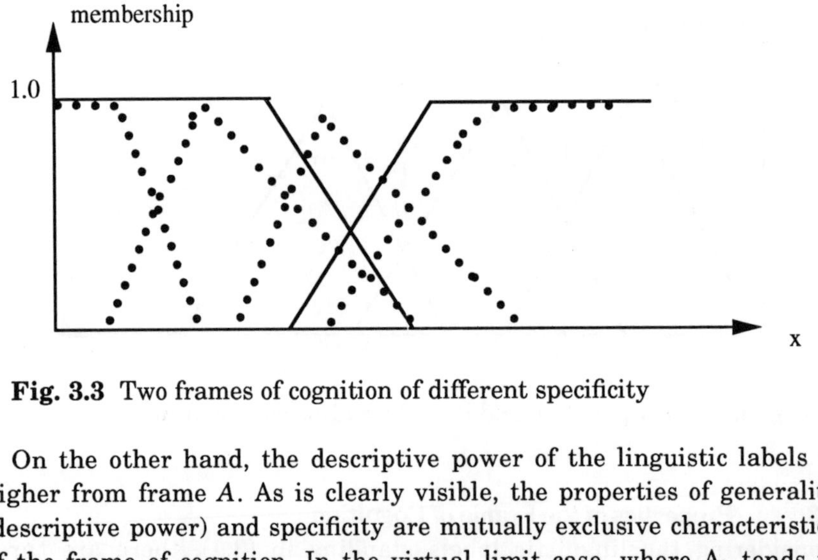

Fig. 3.3 Two frames of cognition of different specificity

On the other hand, the descriptive power of the linguistic labels is higher from frame A. As is clearly visible, the properties of generality (descriptive power) and specificity are mutually exclusive characteristics of the frame of cognition. In the virtual limit case, where A_i tends to shrink to single numerical values, the specificity of A_i increases but its descriptive capabilities decay to zero. Regarding the membership functions of A_i s (assuming that all A_i s are defined in $\mathbf{R}$), we can observe that their spread values are exclusively responsible for the specificity of the elements of their frame of cognition.

Scope of perception

The scope of perception pertains to each linguistic granule. It identifies elements in $\mathbf{X}$ with the highest grades of membership and puts them together. On doing this, our attention is focused on some selected regions of the universe of discourse rather than on single and isolated numerical values. The single numerical elements contributing in this way become nondistinguishable. We can also think of the scope of perception as realising a process of information hiding. For instance, a fuzzy set A_1 with a membership function:

$$A_1(x) = \begin{cases} \text{exponentially increasing over } (-\infty, x_1) \\ 1 \text{ for x in } [x_1, x_2] \\ \text{exponentially decreasing over } (x_2, \infty) \end{cases}$$

$$A_1(-\infty) = 0 \quad , \quad A_1(x_1) = 1 \quad , \quad A_1(x_2) = 1 \quad , \quad A_1(\infty) = 0$$

makes all the elements in $[x_1,x_2]$ equivalent. By defining this membership function of A_1, we selectively hide the information about the elements contained in this interval. In other words, there is no distinction (at the level of the specificity provided by A_1) between a_1 and a_2 as long as both of them are situated in the given interval. The process of information hiding is performed so that all the computational procedures following it can be completed at this predefined conceptual level. This can lead to significant savings in computation.

The concept of information hiding is an inherent property of set theory. Fuzzy sets allow us to add an extra flexibility to the notion by parameterising it by using values of the grades of membership. The resulting λ-cuts are viewed as sets completing the idea of information hiding at several distinct levels. In particular, for the above trapezoidal fuzzy number used as a part of the frame of cognition, the λ-cuts with $\lambda = 1.0$ imply hiding at the highest level.

For the fixed cognitive perspective, information hiding can be accomplished by enhancing the regions of **X** associated with the higher grades of membership. For instance, the operation of contrast intensification (see Chapter 1) amplifies 'high' values of membership (above 0.5) and suppresses those that have already been viewed as insignificant ones.

Information hiding is of a much more general nature. It plays a significant role in software engineering, although the term coined there has a slightly different technical flavour. The higher generality implies a more evident level of information hiding.

Referring to some parametric descriptions of membership functions, we can state that the scope of perception is tied to their modes (modal values). It is also partially implied by the values of the spread values of the membership functions.

Robustness

Fuzzy sets constituting a frame *A* exhibit an interesting and useful property of robustness. Briefly, owing to the smooth transitions in the grades of membership of fuzzy sets, they make it possible to tolerate imprecision in the input information. This tolerance, as will be quantitatively described, is quite significant and surpasses the robustness of the induced Boolean partition. A simple numerical experiment summarises the essence of this feature. We consider the input numerical datum $x \in \mathbf{R}$ that, because of existing noises, is received as x′ and as such becomes processed within this frame. Essentially, this processing realises a mapping within which the levels of activation of A_1, A_2, ..., A_c are formed. For numerical entries x and x′, this process is evident: one invokes A_is to the degree equal to $A_i(x)$. The version corrupted by noise (x′) yields results given as $A_i(x')$. Thus, in fact, the noisy version of x induces:

$$A_1(x'), A_2(x'), \ldots, A_c(x')$$

instead of the original one:

$$A_1(x), A_2(x), \ldots, A_c(x)$$

The lower the difference between $A_i(x)$ and $A_i(x')$, the higher the robustness of the frame with respect to the input disturbances. The overall sum of the absolute differences

$$r(x) = |A_1(x) - A_1(x')| + |A_2(x) - A_2(x')| + \ldots + |A_c(x) - A_c(x')|$$

is used as an indicator of the robustness property of frame *A*. The measure of robustness is an evident function of x. The values of r obtained may vary from point to point. To derive a global robustness performance, we integrate r(x) over the entire universe of discourse:

$$r = \int r(x)\,dx$$

(we assume that the above integral does make sense).

In sets used in the formation of the cognitive perspective, there is a striking regularity, i.e. r(x) achieves zero values if both the corresponding values of $A_i(x)$ and $A_i(x')$ are equal to 1 or 0 (this happens if x and x′ are quite distant from the border points of A_i). If this is not the case, then, even for two close values of x and x′, the corresponding grades of membership may be different. This means that the values of r(x) are profoundly high at the borders of the elements of the induced Boolean partition.

A detailed numerical illustration shows this more clearly.

Example 3.1

We study the frame composed of the three linguistic labels (say, *Small*, *Medium*, and *Large*). The universe of discourse is an interval [0,5]. The membership functions are Gaussian-like and are fully described by their modal values m_i and spreads δ_i, i = 1, 2, 3, say:

$$A_i(x) = \exp\left[-\frac{(x-m_i)^2}{\delta_i}\right]$$

(Fig. 3.4).

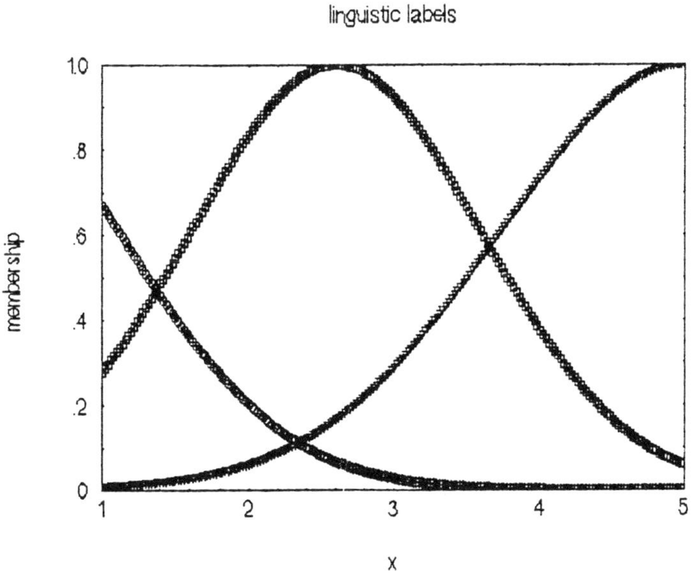

Fig. 3.4 Frame consisting of three Gaussian-like membership functions

The noise affecting the input information x is modelled as an additive random Gaussian variable with a zero mean value and standard deviation σ, namely N(0,σ). The frame is exposed to inputs x as well as their noisy versions x′. The results in terms of r(x) are plotted in Figs. 3.5(a), (b) and (c).

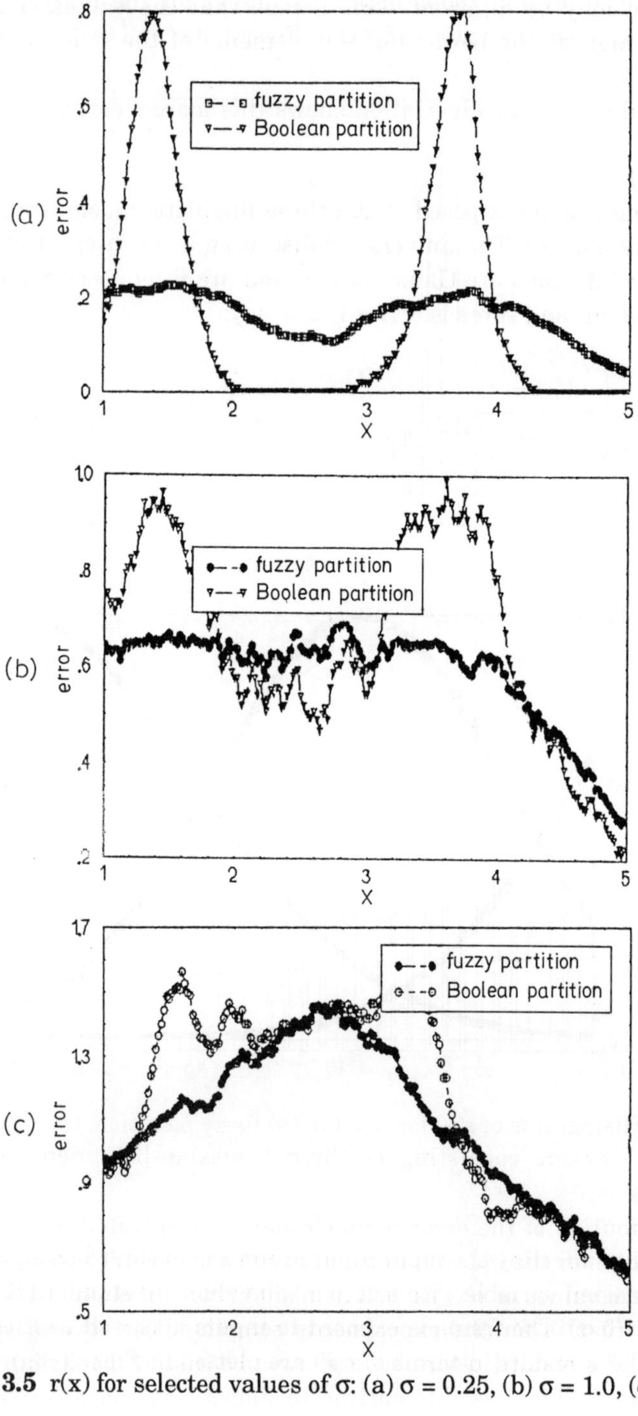

Fig. 3.5 r(x) for selected values of σ: (a) σ = 0.25, (b) σ = 1.0, (c) σ = 3.0

The averaged level of robustness 'r' is higher for the fuzzy partition; see the histogram of the values of r(x) in Figs. 3.6(a) and (b).

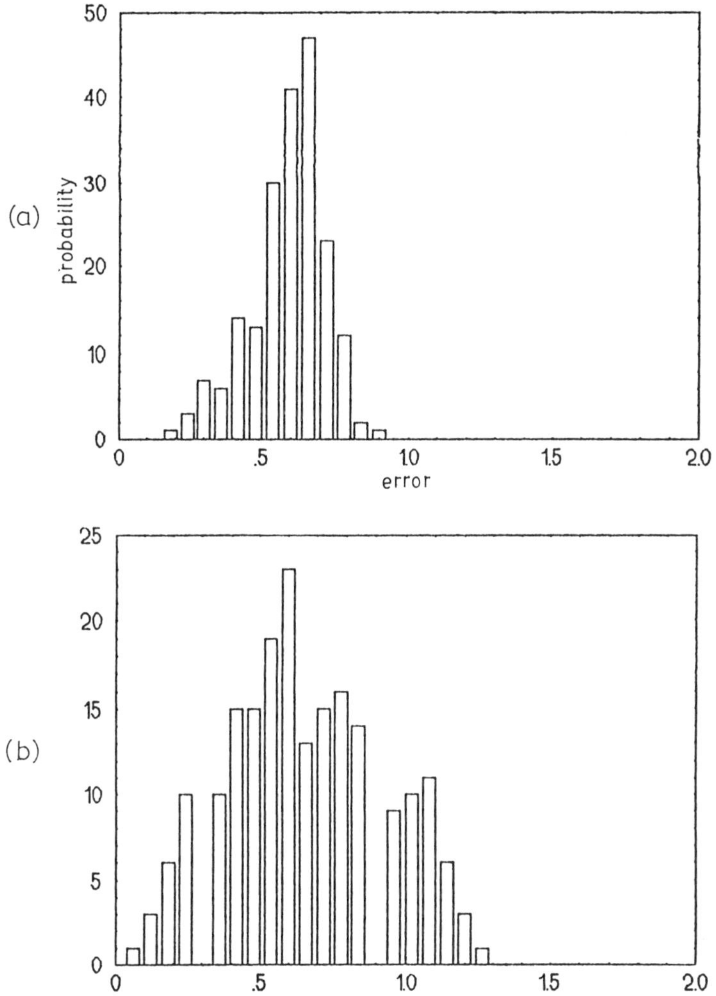

Fig. 3.6 Histograms of r(x) for $\sigma = 1.0$: (a) fuzzy partition, (b) Boolean partition

The distribution of the error is much more concentrated for the fuzzy partition, while for the induced Boolean partition we witness a significant portion of zero values of r(x) as well as a significant fraction of their high values. The results for the experiments completed for several selected values for the standard deviation σ show a consistent behaviour of the fuzzy partition, which, as summarised in Fig. 3.7, is characterised by a higher robustness than its Boolean counterpart.

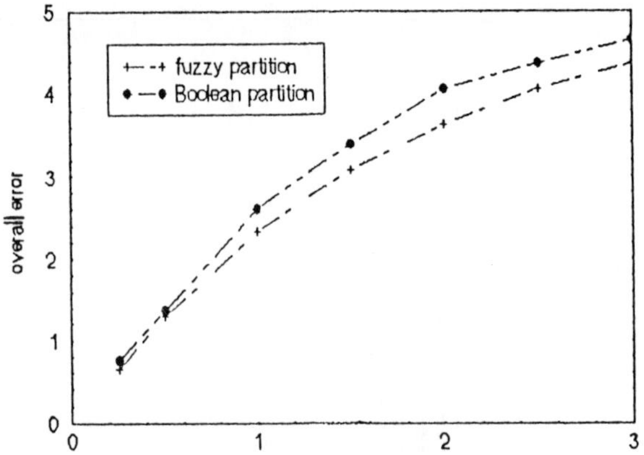

Fig. 3.7 Robustness of the fuzzy partition and its Boolean counterpart versus σ

We also notice that the term of robustness of the frame of cognition has a direct impact on the robustness of any algorithm for which the fuzzy partition plays the role of an interface. Schematically, we put this down thus:

input information — processing algorithm
input error — processing error

Then, any change of the input error, if not absorbed by the fuzzy partition, may have a meaningful effect on the processing error. Thus, the change of the former has a stabilising influence on the obtained level of the error associated with the processing stage.

3.4 Processing Information in the Frame of Cognition

The fuzzy partition plays a primordial role in processing input information that is afterwards utilised by algorithms of identification or control. In addition to the level of robustness achieved, the central issue to be addressed there pertains to matching the input information and the linguistic labels of the frame. The methods of matching are numerous and diverse, in terms of both their methodology and some specialised algorithms; see, for instance, [2,7].

Here, we stress the general point that the input information can be heterogeneous (e.g. can include pointwise numerical values, intervals or fuzzy sets), while the output of the matching phase is always a vector returning the degrees of 'activation' of the individual elements of the frame of cognition, say:

input information — fuzzy partition — results of matching

The results of matching refer exclusively to the linguistic labels A_1, A_2, ..., A_c, respectively.

We study three cases of the input information arranged in an increasing order of its generality. Furthermore, we will restrict ourselves to a single element (label) of A, say A. An additional assumption states that A is defined in **R** (which in many applications holds by default).

The cases are:

(i) a pointwise input information x_0. The linguistic label A becomes activated to the degree $A(x_0)$. In other words, the truth valuc of the predicate $Match(A \mid x_0)$ equals $A(x_0)$. This number is just the value of the possibility measure of x_0 computed with respect to A, cf. [6].

(ii) a collection of input pointwise data $\{x_1, x_2, \ldots, x_n\}$ that are to be processed simultaneously, yielding a single truth value of the predicate $Match(A \mid \{x_1, x_2, \ldots, x_n\})$. We can consider it as resulting from this optimisation problem:

- determine $\lambda \in [0,1]$ such that the expression

$$\sum_{i=1}^{n} [\lambda = A(x_i)]$$

achieves a maximum. The above sum involves the equality index computed for λ and $A(x_i)$, i = 1, 2, ..., n (Chapter 2). The maximum in this expression can be determined iteratively by studying a standard scheme of updating consecutive values of λ, say:

$$\lambda(iter+1) = \lambda(iter) - \xi \frac{\partial}{\partial \lambda} \left\{ \sum_{i=1}^{n} (\lambda(iter) \bullet A(x_i)) \right\} \qquad (3.2)$$

where $\lambda(iter)$ denotes the value of λ obtained in successive iterations, while ξ controls the rate of the changes of λ. The changes themselves are determined through the derivative of the sum computed with respect to λ. This update scheme has an interesting and transparent interpretation for the equality index defined with the aid of the Lukasiewicz implication. Here, we derive:

$$\frac{\partial}{\partial\lambda}(\lambda \equiv A(x_i)) = \begin{cases} +1 \ , & \text{if } \lambda < A(x_i) \\ 0 \ , & \text{if } \lambda = A(x_i) \\ -1 \ , & \text{if } \lambda > A(x_i) \end{cases}$$

Denote by n_1 the number of elements x_i in the above sum for which the condition $\lambda > A(x_i)$ holds. By n_2, we denote the number of elements where the converse inequality holds ($\lambda < A(x_i)$). Then, the scheme reads:

$$\lambda(\text{iter}+1) = \lambda(\text{iter}) - \xi(n_2 - n_1)$$

If, additionally, we set ξ as 1/n, the formula reduces to

$$\lambda(\text{iter}+1) = \lambda(\text{iter}) - \frac{n_2 - n_1}{n}$$

so all the changes in λ are driven by the number of cases reporting the values of $A(x_i)$ greater or lower than the current value of the level of matching .

Finally, we consider an interval-valued variable distributed over a closed interval [a,b]. We are looking for the results of matching obtained with respect to A. The problem is formulated in a similar way by considering the following optimisation (maximisation) problem studied with regard to λ:

$$\int_a^b (\lambda \equiv A(x))\,dx$$

In fact, this constitutes an integral form of the optimisation task formulated in the previous case. The corresponding updates are completed via:

$$\lambda(\text{iter}+1) = \lambda(\text{iter}) - \xi \frac{\partial}{\partial\lambda}\left(\int_a^b (\lambda(\text{iter}) \equiv A(x))\,dx \right) \tag{3.3}$$

With the same equality index (based on the Lukasiewicz implication), some detailed formulas can be obtained. We study two cases arising:

• a single element x_0 satisfying condition $\lambda = A(x_0)$ is contained in the interval [a,b] (Fig. 3.8(a)).

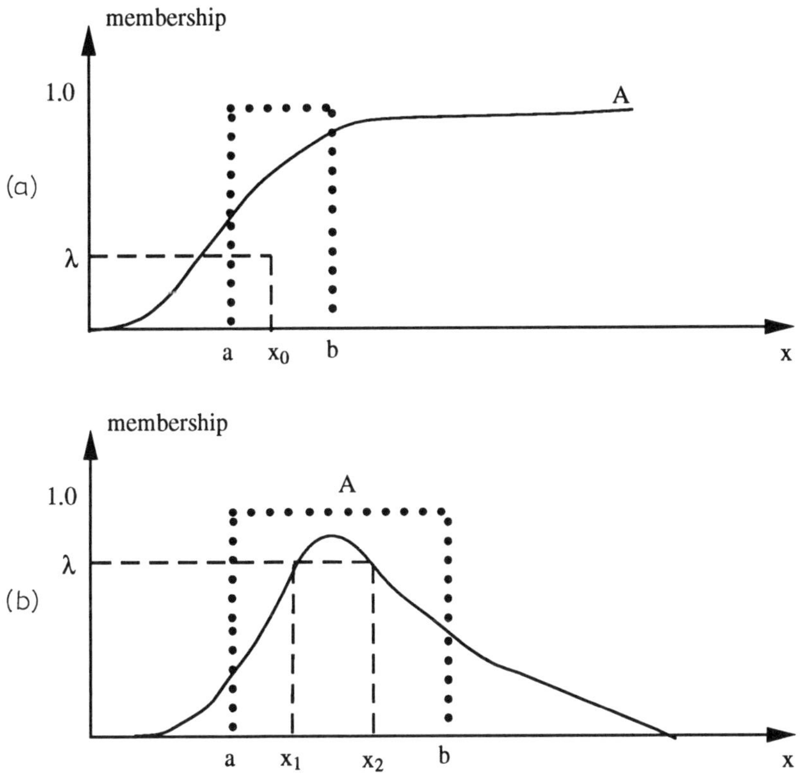

Fig. 3.8 Membership function A(x) and an interval-valued input information: (a) a single point of intersection x_0, (b) two points of intersection x_1 and x_2

One then writes down (3.3) accordingly:

$$\int_a^b (\lambda = A(x))\,dx = \int_a^{x_0} (\lambda = A(x))\,dx + \int_{x_0}^b (\lambda = A(x))\,dx$$

Then, the derivative taken with respect to λ is computed as:

$$\int_{a}^{x_0} (+1)\, dx + \int_{x_0}^{b} (-1)\, dx$$

The zero value of this derivative yields:

$$(x_0 - a) + (b - x_0) = 0 \quad ,$$

i.e.

$$x_0 = \frac{a+b}{2}$$

The result is convincing: x_0 becomes an average value of the ends of the discussed interval [a,b], and this holds despite the form of the membership function (assuming that it satisfies the previous conditions).

The second case includes two values x_1 and x_2 at which the equality index λ coincides with the membership functions, $\lambda = A(x_1)$ and $\lambda = A(x_2)$ (Fig. 3.8(b)).

Simple computations reveal that the derivative equals zero if

$$x_2 - x_1 = \frac{b-a}{2}$$

3.5 Concluding Remarks

The term of the frame of cognition has been developed. Its main properties have been introduced, analysed and illustrated. This notion is central to many applications of fuzzy sets, showing how they can contribute towards an efficient formation of mathematical models incorporating a vast variety of available data.

It is worth stressing that fuzzy sets narrow an existing gap between purely numerical techniques and well known methods of Artificial Intelligence (AI). For an interesting approach on this line, the reader is referred to [1]. In the field of AI, all the information-processing procedures are symbolic — the symbols are manipulated via a collection of specific syntax rules. At the same time, the numerical information is essentially ignored. In the numerical techniques, all the objects are plain numbers (or some of their structures are vectors or arrays). They strive for precision, while their knowledge-representation capabilities are nonexistent.

The AI schemes of knowledge representation are powerful and diversified; however, they do not cope with any numerical information. Fuzzy sets are placed in between. As collections of objects, they are described by symbols (like, for example, *small* or *large*). Simultaneously, these symbols have a certain semantics attached to them that is conveyed by numerical characteristics described by numerical grades of membership. The level of precision as contrasted to generality can easily be modified by changing the number of the linguistic labels and modifying their parameters. As will become obvious later during our studies of fuzzy models, this enhances an ability to implement principles of incompatibility and efficiently express the tradeoffs existing between achievable levels of precision and relevancy.

3.6 References

[1] D'Ambrosio, B. 1989. 'Qualitative Theory Using Linguistic Variables'. Springer-Verlag, Berlin

[2] Dubois, D., & Prade, H. 1988. 'Possibility Theory — An Approach to Computerized Processing of Uncertainty'. Plenum Press, New York

[3] Pedrycz, W. 1990. 'Fuzzy sets framework for development of perception perspective'. *Fuzzy Sets & Syst.*, **37**, pp. 123–137

[4] Pedrycz, W. 1992. 'Selected issues of frame of knowledge representation realized by means of linguistic labels'. *Int. J. Intelligent Syst.*, **7**, pp. 155–170

[5] Zadeh, L.A. 1979. 'Fuzzy sets and information granularity'. *In* 'Advances in Fuzzy Set Theory and Applications' (Eds. M.M. Gupta, R.K. Ragade & R.R. Yager). North Holland, Amsterdam, pp. 3–18

[6] Zadeh, L.A. 1978. 'Fuzzy sets as a basis for a theory of possibility'. *Fuzzy Sets & Syst.*, **1**, pp. 3–28

[7] Zwick, R., Carlstein, E., & Budescu, D.V. 1987. 'Measures of similarity between fuzzy concepts, a comparative analysis'. *Int. J. Approx. Reasoning*, **1**, pp. 221–242

CHAPTER 4

Fuzzy Controllers — Preliminaries and Basic Construction

4.1 Preliminaries

In this Chapter, we introduce fuzzy controllers, discuss the underlying concepts that led to their development, and present their basic concepts. The main design phases are considered, along with four general modes of applications of fuzzy controllers. To put the overall discussion into a broader perspective, we first recall the paradigm of expert control [1]. Also, we give an overview of various applications of fuzzy controllers in control engineering and decision-making, indicating their characteristic features and highlighting their advantages.

4.2 Expert Control and Fuzzy Controller as New Paradigms of Control

Before proceeding to a detailed discussion of the fuzzy controller itself, we must acquire a clear understanding of how everyday tasks, however complex, are completed by us and how a standard way of automation can be worked out in this context. Parking or driving a car, lifting a fragile object, packing a bag at a supermarket, recognising faces: these are only a few among the quite simple tasks that we are faced with in everyday life. Despite their commonality, they create a continuous challenge for robots, whose performance is still far beyond the abilities and skills we demonstrate in solving these problems on an everyday basis. Briefly speaking when a certain task has to be solved (such as, for example, passing an obstacle), we collect all the necessary information available about the current situation (such as, for example, the topology of the terrain and the characteristics of the obstacle articulated in terms of its speed, mobility, size etc.). On that basis and with the aid of situations recalled from our experience, we undertake a series of relevant control actions. Owing to the feedback existing between the system under control and us (Fig. 4.1) a final goal can be achieved within a series of these actions. They form an evident interaction between the elements of the structure.

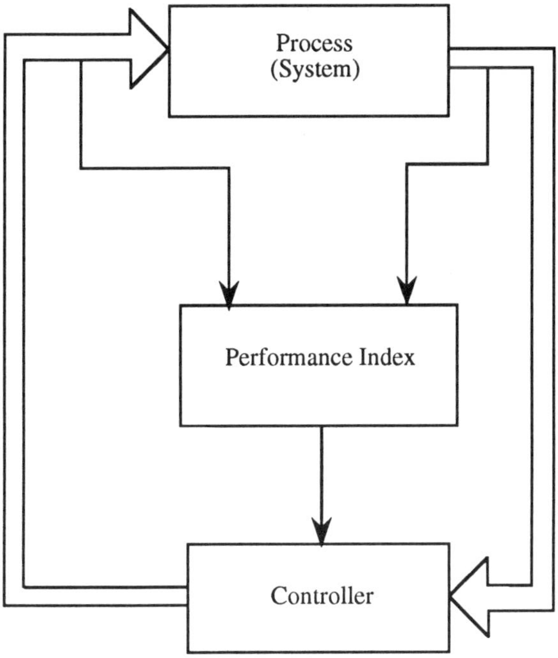

Fig. 4.1 Closed-loop system-controller

If our intent is to replace a human being in this process of control or decision-making, a new module has to be built that is capable of undertaking all the relevant control activities.

The controller *per se* is a mathematical construct and, as such, cannot incorporate a complete control knowledge without the completion of a series of well designed and rationally organised experiments.

The structures of this closed system-controller loop have been the topic of intensive research in control theory and control engineering. Despite the diverse approaches studied there, they are all based on two common and fundamental assumptions:

(i) The system to be controlled has to be known. On this basis, one can predict how it would respond to any given input (control) signal affecting it. To accomplish this prediction task, all the available pieces of knowledge about the system are aggregated into its model. This identification phase is essential for further successful performance of the control algorithms.

(ii) The second strong assumption is that the goal of control has to be specified in terms of concise mathematical formulas directly involving the variables of the system. Once it has been formally articulated, we usually refer to it as a performance index.

For (i) and (ii) established, the structure of the controller as well as its numerical parameters can be computed by some of the existing design methods of control engineering. The model of the system is used to optimise (usually minimise) the given performance index.

Unfortunately, despite the elegant and powerful mathematical framework summarised above, when the complexity of the system increases, both the assumptions (i) and (ii) tend to become infeasible. The model cannot be precisely constructed, owing to the nonlinearities, the nonstationary character of the system and the lack of a representative set of data. Then, the collection of algorithms of system identification, having at its disposal tools based on advanced methods of statistics, experiment design, and multivariable-function optimisation, tends to lose a lot of its key features.

What is more usual these days is that the models constructed are claimed to be more precise, covering the specificity of the system under control, leading to more complex model equations with many parameters to be determined (estimated). Thus, we are near to the so-called 'sin of overmodelling' [28]. Moreover, there is a visible trend towards the 'mathematisation' of control models, as underlined in one of the early papers by Zadeh [39]. In complex decision situations, it is likely that we are unable to specify the relevant performance index that is supposed to be minimised within the control task.

This closely reminds us of a situation in the domain of expert systems where the expert can complete his tasks but cannot comment too much about the optimality of the way in which he has handled the problem. This obstacle seems to be even more severe and difficult to overcome than the first one. This domain of so-called expert control emerged in order to cope not only with quantitative information but also with its qualitative counterpart. It is vigorously advocated in [1].

The key point of the new paradigm of control, this so-called expert control, is the effective use of symbolic computation as a classic application of artificial intelligence in the design process of any control algorithm. This methodology gives a new insight into heuristics (heuristic logic), permitting the design of simple regulators, as well as multivariable controllers with sophisticated control laws. These topics are reviewed by Astrom et al. [1]. A few examples underline the role that heuristics plays in control engineering:

• to obtain a good PID controller, it is necessary to take into account such factors as: (i) operational issue, (ii) operator interfaces, and (iii) effects of nonlinear actuators etc.

• multivariable and self-tuning controllers are based not only on theoretical control laws but also on a significant amount of heuristics [7,11]. In some applications, they may be called supervision safety nets [11].

• heuristics is also applied in any decision-making procedure to achieve 'optimal' (in a certain sense) performance of a controller.

Heuristics plays a key role in the choice of appropriate criteria and in the expression of a compromise between partial criteria of a contradictory character (e.g. when one needs to aggregate a criterion of high accuracy of control and a criterion of low consumption of energy in control actions).

Some instances of expert control can be found in [7] and [8], along with many examples of fuzzy controllers.

In this light, fuzzy sets and fuzzy controllers can be seen as lying in the general stream of expert control, characterised additionally by a suitable mechanism for representing vague human judgments. Studies on the performance of a human being involved in control tasks are described in [3], [9] and [29].

Much research has been carried out on the characteristics of the human operator, treated as the controller in the closed control loop:process — human being. In [29], it is reported that the human being acts as a strongly nonlinear controller, where the parameters of this controller are time-varying functions.

Success in the control of some classes of process by the human being (operator) stimulated further studies to elucidate his control principles. However, a more general opinion is also accepted, based on control protocols containing human strategy, where knowledge of the process controlled as well as of the control goal to be achieved is reflected.

Another fact of significance should be mentioned. If the control protocol is successfully implemented, two primordial problems of control methodology no longer have to be considered:

• first, an elaborate, iterative, error-prone, and time-consuming stage of model-building (identification) is eliminated,

• secondly, a performance index that has to be optimised need not be formulated in an explicit way.

Both are solved in an implicit fashion, assuming that the knowledge of the process and the performance index are hidden in the control protocol itself. Thus, significant benefits can be expected, especially in the formulation of the control task: specification of the proper performance index to be optimised may sometimes be a more complex task than the well known identification procedures. Usually, in existing solutions, such a performance index is a compromise between the real requirements stemming from the process-control needs and the simplicity of the control law that may be used. For instance, for linear models, one often accepts a quadratic form of the performance index from which well known analytical (or semianalytical) formulas are derived [16].

In practice, the human being can take reasonable control actions, even in the time-varying conditions of the process, their nonlinearities, and existing disturbances. This means that knowledge about both the system and the performance index is well represented ('coded') in the control protocol.

All these reasons make this approach an attractive one. A general scheme in which this computerised control protocol, called a fuzzy controller, works is depicted in Fig. 4.2. Here, its inputs are constituted by measures of satisfaction of performance indices that are optimised by the human operator. The structure presents only the main conceptual links existing.

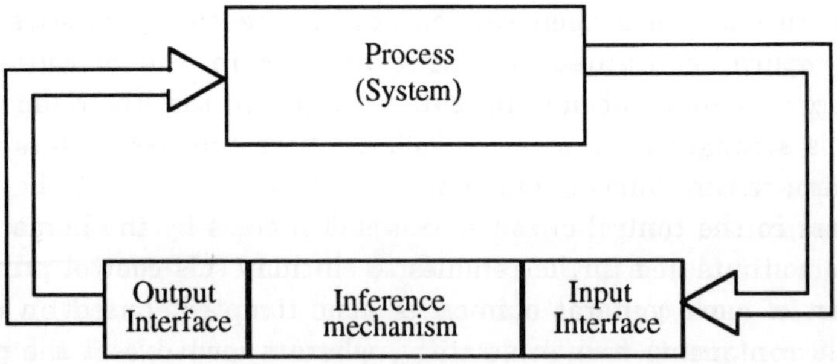

Fig. 4.2 Fuzzy controller in the overall control structure

Nevertheless, several problems in the implementation of this control algorithm have to be solved. First, one needs a control protocol reflecting the main properties of the control strategy. Secondly, for efficiency, one needs a tool that is flexible enough to cope with the linguistic categories in the control protocol and precise enough to allow its computer implementation. Fuzzy sets offer a convenient formal apparatus to meet both requirements.

Before moving into technical details, we concentrate on some general aspects addressing the vital issue of formalising and processing the knowledge available in control tasks.

The fuzzy controller is a tool for processing a fuzzy form of information in a nonfuzzy or fuzzy scheme of reasoning. Here are two applications of fuzzy controllers in the so-called 'soft' sciences (i.e. where the human being plays a central role):

- a model of organisational behaviour using linguistic rules describing relationships at a mining company [35],

- a model of the sociopsychological power theory of Mulder that has been analysed in the framework of fuzzy sets [12].

In control engineering, fuzzy controllers have been extensively studied. Case studies presented their application in various process-control systems previously controlled manually. One has to understand clearly the notion of 'fuzziness' in technical systems (electrical, mechanical, chemical etc.) and its role in information processing. At first glance, it may seem artificial to apply fuzzy sets in such systems. But such a judgment is not valid, for the reasons outlined below.

In a control strategy applied by the operator, we can recognise some concepts that are evidently fuzzy in their nature, for instance the goals of the control (e.g. 'keep *close* to the desired trajectory with a *quite low* control effort' contains the vague expressions *close*, *quite low*). Secondly, fuzziness arises from the way in which knowledge about control is structured.

A main, if not unique, source of knowledge to construct the control algorithm comes from the control protocol of the human operator. This protocol consists of a set of conditional 'if-then' statements, where the first part of each contains a so-called condition (antecedent) while the second (consequent) part deals with an action (control) that has to be taken. Therefore, it conveys the human strategy, expressing which control is to be realised when a certain state of the process controlled is observed.

An example from a textbook for cement-kiln operators [27] deals with the control rules listed below:

Case (no. of rule)	Condition Part	Action to be Realised	Reason
11	BZ OK OX low BE OK	Decrease fuel rate slightly slightly	To raise percentage of oxygen
12	BZ OK OX low BE high	(a) Reduce fuel (b) Reduce ID fan speed	To increase percentage of oxygen to activate (b) to lower back-end temperature and maintain burning-zone temperature
17	BZ OK OX high	Reduce ID speed slightly	To lower percentage of oxygen

The rules indicate how the operator has to adjust the fuel rate, the kiln speed, and the temperature in the burning zone (BZ) of the kiln, the oxygen (OX) percentage in the exhaust gases, and the temperature at the back-end (BE) of the kiln. The remarkable fact is that the rules contain the condition as well as the action part (for instance, OX *high*, OX *low*, reduce ID fan speed *slightly*) of the linguistic terms that reflect the operator's knowledge of the process. Also, they give a clear impression of the level of precision at which one has to perform computations in order to mimic a human's behaviour. The linguistic terms suggest work with sets rather than with single numerical quantities. Hence, it is not surprising that fuzzy sets came into this area of investigation.

While building the linguistic protocols for a process, two main types of question are relevant in the construction of the fuzzy controller [3,34]:

- questions about the operator's own behaviour: e.g. what would you do in such-and-such a situation?

- questions about the behaviour of the process: e.g. why does such-and-such a situation occur?

One should stress that all the terms can be viewed as symbols, implying that the control problem can be considered as a topic of Artificial Intelligence. This subsequently leads to a so-called surface or purely qualitative knowledge of the system. On the other hand, fuzzy sets offer a necessary calibration that produces a so-called deep knowledge.

In general, as stated in [3], a verbal protocol analysis is a fruitful technique for modelling human action in process control. The main aspects to be taken into account are:

- characteristics of human control behaviour,
- development of process-control skills,
- individual differences between process operators,
- task factors affecting performance,
- organisation of the operator's control behaviour.

For a better insight into the control protocol, we focus on a single-input:single-output process in which the control is determined in the light of the satisfaction of two criteria relating to two variables, namely error and change of error, where error is the difference between a specified set point and the actual output value.

For error, change of error, and control, linguistic values are encoded into fuzzy sets, the so-called reference fuzzy sets. These values are used by the operator to describe the control procedure. More precisely, we denote them by fuzzy sets $E_1, E_2, \ldots, E_n$, $DE_1, DE_2, \ldots, DE_m$, and $U_1, U_2, \ldots, U_p$, respectively. They are defined in the universes of discourse **E**, **DE**, and **U**, respectively. Links existing between the linguistic values are, as mentioned above, given in the form of conditional statements (rules):

$$\text{if } (E_i \text{ and } DE_i), \text{ then } U_i \tag{4.1}$$

$i = 1, 2, \ldots, N$, where N denotes the number of the rules. E_i, DE_i, and U_i are fuzzy sets taken from the collection of the reference fuzzy sets already introduced.

More formally, one should write down all the control rules explicitly indicating these reference fuzzy sets, namely:

$$\text{if } \left(E_{i_1(i)} \text{ and } DE_{i_2(i)}\right), \text{ then } U_{i_3(i)}$$

where

$$i_1 : \{1, 2, \ldots, N\} \to \{1, 2, \ldots, n\} \quad ,$$

$$i_2 : \{1, 2, \ldots, N\} \to \{1, 2, \ldots, p\}$$

Since this introduces an evident notation burden, we will use the previous somewhat simplified notation, which, fortunately, does not cause serious misunderstandings.

A set of conditional statements to control a single-input:single-output process is shown in the Table below [14].

Error	Change of Error						
	NB	NM	NS	Z	PS	PM	PB
PB	–	–	NM	NB	NB	NB	NB
PM	–	–	NM	NB	NB	NB	NB
PS	PS	PS	NO	NB	NB	NB	NB
PO	PM	PM	PS	NO	NS	MN	MN
NO	PM	PM	PS	NO	NS	NM	NM
NS	PB	PB	PM	PM	NO	NS	NS
NM	PB	PB	PB	PB	PM	–	–
NB	PB	PB	PB	PB	PM	–	–

where all the entries of the matrix are fuzzy sets of error, change of error, and control. The abbreviations used above mean:

PB : Positive Big
PM : Positive Medium
PS : Positive Small
PO : Positive Zero
NO : Negative Zero
NS : Negative Small
NM : Negative Medium
NB : Negative Big

The Table summarises some general aspects of the control strategy; most of the rules are self-explanatory:

- if error and change of error are zero (in the sense of the linguistic labels, i.e. PO, NO), the control applied is Negative Zero (also in the sense of the respective linguistic label of control)

- if error is Positive Big or Medium and the same as the change of error, we have to use a Negative Big control to reach the state in which error and change of error are reduced to zero

- if error is Negative Big (or Medium) and the same holds for the change of error, the control is set to the greatest positive value (i.e. Positive Big) to ensure that they are reduced significantly.

4.3 Examples of Applications of Fuzzy Controllers

Since 1974, when the first applications of fuzzy controllers were reported, there have been many advances. Now we have a number of valuable theoretical and practical concepts. On the practical aspects of fuzzy controllers, we refer to both the experimental and industrial uses of process control.

We can distinguish two main periods: the first one concentrated on laboratory-scale experiments and prototype developments with relatively rare industrial installations. The second one, starting in the late eighties, brought about a vast number of commercial applications, visible in particular in so-called human electronics. They include a variety of home appliances (vacuum cleaners, washers), camcorders, cameras etc. that are designed to be highly user-friendly devices. Furthermore, more industrial applications of fuzzy controllers started to appear.

Reference [32] discusses fuzzy control of the activated-sludge process, in common use for treating sewage and waste water. This control is difficult because of:

(i) lack of relevant instrumentation,
(ii) poorly understood biological behaviour, and
(iii) poorly formalised control goals for the process model, which is itself hard to identify.

The results with the fuzzy controller were promising; simulation studies clearly indicated a suitable performance of the model.

Another example is the control of water purification [36].

Pappis & Mamdani [25] discuss the use of a fuzzy controller for traffic control at a single intersection of two one-way streets. When a performance index is stated as an average delay of vehicles, the fuzzy controller works better than the conventional vehicle-actuated controller.

Experiments were performed by Lemke & Kickert [13,24] on a warm-water process. Their main aim was to control the temperature of the water leaving a tank for different temperature settings and a given flow of water. The process was subject to existing noises, asymmetries around the set point, and nonlinearities. Comparison of the fuzzy controller with a PI digital controller optimally adjusted showed that the fuzzy controller had a faster response with the same accuracy. Simplified versions of the fuzzy controller (fewer state variables) gave much faster step responses but larger oscillations around the set point.

An autopilot for ships, designed with the aid of fuzzy sets, has been tested on a simulation model [2].

Reference [17] reports use of the fuzzy controller in rotary cement kilns. The fuzzy controller uses the knowledge of the operator of the kiln contained in the control protocol. Two operating situations were studied:

- the kiln is stable as measured by the kiln-drive torque, which shows small variations;

- the kiln-drive torque oscillates greatly: the kiln is unstable.

The fuzzy controller performs well in both modes. Fuzzy control works better than the operator [10]. In general, the fuzzy controller calls for control actions earlier than the human being, and in smaller steps. It also reduces fuel consumption. Dedicated software and hardware are produced by F.L. Smidth & Co.

To characterise the role that fuzzy sets played in the development of some applications, they are described in more detail.

1. Matsushita Electrical recently introduced a fuzzy automatic washing machine. Until now, it has been the machine operator who decides whether a load is small or large and sets the relevant control actions accordingly. Of course, there is a lot of subjectivity in this type of control, e.g. in deciding whether a *small* load with a *large* amount of grease on the clothes needs more washing than a quite *large* load of *slightly* soiled clothes. Fuzzy controllers take care of and process this type of information requiring linguistic values. The washer has additional sensors that can detect and measure the size of the load, the quality and quantity of the dirt (this is done by measuring the murkiness of the wash water) and the type of detergent being used. The results of the measurements (which are not precise, anyway) are afterwards fed into the fuzzy controller as its input information, and a sequence of control decisions is produced.

2. The same company released another product incorporating a fuzzy controller in a vacuum cleaner. It is evident that the suction power of this appliance depends on the amount of dust on the surface. The vacuum cleaner perceives both the condition of the floor and the dust quantity by means of its dust sensors. The fuzzy controller controls the absorption power to the optimum level.

3. Mitsubishi Heavy Industry introduced an air-conditioning system using 25 control rules for heating and 25 others for cooling. An amazing reduction of about 24% in energy consumption in comparison to some previous models was achieved. The development of the entire system was fast: the initial collection of the rules was written in three days; all the membership functions in the rules were collected and improved within one month. The final tuning took no more than three months.

4. Sanyo Fisher's 8mm video camcorder uses fuzzy sets to evaluate focus and lighting conditions. The rules are used to control the diaphragm opening. Another useful application of a fuzzy controller is a camcorder equipped with an electronic image stabiliser developed by Matsushita. This stabiliser helps to correct involuntary camera movement or picture jitter during recording.

We have not mentioned any industrial applications of fuzzy controllers. They are numerous nowadays. We can mention here a cement-production process, a subway train and elevator systems, all controlled by fuzzy controllers. Again, most of them were developed in Japan.

On the basis of the above examples of applications, we can make some general observations:

• Fuzzy controllers are used in an environment where at first the human being operated for a long time, did it successfully, gained a lot of experience and thus can be treated as a source of valuable and meaningful information about the control policies utilised.

• Additional sensors used in these devices to collect more information about the conditions of the environment are usually quite inexpensive. This is because knowledge about the control is provided in terms of fuzzy sets and too-precise current information about the status of the system is not required. In addition, most applications of fuzzy controllers occur in the top-range models of appliances, making the additional cost of sensors fairly negligible.

• The development time is usually lower than for classical control approaches.

Now familiar with the principles of the fuzzy controller and the range of problems handled by it, we take a closer look at the structure of the fuzzy controller itself.

4.4 The Generic Structure of the Fuzzy Controller

The major object of this Section is to clarify the structure of the fuzzy controller, to specify the characteristic mechanisms involved in its construction, and to describe its functioning.

For illustrative purposes, we refer to a control problem where the goal of control is to stabilise temperature in a way that satisfies the goal of *comfortable* temperature.

By their very principles, fuzzy controllers are similar to expert systems. This is reflected, for instance, in their basic elements:

- A collection of control rules (if-then statements), which form the basic scheme of knowledge representation exploited in the fuzzy controller. The rules come with a basic aggregation mechanism explaining how they are put together.

- An inference mechanism.

- An output interface.

Returning to our control example, heating in the system is realised by a heating coil. A heating valve controls the volume of warm water than runs into the coil. From long experience of the system working in diverse conditions, one can summarise the format of the control protocol and specify the fuzzy predicates (fuzzy sets) contributing to the rules. A control action is carried out on the basis of the observed state of the system. The state consists of two variables. One is the difference (error) between the current temperature and the set-point value (target). The second coordinate of the state is expressed as a change in two successive values of error; this gives us a fairly good indicator of the general tendency of the error.

Both error and change-of-error variables are quantified by fuzzy sets. They, in turn, contribute towards the control rules of the controller. We assume that the fuzzy sets are defined in discrete universes of discourse. The dimensionality of the space is determined by the resolution of the available C/A and A/C convertors. For example, a 6-bit convertor yields 128 resolution levels.

As we have said before, the fuzzy sets of error and change of error are used to calibrate the control policy and specify how different fuzzy sets of error and its change complete a partition of the spaces for both input and output variables. For our example, we discuss these three spaces:

$$\mathbf{E} = \{e_1, e_2, \ldots, e_{n_1}\} \quad , \quad \mathbf{DE} = \{de_1, de_2, \ldots, de_{n_2}\} \quad ,$$

$$\mathbf{U} = \{u_1, u_2, \ldots, u_{n_3}\}$$

The basic functional components of the fuzzy controller are described accordingly [2,4,5,8,13,14,15,18,19,20,21,22,23,24,25]:

Knowledge representation:control rules
The aim of this phase is to represent the entire knowledge residing within a collection of control rules:

if E_1 and DE_1, then U_1

or else

if E_2 and DE_2, then U_2

or else

⋮

or else

if E_N and DE_N, then U_N

In the form of the fuzzy relation that combines all the rules:

$$R = R_1 \cup R_2 \cup \ldots \cup R_N = \bigcup_{i=1}^{N} (E_i \times DE_i \times U_i)$$

$$R(e_l, de_j, u_k) = \max_{1 \le i \le N} \left[E_l(e_l) \wedge DE_i(de_j) \wedge U_i(u_k)\right] \tag{4.2}$$

Note that the union of R_i s reflects the modelling of 'or else', the conjunction standing in the control protocol. Also, the Cartesian product is mainly considered to be represented by the minimum operator; however, one can expect any t-norm. In fact, the fuzzy relation summarises associations between fuzzy sets E_i, DE_i, and U_i; note also that, owing to the nature of the composition operator, the fuzzy relation R is made a bidirectional structure; i.e. no particular direction of its further usage is preferred.

Inference mechanism
For any input of the controller, i.e. the output of the process transformed

to error and change of error, E and DE, the fuzzy control is obtained via a compositional rule of inference:

$$U = (E \times DE) \bullet R \tag{4.3}$$

Inserting the values of the membership functions of E and DE as well as R, one has:

$$U(u_k) = \bigvee_{\substack{e \in \mathbf{E} \\ de \in \mathbf{DE}}} \left[E(e_l) \wedge DE(de_j) \wedge R(e_l, de_j, u_k) \right] \tag{4.4}$$

One can also look on (4.4) as a projection operation of R on **E** and **DE** that is focused (specialised) by defining the constraints E and DE in these spaces.

Output interface

The output of the fuzzy controller is a fuzzy set of control. As a process under control requires a nonfuzzy value of control, a 'defuzzification stage' is needed. There are several ways of tackling this problem, reviewed below:

1. *Mean-of-Maxima (MOM) method*. The aim of this is to characterise a fuzzy set of control with the aid of a single representative:

$$\max_{u \in \mathbf{U}} U(u) = U(u_0) \tag{4.5}$$

A difficulty arises when more than one element of **U** possesses this maximal value and thus u_0 is not uniquely determined. For instance, one can take randomly one of the elements with the highest grade of membership. A more systematic approach is to calculate an average value of these maxima (Mean-of-Maxima). Denote by U a set of u_is for which fuzzy control U attains a maximum:

$$\mathcal{U} = \left\{ u_i : \max_{j=1,2,\ldots} U(u_j) = U(u_i) \right\}$$

and assume that its cardinality is equal to 'r'; card(U) = r. Then, u_0 is computed as

$$U_0 = \sum_{u_i \in U} \frac{u_i}{r}$$

The algorithm is based exclusively on a selected, usually fairly small, set of points. It does not include the shape of the membership function as a useful characteristic feature of the fuzzy control obtained.

The second approach includes all the grades of membership by completing their weighted averaging.

2. *Centre of Gravity procedure (COG)*. The centre of gravity of the fuzzy control U treated as its numerical representative is calculated as

$$u_0 = \frac{\int_U u\, U(u)\, du}{\int_U U(u)\, du} \tag{4.6}$$

(we assume that both the integrals exist). Here, u_0 is determined bearing in mind the surface of the fuzzy set U.

In comparison to the MOM procedure, we observe that the resulting input-output characteristic $u = g(\mathbf{x})$ has no switching points (which are characteristic for relay-like curves) but is characterised by a smooth transition between the control values for different inputs of the controller.

A slightly modified version of the COG method treats all the generic fuzzy sets of control standing in the control rules separately. Let $\overline{u}_1$, $\overline{u}_2$, ..., $\overline{u}_N$ denote their numerical representatives (e.g. the centres of gravity of U_1, U_2, ..., U_N, respectively). We note that their levels of activation obtained within the matching phase are equal to λ_1, λ_2, ..., λ_N. Then the pointwise control u_0 is taken as a type of weighted sum of $\overline{u}_i$s:

$$u_0 = \frac{\overline{u}_1 \lambda_1 + \overline{u}_2 \lambda_2 + \ldots + \overline{u}_N \lambda_N}{\lambda_1 + \lambda_2 + \ldots + \lambda_N} \tag{4.7}$$

The modularity of this method is an evident advantage — the control action is computed by separately considering the contributions from the individual rules (Fig. 4.3).

The control actions computed by (4.7) are visualised in Fig. 4.3.

The COG method can be parameterised as follows:

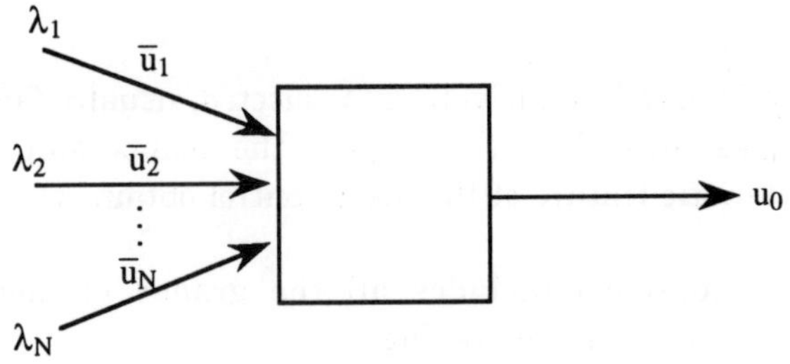

Fig. 4.3 Control action computed by (4.7)

$$u_0 = \frac{\int_{\mathbf{U}(\alpha)} u\, U(u)\, du}{\int_{\mathbf{U}(\alpha)} U(u)\, du}$$

with

$$\mathbf{U}(\alpha) = \left\{ u \in \mathbf{U} \mid U(u) \geq \alpha \right\}$$

Similarly, for the MOM method, we obtain:

$$u_0 = \arg \max_{u \in \mathbf{U}(\alpha)} U(u)$$

Returning to the general presentation of the fuzzy controller, we take a more detailed look in Fig. 4.4. The detailed structure of the fuzzy controller also contains scaling factors, *CE*, *CDE*, and *CU*. They are frequently found in existing implementations of fuzzy controllers. So the input and output variables of the controller are changed in proportion to their values. Hence, the error e′ and the change of error de′ entering the controller are modified accordingly:

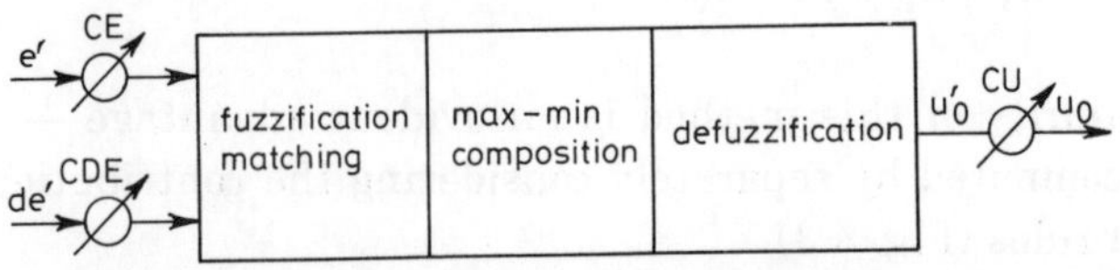

Fig. 4.4 Detailed structure of the fuzzy controller

$$e = CE\, e' \qquad , \qquad de = CDE\, de'$$

and the same relationship holds for the control $u = CU\, u'$. Their role is to tune the fuzzy controller to obtain the desired dynamic properties of the process-controller closed loop. The evident feature of this construct is that the control rules remain the same but the semantics of the generic terms used in the rules is modified. This modification is in accordance with the qualitative nature of the control rules of the protocol, which are supposed to be valid for the broad class of systems under control.

From the structure of the controller, one can easily see that the fuzzy relation requires a lot of memory space to be saved. Recalling n_1, n_2, n_3, the numbers of the elements of the universes of discourse, the fuzzy relation R needs a storage of $n_1 * n_2 * n_3$ elements. Hence, even for a moderate size of **E**, **DE**, and **U**, this may be a problem. For $n_1 = n_2 = n_3 = 10$, the fuzzy relation consists of 1000 elements. Thus, for implementation reasons, it is important to look for another, more prudent, way of calculating the fuzzy control U. This is obtained by avoiding a direct formation of the fuzzy relation of the controller.

Fortunately, for the fuzzy relation and the composition operator specified by (4.3) and (4.4), this problem can be solved easily. The fuzzy relation R does not have to be stored explicitly but can be modularised. We show the result summarised in terms of:

Proposition 4.1

For a fuzzy controller with the fuzzy relation given by (4.3) and the rule of composition specified by (4.4), the fuzzy control U is a fuzzy set in **U** with the membership function:

$$U = \bigcup_{i=1}^{N} (\Lambda_i \cap U_i) \tag{4.8}$$

where Λ_i is a fuzzy set in **U** with a constant membership function equal to:

$$\Lambda_i(u) = \max_{\substack{e \in \mathbf{E} \\ de \in \mathbf{DE}}} \left[E_i(e) \wedge DE_i(de) \wedge E(e) \wedge DE(de) \right] \tag{4.9}$$

In other words, (4.8) becomes:

$$U(u) = \bigvee_{1 \le i \le N} \left[\Pi(E \,|\, E_i) \wedge \Pi(DE \,|\, DE_i) \wedge U_i(u) \right]$$

The proof results from straightforward computations:

$$U(u) = \bigvee_{1 \le i \le N} \left\{ \max_{\substack{e \in \mathbf{E} \\ de \in \mathbf{DE}}} \left[E(e) \wedge DE(de) \wedge E_i(e) \wedge DE_i(de)\right] \wedge U_i(u) \right\} =$$

$$= \bigvee_{1 \le i \le N} \left\{ \max_{\substack{e \in \mathbf{E} \\ de \in \mathbf{DE}}} \left[E(e) \wedge E_i(e) \wedge DE(de) \wedge DE_i(de)\right] \wedge U_i(u) \right\} =$$

$$= \bigvee_{1 \le i \le N} \Pi(E \,|\, E_i) \wedge \Pi(DE \,|\, DE_i) \wedge U_i(u)$$

holding for each $u \in \mathbf{U}$, which completes the proof.

Fuzzy set Λ_i has an interesting interpretation: its value of membership is simply a minimum possibility measure of the input fuzzy sets of error and change of error, taken with respect to the antecedent in the lth rule. Thus, it is viewed as the level of activation of fuzzy sets (E_i , DE_i) by the input information coming to the controller (E , DE).

The proposition states that, instead of storing relation R, one has to store all the linguistic labels — the savings obtained in this way are substantial.

Also, this method of representation of the fuzzy controller has an interesting property of modularity. The fuzzy control derives from the original control actions U_1, U_2, ..., U_N of the rules, where each of them is 'activated' at a certain level matching the actual state of the process and the lth rule. In other words, the ith rule contributes to the fuzzy control at a level equal to the value of the membership function of Λ_i. Finally, the fuzzy set of control is determined by the union of partial results. The two stages of the fuzzy inference (matching and summarising) leading to U are shown in Fig. 4.5.

A simple illustration of how this inference algorithm works is given in Fig. 4.6. Here, the controller utilises one input and generates the fuzzy control; three control rules are available.

The input of the fuzzy controller, say e, is matched to the linguistic categories. This is done immediately, since the matching equals the membership function of the corresponding linguistic category, calculated at point e. Then, as shown on the right of Fig. 4.6, every control rule is activated at the level λ_i; the resulting fuzzy sets are also indicated. Summarising, a union of the fuzzy sets yields the result shown in Fig. 4.6.

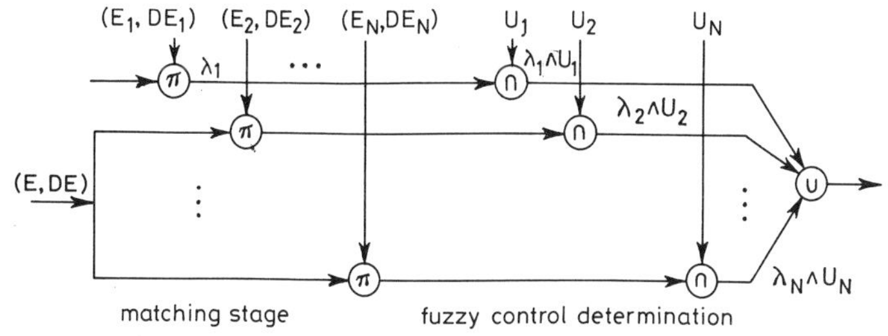

Fig. 4.5 Matching and summarising the inference mechanism

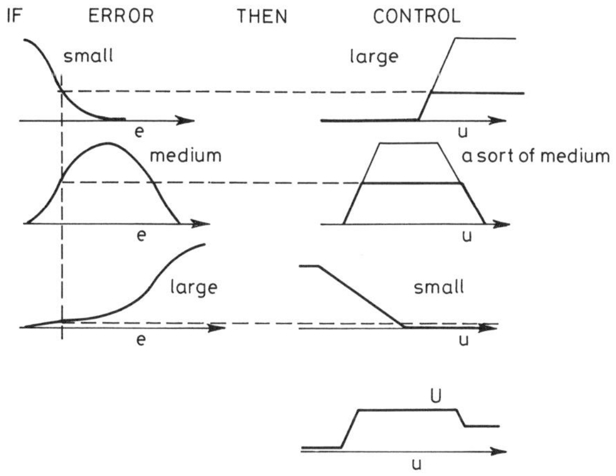

Fig. 4.6 Performance of the fuzzy controller

Some other methods of aggregation (summarising) of the fuzzy relation are also available in the existing literature. They are based on the use of different implication operators, and in this sense are fully unidirectional. The grade of membership R(e,u) denotes the strength (truth value) of implication $E \rightarrow U$ at E(e) and U(u). For instance, cf. [40]:

$$R(e,u) = \max\left[(1 - E(e)), U(u)\right]$$

and

$$R(e,u) = \min\left[1, (1 - E(e) + U(u))\right]$$

The second expression is a well known Lukasiewicz implication operator.

One should also distinguish between the discrete and the continuous schemes of knowledge representation and processing performed in the fuzzy controller.

In essence, in the discrete case, one handles and processes all the individual elements of the discrete universes of discourse for input and control variables.

In its continuous counterpart, the interest is focused on the processing realised at the level of the linguistic labels that are defined as continuous (rather than discrete) membership functions. In this situation, the transformation described by (4.7) is of significant interest since it pertains to the representatives of the fuzzy sets rather than to the overall membership function of the fuzzy control.

4.5 Modes of Operation of the Fuzzy Controller

With respect to the interaction between the fuzzy controller and its environment (namely sensors and actuators), we can distinguish four basic modes of operation.

The taxonomy of these modes is based on the type of information available to the controller and on the way it influences the control environment (Table 4.1).

Table 4.1 Modes of utilisation of the fuzzy controller

input information	output information	
	NUMERICAL	FUZZY
NUMERICAL	closed-loop control systems (1)	open-loop control systems (tutoring systems) (3)
FUZZY	closed-loop control systems (2)	open-loop control systems (intelligent decision-support systems) (4)

1. *Numerical input-numerical output*. This mode has been frequently implemented. Almost exclusively, all the existing realisations of the fuzzy controller available nowadays operate in this mode. They are standard nonlinear elements of closed-loop control systems. The

numerical information is provided by a series of sensors, while the control action is again a single (pointwise) numerical control action affecting the system. The final implementation of the fuzzy controller (using, for example, look-up-table techniques) is computationally efficient. With respect to this mode, one should be aware that, even though the final version of the controller is a nonlinear number-to-number mapping, its design procedure heavily relies on concepts of fuzzy control.

2. Admitting input information to be represented by fuzzy sets (and numerical information, in particular), we can cope with control in the presence of partially known and/or noisy information. Thus, in essence, this mode of control has a prevailing character specific for closed-loop control systems, with an eventual impact from the human being (who may provide some additional linguistic information). Furthermore, some information from sensors can be appropriately discounted for its imprecision. This mode of usage does not allow for the compact implementation that was possible in the first mode.

The two remaining modes refer to open-loop control structures:

3. The input information is of a numerical character, while the fuzzy controller generates a fuzzy set of control. The information represented by it is visualised to a human operator (user), who, based on it, makes a final numerical decision about control actions (numerical values). This mode is also characteristic of tutoring systems, where the user, being exposed to detailed information about the system and its behaviour, is trained how to control it effectively.

4. For the input and output information viewed in general as fuzzy sets, we are again dealing with the open-loop control structure (a sort of expert control or decision-support system). The control information is utilised by the user, who is responsible for issuing the final control action. This mode of use of the fuzzy controller has hardly been exploited; however, it could become a generic mode of operation in the future. In particular, the role of this mode could be essential in intelligent decision-support systems. In these systems, the decision is given in a linguistic form and uses both numerical and linguistic information about the input variables.

An example from this class is a fuzzy controller for determining a safe speed for car drivers. In contrast to pointwise control actions, the output of the controller is a genuine fuzzy set that can be easily interpreted by the user (driver) of this system. The basic object here is to design an intelligent and user-friendly controller capable of advising a car driver about the range of safe speed in current traffic circumstances. Diverse driving conditions, the need to maintain a certain distance between vehicles, and various levels of traffic load are important components of the complex pattern of the everyday traffic any driver has to cope with. A series of variables of the system, involving also weather conditions and the braking characteristics of the vehicle itself, as well as some attitudes and the driving habits of the driver, turn 'safe speed' into a fuzzy notion without clearly defined boundaries. The driver can immediately benefit from the overall knowledge encapsulated within the fuzzy controller and graphically presented by a membership function (this particular display can form an additional indicator in the cluster of indicators traditionally available to the driver). The driver, observing the resulting membership function, can choose one of the speed values (compatible with the highest degree of membership in this fuzzy set). In comparison to existing applications of fuzzy controllers, we can take advantage of information on how the driver accepts the suggestions of the controller. On this specific basis, we can carry out some slight adjustments of the controller to accommodate (personalise) the system's individual characteristics and the habits of the driver and further enhance the safety of driving.

4.6 Summary

This Chapter has considered the concept of fuzzy control and explained research into the formalisation of the heuristic control rules of the human operator, embedded in the paradigm of expert control. The role of fuzzy sets in handling fuzzy labels common in linguistic control protocols has been discussed. The reader may thus gain familiarity with the basic structure and implementation of fuzzy control. Design problems, and the properties of systems under fuzzy control, are studied later, especially in Chapters 6 and 7.

4.7 References

[1] Astrom, K.J., Anton, J.G., & Arzen, K.E. 1986. 'Expert control'. *Automatica*, **13**, pp. 277–286

[2] van Amerongen, J., van Naute Lemke, H.R., & van der Veen, J.C.T. 1977. 'An autopilot for ships designed with fuzzy sets'. Proc. 5th IFAC/IFIP Int. Conf. on Digital Computer Applications to Process Control, The Hague

[3] Bainbridge, L. 1979. 'Verbal reports as evidence of the process operator's knowledge'. *Int. J. Man-Mach. Stud.*, **11**, pp. 411–436

[4] Braae, M., & Rutherford, D.A. 1979. 'Selection of parameters for a fuzzy logic controller'. *Fuzzy Sets & Syst.*, **2**, pp. 185–199

[5] Braae, M., & Rutherford, D.A. 1979. 'Theoretical and linguistic aspects of the fuzzy logic controller'. *Automatica*, **15**, pp. 553–577

[6] Cheng, W.M., Reu, S.J., Ch-Feng Wu & Tsuei, T.H. 1982. 'An expression for fuzzy controller'. *In* 'Fuzzy Information and Decision Processes' (Eds. M.M. Gupta & E. Sanchez). North Holland, Amsterdam, pp. 411–413

[7] Clarke, D.W. 1984. 'Implementation of adaptive controllers'. *In* 'Self-Tuning and Adaptive Control' (Eds. C.J. Harris & S.A. Billings). P. Peregrinus, UK

[8] Daley, S., & Gill, K.F. 1985. 'The fuzzy logic controller: an alternative design scheme'. *Computers in Industry*, **6**, pp. 3–14

[9] Etschmaier, M.M. 1980. 'Fuzzy control for maintenance scheduling in transportation systems'. *Automatica*, **16**, pp. 255–263

[10] Holmblad, L.P., & Ostergaard, J.J. 1982. 'Control of a cement kiln by fuzzy logic'. *In* 'Fuzzy Information and Decision Processes' (Eds. M.M. Gupta & E. Sanchez). North Holland, Amsterdam, pp. 389–299

[11] Iserman, R. 1982. 'Parameter adaptive control algorithms — a tutorial'. *Automatica*, **18**, pp. 513–528

[12] Kickert, W.J.M. 1978. 'Fuzzy Theories on Decision Making'. Martinus Nijhoff, Leiden, Holland

[13] Kickert, W.J.M., & van Naute Lemke, H.R. 1976. 'Application of a fuzzy controller in a warm water plant'. *Automatica*, **12**, pp. 301–308

[14] Kickert, W.J.M., & Mamdani, E.H. 1978. 'Analysis of a fuzzy logic controller'. *Fuzzy Sets & Syst.*, **1**, pp. 29–44

[15] King, P.J., & Mamdani, E.H. 1977. 'The application of fuzzy control systems to industrial processes'. *Automatica*, **13**, pp. 235–242

[16] Kwakernaak, H., & Sivan, R. 1972. 'Linear Optical Control Systems'. J. Wiley, New York, USA

[17] Larsen, P.M. 1980. 'Industrial applications of fuzzy logic control'. *Int. J. Man-Mach. Stud.*, **12**, pp. 3–10

[18] Mamdani, E.H. 1976. 'Applications of fuzzy algorithms for control of simple dynamic plant'. *Proc. IEEE*, **121**, pp. 1585–1588

[19] Mamdani, E.H. 1976. 'Advances in the linguistic synthesis of fuzzy controllers'. *Int. J. Man-Mach. Stud.*, **8**, pp. 669–678

[20] Mamdani, E.H. 1977. 'Application of fuzzy logic to approximate reasoning using linguistic systems'. *IEEE Trans. Comput.*, **26**, pp. 1182–1191

[21] Mamdani, E.H., & Assilian, S. 1981. 'An experiment in linguistic synthesis with a fuzzy logic controller'. *In* 'Fuzzy Reasoning and Its Applications' (Eds. E.H. Mamdani & B.R. Gaines). Academic Press, New York, USA, pp. 311–323

[22] Mamdani, E.H., & Baaklini, N. 1975. 'Prescriptive methods for deriving control policy in a fuzzy logic controller'. *Electron. Lett.*, **11**, pp. 625–626

[23] Mamdani, E.H., Ostergaard, J.J., & Lemblesis, E. 1984. 'Use of fuzzy logic for implementing rule-based control of industrial processes'. *In* 'TIMS/Studies in the Management Sciences', **20**, pp. 429–445

[24] van Naute Lemke, H.R., & Kickert, W.J.M. 1976. 'The application of fuzzy set theory to control a warm water process'. Journal A, **1**, pp. 8–18

[25] Pappis, C.P., & Mamdani, E.H. 1977. 'Fuzzy logic control for a traffic junction'. *IEEE Trans. Syst., Man & Cybern.*, **10**, pp. 707–717

[26] Pedrycz, W. 1984. 'On identification algorithms in fuzzy relational systems'. *Fuzzy Sets & Syst.*, **13**, pp. 153–167

[27] Peray, K.E., & Wadell, J.J. 1972. 'The Rotary Cement Kiln'. Chemical, New York, USA

[28] Schweppe, F.C. 1973. 'Uncertain Dynamic Systems'. Prentice-Hall, Englewood, USA

[29] Sheridan, T.B., & Farrel, W.R. 1974. 'Man-Machine Systems: Information, Control, and Decision Models of Human Performance'. MIT, Cambridge, USA

[30] Sugeno, M. 1985. 'Industrial Applications of Fuzzy Control'. North Holland, Amsterdam, Netherlands

[31] Tong, R.M. 1977. 'A control engineering review of fuzzy systems'. *Automatica*, **13**, pp. 559–569

[32] Tong, R.M., Beck M.B., & Latten, A. 1980. 'Fuzzy control of the activated sludge wastewater treatment process'. *Automatica*, **16**, pp. 695–701

[33] Trankle, T.L., & Markosian, L.Z. 1985. 'An expert system for control system design'. Proc. IEE Int. Conf. Control, Cambridge, UK

[34] Umbers, I.G. 1979. 'Models of the process operator'. *Int. J. Man-Mach. Stud.*, **11**, pp. 263–284

[35] Wenstop, F. 1976. 'Deductive verbal models of organizations'. *Int. J. Man-Mach. Stud.*, **8**, pp. 293–311

[36] Yagishito, O., Itoh, O., & Sugeno, M. 1985. 'Application of fuzzy reasoning to the water purification process'. *In* [30], pp. 19–40

[37] Yasunobu, S., & Miyamoto, S. 1985. 'Automatic train operation system by predictive fuzzy control'. *In* [30], pp. 1–18

[38] Zadeh, L.A. 1973. 'Outline of a new approach to the analysis of complex systems and decision processes'. *IEEE Trans. Syst., Man & Cybern.*, **1**, pp. 28–44

[39] Zadeh, L.A. 1974. 'A rationale for fuzzy control'. *Trans. ASME, Ser. G (USA)*, **94**, pp. 3–4

[40] Zadeh, L.A. 1975. 'Calculus of fuzzy restrictions'. *In* 'Fuzzy Sets and Their Application to Cognitive and Decision Processes' (Eds. L.A. Zadeh , K.S. Fu, K. Tanaka & M. Shimura). Academic Press, New York, USA

CHAPTER 5

Fuzzy Relational Equations

5.1 Introduction

In this Chapter, we present results derived in the field of fuzzy relational equations. In Chapter 1, we saw that fuzzy-relation calculus plays a central role in fuzzy sets, similar to that of function calculus in mathematical analysis. These equations are general and flexible enough to be useful for modelling purposes, e.g. in coping with the complexity of a real-world system.

We look especially at the solutions, starting with the basic relational structures (direct, dual and adjoint) and then discussing a system of equations. We study more specialised forms of equations such as, for example, those with an equality operator or convex combination of equations, and polynomial fuzzy relational equations. We introduce approximate solutions and examine the relevant methodology of solving relational equations in these circumstances. Furthermore, some interpretation aspects of the equations are provided. Multilevel relational equations are introduced as well.

5.2 Fuzzy Relational Equations with sup-t and inf-s Composition

We start with these fuzzy relational equations:

$$Y = X \blacksquare R \tag{5.1}$$

and

$$Y = X \bullet R \tag{5.2}$$

Eq. (5.2) is a dual fuzzy relational equation. Thus, X is a fuzzy set defined in **X**, Y is a fuzzy set expressed in **Y**, and R is a fuzzy relation in **X** × **Y**. In what follows, we restrict ourselves to continuous t-norms.

5.2.1 Interpretation of fuzzy relational equations

Fuzzy relational equations resulting from the relational structures introduced can be interpreted in several ways depending on the nature of the formalism applied. We discuss some of them. We assume that the element of **Y**, say y, is fixed. The fuzzy relation R is then treated as a fuzzy set index by 'y'. To underline this fact, we denote this fuzzy set by R_y.

• The first interpretation stems from the use of set-relation calculus. The intersection completed in terms of t-norms of X and R_y is computed, and the height of the resulting fuzzy set Y is determined as:

$$Y(y) = \mathrm{hgt}(X \cap R_y)$$

• Logic interpretation of fuzzy relational equations. We treat grades of membership as truth values of multivalued logics. Then the supremum operation represents the existential quantifier (∃); note that $\exists(p(x)) = \sup p(x)$ where the supremum is computed over all the possible truth values of predicate 'p'. Subsequently, the t-norm stands for a standard model of the AND (&) connective. Combining these facts together, we consider Y(y) as the truth value of the compound formula:

$$Y(y) = \exists\left[X(x) \,\&\, R_y(x)\right]$$

• OR-AND graph. This interpretation holds for a finite cardinality of **X**, $\mathbf{X} = \{x_1, x_2, \ldots, x_n\}$, and has an interesting illustration encountered, for example, in search techniques. The OR-AND graphs are in common use as one of the standard constructs of Artificial Intelligence. The corresponding graph consists of two types of node distributed at two levels. The AND nodes are situated at the first level, while OR nodes follow this level. Each AND node performs the AND operation for $X(x_i)$ and $R_y(x_i)$. The results are then transmitted to a single OR node that carries out the final OR aggregation. The dual structure (5.2) can be interpreted in a similar fashion (assuming the necessary duality of notions used in the interpretation to those used in the above discussion).

5.2.2 Solving fuzzy relational equations

From (5.1) and (5.2), two problems arise:

- first, to determine R, for a given X and Y,
- secondly, to determine X, for a given R and Y.

To solve these problems, we use the φ- and β-operators. We represent them as residuals [17,18]:

- the φ-operator is a two-site function,

$$\varphi : [0,1] \times [0,1] \to [0,1]$$

so that

(i) $x \,\varphi\, \max(y,z) \geq \max(x \,\varphi\, y\, ,\, x \,\varphi\, z)$,
(ii) $x\, t\, (x \,\varphi\, y) \leq y$,
(iii) $x \,\varphi\, (x\, t\, y) \geq y$.

- the β-operator is a two-site function,

$$\beta : [0,1] \times [0,1] \to [0,1]$$

fulfilling these properties:

(i) $x \,\beta\, \min(y,z) \leq \min(x \,\beta\, y\, ,\, x \,\beta\, z)$,
(ii) $x\, s\, (x \,\beta\, y) \geq y$,
(iii) $x \,\beta\, (x\, s\, y) \leq y$.

The φ- and β-compositions for fuzzy sets and relations are defined accordingly:

- the φ (β) -composition of $X \in F(\mathbf{X})$ and $Y \in F(\mathbf{Y})$ is the fuzzy relation $X \,\varphi\, Y$ ($X \,\beta\, Y$, respectively), with a membership function:

$$(X \,\varphi\, Y)(x,y) = X(x) \,\varphi\, Y(y) \qquad (5.3)$$

$$, \quad x \in X,\ y \in Y$$

$$(X \,\beta\, Y)(x,y) = X(x) \,\beta\, Y(y) \qquad (5.4)$$

- the φ (β) -composition of the fuzzy relation $R \in F(\mathbf{X} \times \mathbf{Y})$ and the fuzzy set $Y \in F(\mathbf{Y})$ is the fuzzy set $R \,\varphi\, Y$ ($R \,\beta\, Y$, respectively), with a membership function:

$$(R\,\varphi\,Y)(x) = \inf_{y \in \mathbf{Y}} [R(x,y)\,\varphi\,Y(y)] \tag{5.5}$$

$$(R\,\beta\,Y)(x) = \sup_{y \in \mathbf{Y}} [R(x,y)\,\beta\,Y(y)] \tag{5.6}$$

We start by proving the following lemmas. They lead us to theorems giving a direct solution to the two problems formulated above.

Lemma 5.1

$$\forall_{X \in F(\mathbf{X})} \forall_{Y \in F(\mathbf{Y})} R \subseteq X\,\varphi\,(X \blacksquare R) \tag{5.7}$$

Proof. Rewriting the right-hand side of (5.7) in terms of the membership functions, we get:

$$(X\,\varphi\,(X \blacksquare R))(x,y) = X(x)\,\varphi\,(X \blacksquare R)(y) =$$

$$= X(x)\,\varphi\,[\sup (X(z)\,t\,R(z,y))] =$$

$$= X(x)\,\varphi\left\{\max\left[\sup_{z \in \mathbf{X}: z \neq x} (X(z)\,t\,R(z,y)), X(x)\,t\,R(x,y)\right]\right\} \geq$$

$$\geq X(x)\,\varphi\,(X(x)\,t\,R(x,y))$$

Applying inequality (iii), we obtain:

$$(X\,\varphi\,(X \blacksquare R))(x,y) \geq R(x,y)$$

which holds for each x and y and completes the proof.

Lemma 5.2

$$\forall_{X \in F(\mathbf{X})} \forall_{Y \in F(\mathbf{Y})} X \blacksquare (X\,\varphi\,Y) \subseteq Y \tag{5.8}$$

Proof. This inequality is a straightforward consequence of (ii).

Lemma 5.3

$$\underset{X\in F(\mathbf{X})}{\forall}\ \underset{R\in F(\mathbf{X}\times\mathbf{Y})}{\forall}\ (R\ \varphi\ Y) \blacksquare R \subseteq Y$$

Proof. Rewriting the above relationship in terms of the membership functions of R and Y, we get:

$$\sup_{x\in\mathbf{X}}\left\{\inf_{y\in\mathbf{Y}}[R(x,y)\ \varphi\ Y(y)]\ t\ R(x,y)\right\} =$$

$$= \sup_{x\in\mathbf{X}}\left\{\min\left[\inf_{z\in\mathbf{Y},z\neq y}(R(x,z)\ \varphi\ Y(z)), R(x,y)\ \varphi\ Y(y)\right] t\ R(x,y)\right\} \leq$$

$$\leq \sup_{x\in\mathbf{X}}\left[(R(x,y)\ \varphi\ Y(y))\ t\ R(x,y)\right]$$

which, taking into account (ii), involves an inequality:

$$\sup_{x\in\mathbf{X}}\left[(R(x,y)\ \varphi\ Y(y))\ t\ R(x,y)\right] \leq Y(y)$$

for every $y \in \mathbf{Y}$, which completes the proof.

Lemma 5.4

$$\underset{X\in F(\mathbf{X})}{\forall}\ \underset{R\in F(\mathbf{X}\times\mathbf{Y})}{\forall}\ X \subseteq R\ \varphi\ (X \blacksquare R)$$

Proof. We have

$$\inf_{y\in\mathbf{Y}}\left[R(x,y)\ \varphi\left(\sup_{x\in\mathbf{X}}(X(x)\ t\ R(x,y))\right)\right] =$$

$$= \inf_{y\in\mathbf{Y}}\left\{R(x,y)\ \varphi\ \max\left[\sup_{z\in\mathbf{X},z\neq x}(X(z)\ t\ R(z,y)), X(x)\ t\ R(x,y)\right]\right\}$$

and (iii) implies that

$$\inf_{y\in\mathbf{Y}}\left[R(x,y)\ \varphi\ (\sup (X(x)\ t\ R(x,y)))\right] \geq \inf_{y\in\mathbf{Y}}\left[R(x,y)\ \varphi\ (X(x)\ t\ R(x,y))\right] X(x)$$

for every $x \in \mathbf{X}$, which completes the proof.

Now we can prove:

Proposition 5.1
(i) If fuzzy sets $X \in F(\mathbf{X})$, $Y \in F(\mathbf{Y})$ fulfil sup-t fuzzy relational equation (5.1), the greatest fuzzy relation $\hat{R} \in F(\mathbf{X} \times \mathbf{Y})$ (in the sense of ordinal-set inclusion) satisfying formula $X \blacksquare \hat{R} = Y$ is given by this formula:

$$\hat{R} = X \varphi Y \tag{5.9}$$

(ii) If a fuzzy set $Y \in F(\mathbf{Y})$ and a fuzzy relation $R \in F(\mathbf{X} \times \mathbf{Y})$ satisfy $X \blacksquare R = Y$, the greatest fuzzy set $\hat{X} \in F(\mathbf{X})$, such that $\hat{X} \blacksquare R = Y$, is given by:

$$\hat{X} = R \varphi Y \tag{5.10}$$

Proof
(i) Lemma 5.1 involves an inequality $R \subseteq \hat{R}$, where $\hat{R} \equiv X \varphi Y$. Moreover, from $\hat{R} \subseteq R$, we get $X \blacksquare \hat{R} \subseteq X \blacksquare R$, i.e. $Y \subseteq X \blacksquare \hat{R}$. Then, from Lemma 5.2 we get $X \blacksquare \hat{R} \subseteq Y$ and, finally, $X \blacksquare \hat{R} = Y$.

(ii) Denoting $\hat{X} = R \varphi Y$, from Lemma 5.3 we obtain the relationship $\hat{X} \blacksquare R \subseteq Y$. The inequality $X \subseteq \hat{X}$ implies $\hat{X} \blacksquare R \subseteq X \blacksquare R$, and from Lemma 5.4 we have $X \subseteq R \varphi Y = \hat{X}$, so $\hat{X} \blacksquare R = Y$.

Returning to the dual fuzzy relational equation, a similar collection of lemmas can be proved. Therefore, they are presented without proof:

Lemma 5.5

$$\underset{X \in F(\mathbf{X})}{\forall} \; \underset{R \in F(\mathbf{X} \times \mathbf{Y})}{\forall} \; X \beta (X \bullet R) \subseteq R$$

Lemma 5.6

$$\underset{X \in F(\mathbf{X})}{\forall} \; \underset{Y \in F(\mathbf{Y})}{\forall} \; Y \subseteq X \bullet (X \beta Y)$$

Lemma 5.7

$$\underset{R\in F(\mathbf{X}\times\mathbf{Y})}{\forall}\ \underset{Y\in F(\mathbf{Y})}{\forall}\ Y \subseteq (R\,\beta\,Y) \bullet R$$

Lemma 5.8

$$\underset{X\in F(\mathbf{X})}{\forall}\ \underset{R\in F(\mathbf{X}\times\mathbf{Y})}{\forall}\ R\,\beta\,(X \bullet R) \subseteq X$$

Proposition 5.2 is a straightforward outcome of these lemmas.

Proposition 5.2

(i) If the set of solutions of equation $X \bullet R = Y$ with respect to R for known X and Y is nonempty, $\mathcal{R} = \{R \in F(\mathbf{X}\times\mathbf{Y}) \mid X \bullet R = Y\} = \varnothing$, then the least element of $\mathcal{R}$ is equal to:

$$\check{R} = X\,\beta\,Y$$

(ii) If the set of solutions of Proposition 5.2 with respect to X is nonempty, $\mathcal{X} = \{X \in F(\mathbf{X}\times\mathbf{Y}) \mid X \bullet R = Y\}$, $\mathcal{X} = \varnothing$, then the least element of $\mathcal{X}$ is equal to:

$$\check{X} = R\,\beta\,Y$$

5.3 Interaction in Fuzzy Relational Equations

The operations involved in fuzzy relational equations can be characterised with respect to their level of interaction. The property of interaction pertains to the way in which fuzzy sets and fuzzy relations interact within the process of composition. Two levels of interaction (hierarchy of interaction) are distinguished. The first one, called a local one, deals with the interaction existing between the two grades of membership associated with the same element of the universe of discourse. The global interaction holds at the level of all the elements of the universe of discourse. The local level pertains to the level of interaction that is characteristic of a certain logical operation (such as AND or OR) utilised in the composition operator. The global level refers to the entire space. A few examples below clarify these notions.

The sup-t (or inf-s, respectively) composition exhibits a local level of interaction:

$$Y(y) = \sup_{x \in \mathbf{X}} [X(x)\, t\, R(x, y)]$$

Denote by X a set of elements x for which the expression $X(x)\, t\, R(x,y)$ attains a supremum. Then Y(y) depends only on the values obtained on X, while all the remaining elements of **X** are ignored in the computations and do not influence the result:

$$Y(y) = X(x_0)\, t\, R(x_0, y) \quad , \quad x_0 \in \mathbf{X}$$

In particular, if card(X) = 1, the degree of membership Y(y) depends solely on a single pair of arguments $X(x_0)\, t\, R(x_0,y)$, and only $X(x_0)$ and $R(x_0,y)$ contribute to this result. Note that the sup-min composition exhibits a limited property of the local interaction, since the membership value Y(y) depends only on $X(x_0)$ or $R(x_0,y)$. The global interaction occurs for the s-t and t-s composition operators, where sup (inf, respectively) are generalised to any triangular norm:

$$Y(y) = \underset{x \in \mathbf{X}}{\mathrm{S}} [X(x)\, t\, R(x, y)]$$

and

$$Y(y) = \underset{x \in \mathbf{X}}{\mathrm{T}} [X(x)\, s\, R(x, y)]$$

This type of composition forms the most general operation on X and R. The s- or t-norm taken over elements of **X** causes a global level of interaction, leading to an overall combination of 'local' contributions computed at individual elements of **X**. The monotonicity property holds:

$$\underset{x \in \mathbf{X}_1}{\mathrm{S}} [X(x)\, t\, R(x, y)] \le \underset{x \in \mathbf{X}_2}{\mathrm{S}} [X(x)\, t\, R(x, y)]$$

for $\mathbf{X}_1 \subset \mathbf{X}_2$. Generally, in virtue of the 'distributed' contributions of the elements of **X**, one cannot find a subset $\mathbf{X}_0 \subset \mathbf{X}$ such that this equality is satisfied:

$$\underset{x \in \mathbf{X}_0}{\mathrm{S}} [X(x)\, t\, R(x, y)] = \underset{x \in \mathbf{X}}{\mathrm{S}} [X(x)\, t\, R(x, y)]$$

5.4 Adjoint Fuzzy Relational Equations and their Solutions

An adjoint fuzzy relational equation takes the form:

$$Y = X \,\varphi\, R \tag{5.11}$$

where Y has the membership function:

$$Y(y) = \inf_{x \in \mathbf{X}} \left[X(x) \,\varphi\, R(x,y)\right] \quad , \quad y \in \mathbf{Y}$$

In [10], this equation is solved using logic (Gentzen sequent calculus). However, it can be solved by the above method. The final result is formulated in Proposition 5.3.

Proposition 5.3

If fuzzy sets A and Y fulfil the adjoint fuzzy relational equation (5.11), the least element of the set of solutions (i.e. a family $\{R \in F(\mathbf{X} \times \mathbf{Y}) \mid X \,\varphi\, R = Y\}$) is given by:

$$\check{R} = X \;t\; Y \tag{5.12}$$

Thus

$$\check{R}(x,y) = X(x) \;t\; Y(y) \quad , \quad x \in \mathbf{X}, y \in \mathbf{Y}$$

Proposition 5.4

If a fuzzy set Y and a fuzzy relation R fulfil (5.11), the greatest element of $\{X \in F(\mathbf{X}) \mid X \,\varphi\, R = Y\}$ is given by:

$$\hat{X} = R \,\varphi\, Y$$

$$\hat{X}(x) = \inf_{x \in \mathbf{X}} \left[R(x,y) \,\varphi\, Y(y)\right] \tag{5.13}$$

Summarising, we can formulate the necessary and sufficient conditions for the corresponding equations to have a solution with respect to *R*:

- for sup-t composition, we get:

$$\forall_{y \in \mathbf{Y}} \exists_{x \in \mathbf{X}} X(x) \geq Y(y)$$

- for the dual fuzzy relational equation:

$$\forall_{y \in \mathbf{Y}} \exists_{x \in \mathbf{X}} X(x) \leq Y(y)$$

The adjoint fuzzy relational equation is not as restrictive as the two equations above.

5.5 Solving Systems of Fuzzy Relational Equations

We next consider this set of fuzzy relational equations:

$$X_i \blacksquare R = Y_i \tag{5.14}$$

$$X_i \bullet R = Y_i \tag{5.15}$$

$$X_i \varphi R = Y_i \tag{5.16}$$

where i = 1, 2, ..., N_i, treating X_i and Y_i as given fuzzy sets, $X_i \in F(\mathbf{X})$, $Y_i \in F(\mathbf{Y})$, seeking the fuzzy relation R that ties all the sets together. The solutions of equations with sup-t composition are derived [19] from Proposition 5.5.

Proposition 5.5

Let $\hat{R}$ be the fuzzy relation:

$$\hat{R} = \bigcap_{i=1}^{N} (X_i \varphi Y_i) \tag{5.17}$$

Denote by R_i a family of fuzzy relations satisfying the ith equation in (5.14).

If their intersection

$$R = \bigcap_{i=1}^{N} R_i$$

is nonempty, $\hat{R}$ belongs to $\mathcal{R}$.

Proof. Consider a fuzzy relation R that is any element of $\mathcal{R}$. For each i, i = 1, 2, ..., N, we get

$$X_i \blacksquare R = Y_i$$

and

$$R \subseteq X_i \varphi Y_i = \hat{R}_i$$

From the intersection of the second relations, this inequality is derived:

$$R \subseteq \bigcap_{i=1}^{N} \hat{R}_i = \hat{R}$$

The monotonicity of the sup-t composition gives a converse situation:

$$Y_i = X_i \blacksquare R \subseteq X_i \blacksquare R \subseteq X_i \blacksquare \hat{R} \subseteq X_i \blacksquare R_i = Y_i$$

Combining these, $\hat{R}$ belongs to $\mathcal{R}$. For the family of adjoint fuzzy relational equations, Proposition 5.6 holds.

Proposition 5.6
Assuming that a family of fuzzy relations

$$\mathcal{R} = \bigcap_{i=1}^{N} \mathcal{R}_i$$

with $\mathcal{R}_i$ forming a set of solutions of the ith fuzzy relational equation (5.16), is nonempty, a fuzzy relation:

$$\check{R} = \bigcup_{i=1}^{N} (X_i \, t \, Y_i)$$

is the solution of (5.16).

Proof. Notice that a simple inequality is preserved:

$$X \varphi S \subseteq X \varphi T$$

for $S \subset T$ as fuzzy relations in $\mathbf{X} \times \mathbf{Y}$ (this property derives from the monotonicity of the φ-composition with respect to the second argument, i.e. $a \varphi b \le a \varphi c$, if $b \le c$, $a, b, c \in [0,1]$). Bearing in mind the assumption of Proposition 5.6, we have:

$$X_i \varphi R = Y_i \qquad \text{and} \qquad \check{R}_i = X_1 \times Y_i \subset R$$

which implies:

$$\check{R} = \bigcup_{i=1}^{N} \check{R}_i \subset R$$

Also, for each index 'i', we have:

$$Y_i = X_i \varphi \check{R}_i \subseteq X_i \varphi \check{R} \subseteq X_i \varphi R = Y_i$$

which completes the proof.

Analogously, a similar proposition can be stated for dual fuzzy relational equations.

Proposition 5.7

Assuming that a family of fuzzy relations

$$\mathcal{R} = \bigcap_{i=1}^{N} \mathcal{R}_i$$

with $\mathcal{R}_i = \{R \mid X_i \bullet R = Y_i\}$ is nonempty, the fuzzy relation

$$\check{R}' = \bigcup_{i=1}^{N} (X_i \beta Y_i)$$

is the solution of (5.15), $R' \in \mathcal{R}$.

5.6 Solving Extended Versions of Fuzzy Relational Equations

Section 5.2 resolved the equation with the simplest structure. We now pass to equations with this structure:

$$Y = X_1 \blacksquare X_2 \blacksquare \ldots \blacksquare X_n \blacksquare R \tag{5.18}$$

$$Y = X_1 \bullet X_2 \bullet \ldots \bullet X_n \bullet R \tag{5.19}$$

and the composite fuzzy relational equations:

$$T = S \blacksquare W \tag{5.20}$$

$$T = S \bullet W \tag{5.21}$$

where

$$X_i \in F(\mathbf{X}_i) \quad , \quad i = 1, 2, \ldots, N \quad , \quad T \in F(\mathbf{Y}) \quad , \quad R \in F\left(\underset{i=1}{\overset{N}{\mathsf{X}}} \mathbf{X}_i \times \mathbf{Y}\right)$$

and

$$S \in F(\mathbf{X} \times \mathbf{Z}) \quad , \quad W \in F(\mathbf{Z} \times \mathbf{Y}) \quad , \quad T \in F(\mathbf{X} \times \mathbf{Y})$$

In other words, fuzzy set Y and fuzzy relation T equal:

$$Y(y) = \sup_{\substack{x_1 \in F(\mathbf{X}_1) \\ x_2 \in F(\mathbf{X}_2) \\ \vdots \\ x_n \in F(\mathbf{X}_n)}} \left[X_1(x_1)\ t\ X_2(x_2)\ t \ldots t\ X_n(x_n)\ t\ R(x_1, x_2, \ldots, x_n, y)\right]$$

$$Y(y) = \inf_{\substack{x_1 \in F(\mathbf{X}_1) \\ x_2 \in F(\mathbf{X}_2) \\ \vdots \\ x_n \in F(\mathbf{X}_n)}} \left[X_1(x_1)\ s\ X_2(x_2)\ s \ldots s\ X_n(x_n)\ s\ R(x_1, x_2, \ldots, x_n, y)\right]$$

$$y \in \mathbf{Y}$$

$$T(x, y) = \sup_{z \in F(\mathbf{Z})} \left[S(x, z)\ t\ W(z, y)\right]$$

$$T(x, y) = \inf_{z \in F(\mathbf{Z})} \left[S(x, z)\ s\ W(z, y)\right]$$

Now the problem for (5.18) and (5.19) can be formulated thus:

- let $X_1, X_2, \ldots, X_n$ be given, and determine the fuzzy relation R,

- let $R, X_1, X_2, \ldots, X_{j-1}, X_{j+1}, \ldots, X_n, Y$ be given, and determine X_j.

In the first problem, notice that (5.18) can be arranged so that:

$$Y(y) = \sup \left[\underline{X}(\mathbf{x})\ t\ R(\mathbf{x}, y)\right]$$

where $\mathbf{x}$ indicates an element of the Cartesian product of $\mathbf{X}_i$s. The solution of the first problem is thus:

$$\hat{R} = \underline{X}\ \varphi\ Y$$

In the second problem, we get:

$$Y = X_j \blacksquare \left(X_1 \blacksquare X_2 \blacksquare \ldots \blacksquare X_{j-1} \blacksquare X_{j+1} \blacksquare \ldots \blacksquare X_n\right) \blacksquare R$$

i.e.

$$Y = X_j \blacksquare G$$

$$G = X_1 \blacksquare X_2 \blacksquare \ldots \blacksquare X_{j-1} \blacksquare X_{j+1} \blacksquare \ldots \blacksquare X_n \blacksquare R$$

with

$$G \in F\left(\mathop{\mathsf{X}}_{\substack{i=1 \\ i=j}}^{n} \mathbf{X}_i \times \mathbf{Y}\right)$$

Therefore

$$\hat{X}_j \neq G\ \varphi\ Y$$

For inf-s composition, we derive results in an analogous manner.

Returning to the composite fuzzy relational equations (5.20) and (5.21), we formulate the tasks:

- S, T are provided, determine W,
- W, T are provided, obtain S.

Avoiding tedious calculation, we recall the fundamental results derived in the literature [2,6,12,18,23]. Assuming that the relevant families of solutions:

$$\mathcal{W} = \{W \mid T = S \blacksquare W\} \qquad \text{and} \qquad \mathcal{S} = \{S \mid T = S \blacksquare W\}$$

do exist, their maximal elements are:

$$\hat{W} = S^T \varphi T$$

i.e.

$$\hat{W}(z,y) = \inf_{x \in \mathbf{X}} \left[S^T(z,x) \varphi T(x,y)\right]$$

and

$$\hat{S} = \left(W \varphi T^T\right)^T$$

i.e.

$$\hat{S}(x,z) = \left\{\inf_{y \in \mathbf{Y}} \left[W(z,y) \varphi T^T(y,x)\right]\right\}^T$$

where T denotes the transpose of the fuzzy relation:

$$T(x,y) = T^T(y,x)$$

The extremal elements of the solutions of (5.21) are described by these formulas (of course, one has to assume that the families of the solutions are nonempty):

- for S and T given, the least element of the family of solutions is given by:

$$\check{W} = S^T \beta T$$

- for W and T known, we get:

$$\check{S} = \left(W \, B \, T^T\right)^T$$

The solutions of (5.20) and (5.21) can be obtained differently as follows. Take (5.20). For any fixed element of the space **X**, (5.20) reduces to the equation with relation and fuzzy sets discussed in Section 5.2. To focus this, only one element of **Y** is considered. We denote $T_x(y)$ instead of T(x,y) and $S_x(z)$ instead of S(x,z). Then (5.20) reads:

$$T_x(y) = \sup_{z \in F(\mathbf{Z})} [S_x(z) \; t \; W(z,y)] \quad , \quad y \in \mathbf{Y}, x \in \mathbf{X}$$

Recalling (5.17) (Proposition 5.5), the fuzzy relation W we are looking for is:

$$\hat{W}_x = S_x \; \varphi \; T_x$$

where $\hat{W}_x$ shows that this solution is valid for a certain value of x. Taking all these partial results, i.e. intersecting them with respect to the elements of **X**, we obtain:

$$\hat{W} = \bigcap_{x \in \mathbf{X}} \hat{W}_x$$

and get the same result as derived above.

5.7 Solving Polynomial Fuzzy Relational Equations

A polynomial fuzzy relational equation [14] is of this type:

$$\bigcup_{i=1}^{I} \left(A^{(i)} \blacksquare X \blacksquare B^{(i)} \right) = Y \tag{5.22}$$

where $A^{(i)} \in F(\mathbf{X} \times \mathbf{Z})$, $X \in F(\mathbf{Z} \times \mathbf{W})$, $B^{(i)} \in F(\mathbf{W} \times \mathbf{Y})$, $Y \in F(\mathbf{X} \times \mathbf{Y})$. Assuming all the universes of discourse are finite, $\mathbf{X} = \{x_1, x_2, \ldots, x_n\}$, $\mathbf{Z} = \{z_1, z_2, \ldots, z_r\}$, $\mathbf{W} = \{w_1, w_2, \ldots, w_p\}$, $\mathbf{Y} = \{y_1, y_2, \ldots, y_m\}$, it is convenient to express this equation thus:

$$\mathbf{Y} = \mathbf{X} \blacksquare \mathbf{A} \tag{5.23}$$

where **A** is a fuzzy relation derived from $A^{(i)}$ and $B^{(i)}$ in (5.22), i = 1, 2, ..., I. In fact, **A** is a tensor product of these relations, so that:

$$\mathbf{A} = \bigcup_{i=1}^{I} \left(A^{(i)}\ t\ (B^{(i)})^T \right)^T$$

Thus

$$A^{(i)}\ t\ (B^{(i)})^T = \left[A^{(i)}(x_j, z_k)\ t\ (B^{(i)})^T \right]_{\substack{j=1,2,\ldots,n \\ k=1,2,\ldots,r}}$$

where

$$A^{(i)}(x_s, z_v)\ t\ (B^{(i)})^T = \left[A^{(i)}(x_s, z_v)\ t\ B^{(i)}(w_j, y_k) \right]_{\substack{j=1,2,\ldots,p \\ k=1,2,\ldots,m}}$$

with s = 1, 2, ..., n, v = 1, 2, ..., r. The fuzzy sets **X** and **Y** are vector representations of the fuzzy relations X and Y in (5.22):

$$\mathbf{X} = \left[X(z_1, w_1)\ X(z_1, w_2) \ldots X(z_1, w_p) \ldots X(z_r, w_p) \right]$$

$$\mathbf{Y} = \left[Y(x_1, y_1)\ Y(x_1, y_2) \ldots Y(x_1, y_m) \ldots Y(x_n, y_m) \right]$$

The transformation of (5.22) into (5.23) enables us to derive a basic fuzzy relational equation.

5.8 Fuzzy Relational Equations with Equality and Difference Composition Operators

In measuring the similarity of two fuzzy sets, we have two operations: a degree of equality of the grades of the membership function, say a and b, denoted by a & b:

$$a \mathbin{\&} b = (a\ \alpha\ b) \wedge (b\ \alpha\ a)$$

and a degree of difference of a and b:

$$a \mathbin{\hat{\&}} b = (a\ \varepsilon\ b) \vee (b\ \varepsilon\ a)$$

with α and ε defined as

$$a\ \alpha\ b = \begin{cases} 1\ , & \text{if } a \le b \\ b\ , & \text{otherwise} \end{cases}$$

$$a\,\varepsilon\,b = \begin{cases} b & , \text{ if } a < b \\ 0 & , \text{ otherwise} \end{cases}$$

Taking into account the fuzzy set A defined in **X** and the fuzzy relation R expressed in $\mathbf{X} \times \mathbf{Y}$, two new compositions are:

$$B(y) = (A \;\&\; R)(y) = \sup_{x \in \mathbf{X}} [A(x) \;\&\; R(x,y)] \tag{5.24}$$

and

$$B(y) = (A \;\hat{\&}\; R)(y) = \inf_{x \in \mathbf{X}} \left[A(x) \;\hat{\&}\; R(x,y)\right] \tag{5.25}$$

In (5.24), B(y) expresses an 'optimistic' degree of the equality of fuzzy set A and fuzzy relation R (at the specified value of y); while (5.25) is a 'pessimistic' degree of the difference of A and R.

For a solution of (5.24), denote by *R* a family of all fuzzy relations satisfying the equation for A and B. Assume that the universes of discourse **X** and **Y** are finite. Introduce also:

$$\xi : [0,1] \times [0,1] \rightarrow [0,1] \tag{5.26}$$

defined pointwise as

$$a\,\xi\,b = \begin{cases} a\,\alpha\,b & , \text{ if } b < 1 \\ a & , \text{ otherwise} \end{cases}$$

$a, b \in [0,1]$. Computations show these properties:

(i) $a\,\xi\,\max(b,c) \geq \max(a\,\xi\,b\,,\,a\,\xi\,c)$, if $\max(b,c) < 1$
(ii) $a\,\xi\,(a \;\&\; b) \geq b$
(iii) $(a \;\&\; b) \leq (a \;\&\; c)$ if $a \;\&\; b < 1$ and $b \leq c$
(iv) $a \;\&\; (a\,\xi\,b) \leq b$,

$a, b, c \in [0,1]$. The conditions imposed cannot be skipped. Note that, if $\max(b,c) = 1$, then (i) is not valid. If $a \leq c < b = 1$, then:

$$a\,\xi\,\max(b,c) = a\,\xi\,1 = a < 1 = \max(a,1) = \max(a\,\xi\,b\,,\,a\,\xi\,c)$$

A similar case holds for (iii). If $a \;\&\; b = 1$, then, for $a = b < c$, one gets:

$$a \,\&\, c = (a \,\alpha\, c) \wedge (c \,\alpha\, a) = 1 \wedge a = a < 1 = a \,\&\, b$$

On this definition of the ξ-operator, the fuzzy relation $A \,\xi\, B$ for $A \in F(\mathbf{X})$ and $B \in F(\mathbf{Y})$ is:

$$(A \,\xi\, B)(x,y) = A(x) \,\xi\, B(y) \tag{5.27}$$

Then we can prove these lemmas:

Lemma 5.9

$$\underset{A\in F(\mathbf{X})}{\forall} \; \underset{B\in F(\mathbf{Y})}{\forall} \; (A \,\xi\, B) \,\&\, A \subseteq B \tag{5.28}$$

Proof. The left-hand side of (5.28):

$$\left[(A \,\xi\, B) \,\&\, A\right](y) = \sup_{x\in\mathbf{X}} \left\{A(x) \,\&\, \left[A(x) \,\xi\, B(y)\right]\right\} \le \sup_{x\in\mathbf{X}} B(y) = B(y)$$

for any $y \in \mathbf{Y}$, which completes the proof.

Lemma 5.10

$$\underset{A\in F(\mathbf{X})}{\forall} \; \underset{R\in F(\mathbf{X}\times\mathbf{Y})}{\forall} \; \underset{y'\in\mathbf{Y}:(A\&R)(y')<1}{\forall} \; R(x,y') \le \left[A \,\xi\, (A \,\&\, R)\right](x,y')$$

for any $x \in \mathbf{X}$.

Proof. In virtue of properties (i) and (ii), we have:

$$\left[A \,\xi\, (A \,\&\, R)\right](x,y') = A(x) \,\xi\, \left[(A \,\&\, R)(y')\right] =$$

$$= A(x) \,\xi\, \left\{\sup_{z\in\mathbf{X}} \left[A(z) \,\&\, R(z,y')\right]\right\} =$$

$$= A(x) \,\xi\, \left\{\sup_{\substack{z\in\mathbf{X}\\ z\ne x}} \left[A(z) \,\&\, R(z,y')\right] \vee \left[A(x) \,\&\, R(x,y')\right]\right\} \ge$$

$$\ge A(x) \,\xi\, \left[A(x) \,\&\, R(x,y')\right] \ge R(x,y')$$

for each $x \in \mathbf{X}$.

Lemma 5.11
If $R(x,y') \le R'(x,y')$ for any $x \in \mathbf{X}$ and for some $y' \in \mathbf{Y}$:

$$(A \,\&\, R)(y') \le (A \,\&\, R')(y')$$

provided that $(A \,\&\, R)(y') < 1$.

Proof. Since **X** is finite,

$$1 > (A \,\&\, R)(y') = \sup_{x \in \mathbf{X}} [A(x) \,\&\, R(x,y')] = A(z) \,\&\, R(z,y')$$

for $z \in \mathbf{X}$. In virtue of (iii), we deduce:

$$[A(z) \,\&\, R(z,y')] \le [A(z) \,\&\, R'(z,y')]$$

since $R(z,y') \le R'(z,y')$. Then

$$(A \,\&\, R)(y') = \sup_{x \in \mathbf{X}} [A(x) \,\&\, R(x,y')] = A(z) \,\&\, R(z,y') \le$$

$$\le A(z) \,\&\, R'(z,y') \le \sup_{x \in \mathbf{X}} [A(x) \,\&\, R'(x,y')] = (A \,\&\, R')(y')$$

and hence the proof.

From these lemmas, a proposition enables us to solve (5.24):

Proposition 5.8
$R \neq \varnothing$ if and only if the fuzzy relation $(A \,\xi\, B)$ defined in (5.27) belongs to $\mathcal{R}$.

Proof. We prove only the nontrivial implication: if $\mathcal{R} \neq \varnothing$, $A \,\xi\, B$ belongs to $\mathcal{R}$. Let $y' \in \mathbf{Y}$ so that $B(y') < 1$, and let $R \in \mathcal{R}$. Then

$$B(y') = (A \,\&\, R)(y') < 1$$

From Lemma 5.10:

$$R(x,y') \le (A \,\xi\, B)(x,y')$$

for any $x \in \mathbf{X}$. From Lemma 5.10 also:

$$[A \,\&\, (A \,\xi\, B)](y') \geq (A \,\&\, R)(y') = B(y') \tag{5.29}$$

From Lemma 5.9:

$$[A \,\&\, (A \,\xi\, B)](y') \leq B(y') \tag{5.30}$$

The inequalities (5.29) and (5.30) imply that

$$[(A \,\xi\, B) \,\&\, A](y') = B(y')$$

Now let $y'' \in \mathbf{Y}$ so that $B(y'') = 1.0$. We derive:

$$[A \,\&\, (A \,\xi\, B)](y'') = \sup_{x \in \mathbf{X}} \left\{A(x) \,\&\, [(A \,\xi\, B)(x, y'')]\right\} =$$

$$= \sup_{x \in \mathbf{X}} [A(x) \,\&\, A(x)] = 1 = B(y'')$$

which proves the Proposition.

Unfortunately, we cannot claim $A \,\xi\, B$ as the greatest element of R. To make that statement, another assumption is required. For this, we define these sets:

$$\Omega(A) = \{x \in \mathbf{X} \mid A(x) = 1\}$$

$$\Omega(B) = \{y \in \mathbf{Y} \mid B(y) = 1\}$$

Then this result is derived [4]:

Proposition 5.9
Let $R \neq \emptyset$:

(i) if $\Omega(B) = \emptyset$, the fuzzy relation $A \,\xi\, B$ is the greatest element of R,

(ii) if $\Omega(A) \neq \emptyset$, the fuzzy relation R defined pointwise:

$$R(x, y) = \begin{cases} A(x) \,\alpha\, B(y) & , \text{ if } y \notin \Omega(B) \\ 1 & , \text{ otherwise} \end{cases}$$

$x \in \mathbf{X}$, forms the greatest element of R, if $\Omega(A) \neq \emptyset$,

(iii) if $\Omega(A) = \varnothing$, the fuzzy relation R defined pointwise:

$$R(x,y) = \begin{cases} A(x)\,\alpha\,B(y) & , \text{ if } x \in \mathbf{X},\, y \notin \Omega(B) \\ 1 & , \text{ if } x \neq x',\, y \in \Omega(B) \\ A(x') & , \text{ if } x = x',\, y \in \Omega(B) \end{cases}$$

for every $x' \in \mathbf{X}$, forms a subfamily M of all the solutions R.

5.9 Convex Combination of Fuzzy Relational Equations

We defined above the conditions in which the relevant fuzzy relational equations have solutions. A generalisation of the fuzzy relational equation proposed in [16] allows us to get equations that are always solvable with respect to R. With fuzzy sets X and Y given, fuzzy relations exist that fulfil the equation formed as a convex combination of the two 'pure' forms. The convex combination introduced in [15] is given by:

$$Y = \Lambda\,(X \otimes R) + \overline{\Lambda}\,(X \bullet G) \tag{5.31}$$

i.e.

$$Y(y) = \Lambda(y) \inf_{x \in \mathbf{X}} [X(x) \vee R(x,y)] + (1 - \Lambda(y)) \sup_{x \in \mathbf{X}} [X(x) \wedge G(x,y)]$$

Λ is a fuzzy set defined in **Y** that acts as a parameter controlling the contribution of each 'pure' equation to the convex combination. In this context, the second part of (5.31) is an optimistic form of composition of X and R, while the first one with inf-max composition is a pessimistic form of composition. Moving with Λ from an empty fuzzy set to a set with membership function identical to 1.0, we pass from an inf-max composition to a sup-min one. Fuzzy set Λ gives us an extra degree of freedom, ensuring that its appropriate choice gives a solution of (5.31) with respect to R and G whatever the value of X and Y.

We can state the problem as follows:

- for (5.31) with X and Y known, determine fuzzy relations R and G and a fuzzy set Λ.

An algorithm for this is given in [15]:

1. Determine a family of fuzzy sets $\Lambda : \mathbf{Y} \rightarrow [0,1]$, with the membership functions bounded as:

$$1-\left\{[1-Y(y)]\#\left[1-\sup_{x\in\mathbf{X}} X(x)\right]\right\} \le \Lambda(y) \le Y(y)\#\inf_{x\in\mathbf{X}} X(x)$$

y ∈ **Y**, where a bounded division # is defined pointwise as:

$$a \# b = \min(1, a/b) \quad , \quad a, b \in [0,1]$$

Note that the bounded division is merely the φ-operator induced by the product. Denote by L(X,Y) a family of all fuzzy sets Λ satisfying the above inequalities.

2. Choose any fuzzy set from L(X,Y) and calculate the fuzzy relations R and G from:

$$R(x,y) = \begin{cases} X(x)\,\varepsilon\, Y(y) \ , & \text{if } \inf_{x\in\mathbf{X}} X(x) \le Y(y) \le \sup_{x\in\mathbf{X}} X(x) \\ X(x)\,\varepsilon\left[Y(y) + \dfrac{1-\Lambda(y)}{\Lambda(y)}\left(Y(y) - \sup_{x\in\mathbf{X}} X(x)\right)\right] \ , & \text{otherwise} \end{cases}$$

where ε is an operator used in solving dual fuzzy relational equations (with inf-max composition) and the fuzzy relation G is given by:

$$G(x,y) = \begin{cases} X(x)\,\alpha\left[Y(y) - \dfrac{\Lambda(y)}{1-\Lambda(y)}\left(\inf_{x\in\mathbf{X}} X(x) - Y(y)\right)\right] \ , \\ \qquad\qquad \text{if } \inf_{x\in\mathbf{X}} X(x) < Y(y) < \sup_{x\in\mathbf{X}} X(x) \\ X(x)\,\alpha\, Y(y) \ , \quad \text{otherwise} \end{cases}$$

The solution of (5.31) gives us a family of fuzzy relations R and G with parameter Λ ∈ L(X,Y). We see that:

- there is no uniqueness of the greatest or the least solution, as it has been obtained for 'pure' equations. Fuzzy relations R and G depend heavily on the choice of fuzzy set Λ from L(X,Y),

- we can determine the greatest or the least fuzzy set Y from all the solutions L(X,Y).

We use this notation:

$$r(y) = \inf_{x \in \mathbf{X}} [X(x) \vee R(x, y)]$$

and

$$g(y) = \sup_{x \in \mathbf{X}} [X(x) \wedge G(x, y)]$$

$y \in \mathbf{Y}$. Taking for each y the bounds in the family $L(X,Y)$, say $\lambda_-(y)$ and $\lambda_+(y)$, equal to:

$$\lambda_-(y) = \inf_{\Lambda(y) \in L(X,Y)} \Lambda(y) \quad , \quad \lambda_+(y) = \sup_{\Lambda(y) \in L(X,Y)} \Lambda(y)$$

the intervals of a generated fuzzy set $[Y_- \; Y_+]$ are:

$$Y_-(y) = \begin{cases} \lambda_-(y)(r(y) - g(y)) + g(y) & , \text{ if } r(y) - g(y) > 0 \\ \lambda_+(y)(r(y) - g(y)) + g(y) & , \text{ otherwise} \end{cases}$$

$$Y_+(y) = \begin{cases} \lambda_+(y)(r(y) - g(y)) + g(y) & , \text{ if } r(y) - g(y) > 0 \\ \lambda_-(y)(r(y) - g(y)) + g(y) & , \text{ otherwise} \end{cases}$$

The convex combination enables us to solve a single equation with no extra assumptions. However, precision is sacrificed, and only an interval-valued fuzzy set Y is available (because we are dealing with a family of fuzzy sets covered by $L(X,Y)$). Can we solve a system of fuzzy relational equations with the convex combination? There is no guarantee that fuzzy set Λ can produce a solution without extra assumptions.

5.10 Multilevel Structures of Fuzzy Relational Equations

So far, the fuzzy relational equations discussed possess an evident single-level structure. This means that the values of the membership function of fuzzy set Y are always determined independently for each element of **Y**. The natural extension is to introduce some intermediate levels of processing governed by their own relation operations, so that the structure obtained becomes multilevel. First, we introduce some useful notation and clarify the origin of some of these structures. The membership values of X, say $X(x_1)$, $X(x_2)$, ..., $X(x_n)$, can be viewed as values of the arguments of a certain multivalued function (fuzzy function).

By a minterm term, we mean an AND combination of $X(x_i)$s, as well as their complementary (negated) values $\overline{X}(x_i)$:

$$X(x_1)\ \mathrm{AND}\ X(x_2)\ \mathrm{AND}\ \ldots\ \mathrm{AND}\ X(x_n)\ \mathrm{AND}$$
$$\mathrm{AND}\ \overline{X}(x_1)\ \mathrm{AND}\ \overline{X}(x_2)\ \mathrm{AND}\ \ldots\ \mathrm{AND}\ \overline{X}(x_n)$$

where the AND operation is realised by any t-norm.

The generalisation of this expression includes additional parameters (one can think of them as individual threshold values associated with $X(x_i)$s and their complements). Denote those values by $W(x_1,z_k)$, $W(x_2,z_k)$, ..., $W(x_n,z_k)$, respectively. The generalised minterm reads:

$$Z(z_k) = (X(x_1)\ \mathrm{OR}\ W(x_1,z_k))\ \mathrm{AND}\ (X(x_2)\ \mathrm{OR}\ W(x_2,z_k))\ \mathrm{AND}\ \ldots$$
$$\mathrm{AND}\left(\overline{X}(x_1)\ \mathrm{OR}\ W'(x_1,z_k)\right)\mathrm{AND}\ \ldots\ \mathrm{AND}\left(\overline{X}(x_n)\ \mathrm{OR}\ W'(x_n,z_k)\right)$$

which can be treated as an example of the t-s composition of X (and its complementary version) and a 2n-dimensional vector of thresholds. By augmenting the original space **X** by new elements with the complementary values of $X(x_i)$s, we obtain this form of the fuzzy relational equation:

$$Z = X \bullet W \tag{5.32}$$

We recall that any two-valued (Boolean) function has two equivalent canonical representations, namely a sum of minterms (called also a conjunctive normal form, CNF) and a produce of maxterms (disjunctive normal form, DNF) [24]. We can do the same thing in representing fuzzy functions. Note, though, that neither is this canonical representation (including even the generalised forms of minterms and maxterms) sufficient to fully represent (approximate) multivalued functions nor are both the CNF and DNF equivalent (the latter arising from the existence of overlap and underlap properties in fuzzy sets). Considering (5.32) as a generalised minterm, the logical sum of some of them is given by this relationship:

$$Y = Z \blacksquare T \tag{5.33}$$

where T is a fuzzy relation transforming Z into Y.

Putting (5.32) and (5.33) together, we arrive at this two-level fuzzy relational structure:

$$\begin{cases} Z = X \bullet W \\ Y = Z \blacksquare T \end{cases} \tag{5.34}$$

One can treat it as a two-level fuzzy relational equation. For instance, for X and Y provided, the fuzzy relations W and T have to be determined. The dimension of the space in which fuzzy set Z is defined adds an additional degree of freedom to the problem and makes it even more unconventional than the fuzzy relational equations studied in this Chapter.

There may be other multilevel fuzzy relational structures, like this one:

$$Z = X \,\&\, W_1$$

$$V = Z \bullet W_2$$

$$Y = V \blacksquare W_3$$

Their aim is to grasp the nature of the problem being discussed and encapsulate these specific components in the structure. Usually, our ability to cope with these equations in a purely analytical manner is quite limited, and some optimisation methods are indispensable. We discuss them in depth by introducing fuzzy neural networks and study various schemes of learning available there. Nevertheless, in general, one should be aware that they originated from the realm of fuzzy relational structures and multilevel fuzzy relational equations.

5.11 Solvability of Fuzzy Relational Equations

If a fuzzy relational equation or a system of equations has a solution or a set of solutions, we can get it from the corresponding formulas. If not, we need to know how far the analytical formulas are quantitatively valid for [16].

Focusing on a single equation $X \blacksquare R = Y$ and then on a system of equations $X_i \blacksquare R = Y_i$, i = 1, 2, ..., N, we recall that the equality index of two fuzzy sets, say C and D, defined in the same space, is:

$$\| C = D \| = \inf_{x \in \mathbf{X}} \left[(C(x) \,\varphi\, D(x)) \, t \, (D(x) \,\varphi\, C(x)) \right]$$

It is instructive to examine the 'solvability indices':

- for a single equation:

$$\mathbf{s} = \sup_{R \in F(\mathbf{X} \times \mathbf{Y})} \| X \blacksquare R = Y \|$$

- for a system of equations:

$$\mathbf{s} = \sup_{R \in F(\mathbf{X} \times \mathbf{Y})} \mathop{\mathbf{T}}_{i=1}^{N} \| X_i \blacksquare R = Y_i \|$$

with **T** a finite iteration of t-norms:

$$\mathop{\mathbf{T}}_{i=1}^{1} v_i = v_1 \quad , \quad \mathop{\mathbf{T}}_{i=1}^{n+1} v_i = \left(\mathop{\mathbf{T}}_{i=1}^{n} v_i \right) t \; v_{n+1}$$

This index offers a multivalued evaluation of the truth of these statements about fuzzy relational equations:

'A fuzzy relational equation has a solution'

or

'A system of fuzzy relational equations has a solution'

If a solution of the equation (or the system of equations) exists, **s** is 1.0. Conversely, if the t-norm is lower semicontinuous, then from **s** = 1.0 we can deduce that the equation (or system of equations) has a solution.

As we are working with continuous t-norms, these solvability indices can be derived [7]:

- for a single equation:

$$\mathbf{s} = \mathrm{hgt}(X) \; \varphi \; \mathrm{hgt}(Y)$$

- for a system of equations:

$$\mathbf{s} \geq \mathop{\mathbf{T}}_{i=1}^{N} \sup_{y \in \mathbf{Y}} \left\{ Y_i(y) \; \varphi \left[\mathrm{hgt}(X_i) \; t \left(\bigcap_{i=1}^{N} Y_i \right)(y) \right] \right\}$$

We can estimate the solvability index without solving the relevant equation(s). Subsequently, we can seek ways to improve the index. It is a useful passive mechanism for evaluating the solutions of the equations, allowing us to decide whether to expect a fairly good solution or to conclude that the value of the index makes the task hopeless. Nevertheless, no suggestion is made concerning a slight modification (of fuzzy sets or of the fuzzy relational equation) by which a higher solvability index might be achieved.

Two plausible ways of achieving higher values of the solvability indices are:

(i) modification of the fuzzy sets in the equation(s) (X,Y or X_i, Y_i),
(ii) modification of the fuzzy relational equation.

The first modification may involve systematic changes in the membership functions of the fuzzy sets or choosing some representatives from a system of equations [9]. In [7], a modification of X_i and Y_i is proposed that leads to their 'fuzzification'; i.e. they are replaced by X_i^α and Y_i^α:

$$X_i^\alpha = X_i \cup \alpha \qquad \text{and} \qquad Y_i^\alpha = Y_i \cup \alpha$$

where α is treated as a fuzzy set with a membership function equal to α, $\alpha \in [0,1]$.

The second modification replaces a 'pure' fuzzy relational equation by a convex combination. If there is a sup-t composition in the equation, an extra parameter Λ in the convex combination enables us to control the solvability index. The smallest fuzzy set Λ that gives an acceptable level of **s** imposes the least distortion on the original fuzzy relational equation.

Decoupled fuzzy sets [20] give higher solvability indices and are particularly useful in a system of equations. If each fuzzy relational equation is solvable, then adding a family of decoupled sets, defined in additional space, allows us to get a solution for the entire system of equations (see Chapter 10). More detailed methods are provided in subsequent Chapters.

5.12 Conclusions

We have tackled the procedures for solving the different equations. Their diversity allows us to find a model that is suitable for coping with real-world systems.

Solving equations having their own theoretical values is valuable in formulating tasks arising in these models, especially identification and control. The solvability property of a single equation or a system of equations is significant for modelling purposes, where the relevance of the model constructed determines its range of application.

We indicate some trends in research. First results in this area were obtained in many-valued logic and were studied in the framework of the analysis of digital circuits. Fuzzy-set equations covered in [23] start with sup-min composition defined in [0,1] and evolve towards the composition of L-fuzzy sets [5], and their algebraic characterisation. It is of interest to determine in families of solutions those possessing extremal properties (e.g. equations with maximal or minimal entropy or consensus). We refer the interested reader to [1], [2], [3], [8], [13], [21] and [22]. Yet another trend relies on keeping the unit interval unchanged, introducing various composition operators, and studying the different structures of the equations.

The multilevel relational structures that have been introduced open a new field of investigation. Owing to their generality, they are definitely worth studying.

5.13 References

[1] Di Nola, A., & Pedrycz, W. 1982. 'Entropy and energy measures: characterization of resolution of some fuzzy relational equations'. *BUSEFAL*, **10**, pp. 44–53

[2] Di Nola, A., Pedrycz, W., Sessa, S., & Pei-Zhuang, W. 1984. 'Fuzzy relation equations under a class of triangular norms: a survey and new results'. *Stochastica*, **2**, pp. 99–145

[3] Di Nola, A., Pedrycz, W., & Sessa, S. 1985. 'On measures of fuzziness of solutions of fuzzy relation equations with generalized connectives'. *J. Math. Anal. & Appl.*, **106**, pp. 443–453

[4] Di Nola, A., Pedrycz, W., & Sessa, S. 1988. 'Fuzzy relational equations with equality and difference operators'. *Fuzzy Sets & Syst.*, **25**, pp. 205–216

[5] Goguen, J.A. 1967. 'L-fuzzy sets'. *J. Math. Anal. & Appl.*, **18**, pp. 145–174

[6] Gottwald, S. 1986. 'Characterization of the solvability of fuzzy equations'. *Elektron. Informationsverarb. Kybern.*, **22**, pp. 67–91

[7] Gottwald, S., & Pedrycz, W. 1986. 'Solvability of fuzzy relational equations and manipulation of fuzzy data'. *Fuzzy Sets & Syst.*, **18**, pp. 45–65

[8] Higashi, M., Di Nola, A., Pedrycz, W., & Sessa, S. 1984. 'Ordering fuzzy sets by consensus concept and fuzzy relation equations'. *Int. J. Gen. Syst.*, **10**, pp. 47–56

[9] Hirota, K., & Pedrycz, W. 1983. 'Analysis and synthesis of fuzzy systems by the use of probabilistic sets'. *Fuzzy Sets & Syst.*, **10**, pp. 1–13

[10] Izumi, K., Tanaka, H., & Asai, K. 1986. 'Adjointness of fuzzy systems'. *Fuzzy Sets & Syst.*, **20**, pp. 211–231

[11] Ledley, R.S. 1960. 'Digital Computer and Control Engineering'. McGraw-Hill, New York, USA

[12] Miyakoshi, M., & Shimbo, M. 1985. 'Solutions of composite fuzzy relational equations with triangular norms'. *Fuzzy Sets & Syst.*, **16**, pp. 53–63

[13] Miyakoshi, M., & Shimbo, M. 1986. 'Lower solutions of systems of fuzzy equations'. *Fuzzy Sets & Syst.*, **19**, pp. 37–46

[14] Ohsato, A., & Sekiguchi, T. 1981. 'Methods of solving the polynomial form of composite fuzzy relational equations'. *Trans. Soc. Instrum. Contr. Eng. (Japan)*, **7**, pp. 56–63

[15] Ohsato, A., & Sekiguchi, T. 1984. 'Convexly combined form of fuzzy relational equations and its application to knowledge representation'. *Proc. Int. Conf. Syst., Man & Cybern.*, Bombay, pp. 294–299

[16] Pedrycz, W. 1982. 'Numerical and applicational aspects of fuzzy relational equations'. *Fuzzy Sets & Syst.*, **11**, pp. 1–18

[17] Pedrycz, W. 1983. 'Fuzzy relational equations with generalized connectives and their applications'. *Fuzzy Sets & Syst.*, **10**, pp. 185–201

[18] Pedrycz, W. 1985. 'On generalized fuzzy relational equations and their applications'. *J. Math. Anal. & Appl.*, **107**, pp. 520–536

[19] Pedrycz, W. 1985. 'Applications of fuzzy relational equations for methods of reasoning in presence of fuzzy data'. *Fuzzy Sets & Syst.*, **16**, pp. 163–175

[20] Pedrycz, W. 1988. 'Approximate solutions of fuzzy relational equations'. *Fuzzy Sets & Syst.*. **28**, pp. 183–202

[21] Pei-Zhuang, W., Sessa, S., Di Nola, A., & Pedrycz, W. 1984. 'How many lower solutions does a fuzzy relation equation have?' *BUSEFAL*, **18**, pp. 67–74

[22] Prevot, M. 1981. 'Algorithm for solution of fuzzy relations'. *Fuzzy Sets & Syst.*, **5**, pp. 103–117

[23] Sanchez, E. 1976. 'Resolution of composite fuzzy relation equations'. *Inf. & Contr.*, **34**, pp. 38–48

[24] Schneeweiss, W.G. 1989. 'Boolean Functions with Engineering Applications'. Springer-Verlag, Berlin

CHAPTER 6

Design Aspects of Fuzzy Controllers

6.1 Introduction

This Chapter covers the fundamental design aspects of fuzzy controllers. Two streams of design are distinguished. Static analysis gives us an extensive look at the logical properties of the fuzzy controller. This type of analysis is aimed at studying the open-loop behaviour of the controller. It includes several aspects of the control protocol itself (such as its completeness, its consistency and the interaction between the rules) and includes the investigation of other interesting properties such as robustness, which has a direct impact on the closed-loop performance of the fuzzy controller. Relay-like properties also fall into the category of static analysis. In the second stream, we study the dynamic aspects of the controller including its adaptive mechanisms.

We start with a short discussion of the properties of information processing in the fuzzy controller studied in Chapter 4.

6.2 Properties of Information Processing in the Generic Structure of the Fuzzy Controller

We investigate the basic properties of the fuzzy controller and comment on some of its features. As before, the control protocol involves N control rules, where each of them consists of the two antecedents:

if E_k and DE_k, then U_k

k = 1, 2, ..., N. These pieces of knowledge are summarised in this fuzzy relation:

$$R = \bigcup_{k=1}^{N} (E_k \times DE_k \times U_k) = \bigcup_{k=1}^{N} R_k$$

namely

$$R(e,de,u) = \max_{k=1,2,\ldots,N} R_k(e,de,u)$$

The rule of knowledge combination applied here can be viewed as a fuzzy-set analogue of the well known rule of Hebbian learning (correlational learning). In this context, the fuzzy relation is treated as a matrix containing connections between the input nodes (elements of **E** and **DE**) and output nodes (elements of **U**). The minimum operation is identified by the product, while the maximum is replaced by a standard sum. Then the corresponding formula reads:

$$R = \sum_{k=1}^{N} E_k \, DE_k \, U_k$$

i.e.

$$R(e,de,u) = \sum_{k=1}^{N} E_k(e) \, DE_k(de) \, U_k(u)$$

We show that this Hebbian-like learning has far-reaching consequences for the way in which information processing is realised in the fuzzy controller. We also recall that, in the max-min composition, the fuzzy set of control is:

$$U(u) = \max_{k=1,2,\ldots,N} \left[\min\left(\lambda_k, U_k(u)\right)\right]$$

where the coefficient λ_k:

$$\lambda_k = \min\left\{\min_{e \in \mathbf{E}} \left[E(e), E_k(e)\right], \min_{de \in \mathbf{DE}} \left[DE(de), DE_k(de)\right]\right\}$$

has a clear interpretation as the degree of 'firing' (activation) of the kth control rule caused by the input datum (E and DE).

Selected properties of information processing in the fuzzy controller are given below:

- *boundary conditions*

If an empty information is provided, E = 0, DE = 0 (membership functions identically equal to zero), the controller infers an empty fuzzy set of control, U = 0.

If nothing is known about E and DE, i.e. all the elements of **E** and **DE** are accepted to the highest degree of membership (the membership functions of E and DE are identically equal to 1), the fuzzy set of control is

$$U(u) = \max_{k=1,2,\ldots,N} U_k(u)$$

i.e. U becomes a union of U_ks, since all the levels of activation λ_k are set to 1 (assuming, of course, that E_k and DE_k are normal fuzzy sets). In other words, owing to the evident lack of specificity in this input information, all the rules are fired in parallel, yielding a union of fuzzy sets of control standing in the control rules. Observe also that U(u) results from a straightforward projection of R on **U**.

• *monotonicity*

If $E \subset E'$ and $DE \subset DE'$, then $U \subset U'$; the more specific the input information, the more significant the derived conclusion (fuzzy control).

• *crosstalk*

The crosstalk or interaction between the rules is the most evident and striking consequence of the Hebbian learning used in the controller. Its essence is that, even when the input information coincides fully with the antecedents of one of the rules, say $E = E_j$, $DE = DE_j$, the inferred fuzzy set of control is not equal to U_j but becomes corrupted by residual contributions coming from the other rules of the protocol:

$$U = U_j \cup \Delta U$$

where

$$\Delta U = \bigcup_{k \neq j}^{N} (\lambda_k \wedge U_k)$$

The additional component ΔU represents crosstalk. Its occurrence is due to the overlap between the fuzzy sets of the antecedents of the rules. We study its quantitative characterisation later.

Changing our notation, X is an input of the fuzzy controller treated as a fuzzy relation, defined in input spaces of the controller. In Chapter 4, X comprised the two coordinates, error and change of error, $X = E \times DE$.

There are four important aspects of design:

- completeness of the control rules,
- consistency of the control rules,
- interaction of the control rules, and
- robustness of the controller.

6.2.1 Completeness of the control rules

By completeness, we mean that the controller can generate control for any input fuzzy state X. Recalling the basic equation of the controller with sup-min composition applied, an empty fuzzy set of control corresponding to a specified nonempty fuzzy set of input cannot be accepted. More formally a set of control rules:

$$\text{if } X_i, \text{ then } U_i \quad , \quad i = 1, 2, \ldots, N$$

is complete in this condition:

$$\underset{x \in \mathbf{X}}{\forall} \; \underset{1 \le i \le N}{\exists} \; X_i(x) > \varepsilon \tag{6.1}$$

where $\varepsilon \in (0,1]$.

All the fuzzy controllers studied experimentally satisfy the condition of completeness. In other words, (6.1) requires that the union of fuzzy relations X_i be greater than zero for all $x \in \mathbf{X}$, i.e.

$$\underset{x \in \mathbf{X}}{\forall} \left(\bigcup_{i=1}^{N} X_i(x) \right) > \varepsilon \tag{6.2}$$

This relationship is not difficult to comply with. The requirements seem to be natural, since we are dealing with fuzzy labels (categories) that usually overlap. The parameter ε describes this overlap. The violation of inequality (6.2) occurs only if some labels are missing, which could be caused by the omission of a certain relevant condition-control pair.

6.2.2 Interaction of the control rules

As discussed in Section 6.2, interaction between the control rules holds if

$$\underset{1 \le i \le N}{\exists} \; X_i \bullet R \neq U_i \tag{6.3}$$

i.e.

$$\underset{1\le i\le N}{\exists}\ \underset{u\in U}{\exists}\ (X_i \bullet R)(u) \neq U_i(u)$$

Thus, the constructed relation R and the composition operator applied modify the original fuzzy set of control U_i: instead of U being equal to U_i, a deformed fuzzy set of control appears. As shown for the normal fuzzy relations X_i, one gets:

$$\underset{1\le i\le N}{\forall}\ U_i \subseteq X_i \bullet R$$

Proposition 6.1 specifies the conditions ensuring the nonexistence of interaction between the control rules.

Proposition 6.1
If the fuzzy relations X_j, j = 1, 2, ..., N, are pairwise disjoint, i.e.

$$X_i \cap X_j = \varnothing \quad , \quad \text{for } i = j,\ i,j = 1, 2, \ldots, N$$

i.e.

$$\min\left(X_i(x), X_j(x)\right) = 0$$

and they are normal, the fuzzy controller produces the original fuzzy set of control:

$$\underset{1\le i\le N}{\forall}\ X_i \bullet R = U_i$$

Proof. Denote by $\mathrm{supp}(X_j)$ a support of the fuzzy relation X_j and calculate the sup-min composition of X_j and the relation of the controller, $X_j \bullet R$:

$$(X_j \bullet R)(u) = \sup_{x\in X}\left\{\min\left[X_j(x), R(x,u)\right]\right\} =$$

$$= \max\left\{\sup_{x\in \mathrm{supp}(X_j)}\left\{\min\left[X_j(x), R(x,u)\right]\right\},\ \sup_{x\notin \mathrm{supp}(X_j)}\left\{\min\left[X_j(x), R(x,u)\right]\right\}\right\} =$$

$$= \sup_{x\in \mathrm{supp}(X_j)}\left\{\min\left\{X_j(x), \max\left[\min\left[X_j(x), U_j(u)\right]\right]\right\}\right\} = U_j(u)$$

However, the assumptions of the above Proposition may be significantly relaxed:

Proposition 6.2

Consider a fuzzy controller with a fuzzy relation and composition operator specified as in Chapter 4. If:

(i) X_i are normal fuzzy relations, $\mathrm{hgt}(X_i) = 1$,

(ii) $\underset{u \in \mathbf{U}}{\forall} \underset{\substack{1 \le k \le N \\ k \ne i}}{\forall} \Pi\left(X_k \mid X_i\right) \le U_k(u)$

there is no interaction between the control rules.

Proof. Rewrite the membership function of the fuzzy control U accordingly (here X is put equal to X_k):

$$U(u) = \max_{1 \le i \le N} \left\{ \left[\sup_{x \in \mathbf{X}} \left(X_k(x) \wedge X_i(x)\right) \right] \wedge U_i(u) \right\} =$$

$$= \max \left\{ \max_{\substack{1 \le i \le N \\ i \ne k}} \left\{ \left[\sup_{x \in \mathbf{X}} \left(X_k(x) \wedge X_i(x)\right) \right] \wedge U(u) \right\}, \sup_{x \in \mathbf{X}} \left[X_k(x) \wedge X_k(x)\right] \wedge U_k(u) \right\}$$

In virtue of the normality of X_k, this becomes:

$$U(u) = \max_{\substack{1 \le i \le N \\ i \ne k}} \left[\Pi\left(X_k \mid X_i\right) \wedge U_i(u) \right] \vee U_k(u)$$

Moreover, this inequality is preserved:

$$\max_{\substack{1 \le i \le N \\ i \ne k}} \left[\Pi\left(X_k \mid X_i\right) \wedge U_i(u) \right] \le \max_{\substack{1 \le i \le N \\ i \ne k}} \Pi\left(X_k \mid X_i\right)$$

which, from (ii), implies that:

$$U(u) = U_k(u) \qquad \text{for all } u \in \mathbf{U}$$

From Proposition 6.2, we can infer:

if $\Pi(X_k \mid X_i) \neq 0$ for any pair of indices i and k, $i \neq k$, then, to ensure noninteraction of the control rules, the fuzzy sets of control must not be too precise and must cover the entire space of control **U**.

In general, it is not so obvious how to evaluate the degree of interaction shared by the control rules. Therefore, we offer the following construction simply as a candidate for this task. Denote by $R^{(0)}$ the fuzzy relation of the controller and calculate for each fuzzy relation of the input from the control rules the relevant fuzzy set of control. Thus:

$$U_i^{(0)} = X_i \bullet R^{(0)}$$

i = 1, 2, ..., n. Assume that each X_i is a normal relation. Then, from Proposition 6.1, the inclusion of $U_i \subseteq U_i^{(1)}$ is valid. We next construct the fuzzy relation:

$$R^{(1)} = R^{(0)} \cup \bigcup_{i=1}^{N} \left[X_i \times \left(U_i \cup U^{(1)}\right)\right]$$

This construction bears out that the fuzzy relation of the controller results from the fuzzy data from the equation of the controller itself. Thus, fuzzy relations $R^{(2)}$, $R^{(3)}$, ... can be obtained. From the above scheme:

$$R^{(2)} = R^{(1)} \cup \bigcup_{i=1}^{N} \left[X_i \times \left(U_i \times U_i^{(2)}\right)\right]$$

with

$$U_i^{(2)} = X_i \bullet R^{(1)}$$

and

$$R^{(3)} = R^{(2)} \cup \bigcup_{i=1}^{N} \left[X_i \times \left(U_i \times U_i^{(3)}\right)\right]$$

where

$$U_i^{(3)} = X_i \bullet R^{(2)}$$

Uniting these relations, say $R^{(0)}$, $R^{(1)}$, ..., we get:

$$\tilde{R} = \bigcup_{k=1}^{\infty} \bigcup_{i=1}^{N} \left[X_i \times \left(U_i \cup U_i^{(k)}\right)\right] = \bigcup_{k=0}^{\infty} \bigcup_{i=1}^{N} \left[X_i \times \left(U_i \cup U_i^{(k)}\right)\right]$$

where we use the notation $U_i^{(0)} = U_i$. Its membership function is:

$$\tilde{R}(x,u) = \sup_{\substack{0 \le k \le \infty \\ 1 \le i \le N}} \left[X_i(x) \wedge U_i^{(k)}(u)\right]$$

Obviously, we can bound R and drop the second component in the above expression:

$$\tilde{R}(x,u) \le \max_{1 \le i \le N} X_i(x) = c(x)$$

Hence, we can bound R by uniting the fuzzy relations in the control rules. In this light, a number:

$$\delta(x,u) = c(x) - \tilde{R}(x,u)$$

can be treated as a measure of the interaction of the control rules. This depends on the Cartesian product of the relevant spaces; thus the fuzzy controller may have a quite high interaction in one region of $\mathbf{X} \times \mathbf{U}$, and in the other no interaction at all. For a global evaluation of this property for a given controller, $\delta(x,u)$ can be averaged or its maximal value chosen.

The controller's interaction derives from its logical construction. But not when the fuzzy-relational approach is used. Nevertheless, if the values of the interaction are too high and unacceptable, the fuzzy relations X_i should be derived with more refinement, to avoid significant overlap between them. Another useful property of the fuzzy controller is that its precision is closely related to the energy of the input and output information. This feature gives us a clear picture of the quality of the fuzzy control, which may allow us to determine whether the fuzzy relation X entering the controller is precise enough to enable us to infer the fuzzy control. Recalling the basic construction of the controller, we know that, for X being a singleton, the fuzzy control constitutes the relevant row (or column) of the relation of the controller. As X becomes more and more 'fuzzy' (still assuming that X is a normal fuzzy relation), fuzzy set U also becomes more and more 'fuzzy', in the limit becoming a fuzzy set with almost constant membership function. This means that fuzzy control U conveys no valuable information. Specifying the fuzziness of the fuzzy sets in terms of energy precision is done thus:

- the precision of fuzzy relation X is acceptable if this holds:

  if $D(X) \leq \rho$, then $D(U) \leq \varepsilon$

where ρ and ε are non-negative constants. The second term depends on ρ and is the feasible precision of the fuzzy control acceptable for the input fuzzy information. D(.) is the energy of the fuzzy relation.

The interaction of the control rules results from the above construction of the controller: the fuzzy relation is a union of all subrelations, and the inference scheme of the fuzzy control uses sup-min composition. No evident links exist between these two stages. Other ways of computing the fuzzy relation of a controller are based on logical studies of implication operators [8]. We show how a fuzzy controller can be built from fuzzy-relational equations: both stages, namely the construction of the relation of the controller and the inference mechanism for control calculation, are related. In this way, the problem of interaction of the control rules is avoided.

6.2.3 Consistency of the control rules

A fundamental problem is the consistency of the control protocols. Implemental and/or contradictory information in the protocols may lead to unexpected and unsatisfactory results.

Where a set of control rules is available, inconsistency is evident if, for a given input of the controller, the resulting fuzzy set of control is multimodal. This occurs when there are two control rules with almost the same state (condition part) and diverse actions are suggested. It may happen if the rules formulated by the operator are articulated in terms of two contradictory criteria. High accuracy and low fuel consumption considered in the same control problem often become contradictory. Remarkably, multimodality disappears on defuzzification, using for example a mean-of-maxima procedure. Nevertheless, the control obtained may not be effective on either of the two criteria mentioned above. We should study the control rules to get a deeper insight into their interrelationships. We must then try to eliminate or replace the main contradictory control rules.

To express the structure of the fuzzy controller, we recall the measures of possibility and necessity. The possibility of X with respect to X_l, $\Pi(X | X_l)$, expresses the degree to which X and X_l overlap, while the necessity $N(X | X_l)$ is the degree to which X is contained in X_l.

The goal is to characterise interaction between X and X_l. We will express it numerically by providing activation levels of X_l produced by the datum. The level of interaction is bounded by the values of the possibility and necessity measures.

Definition 6.1

The upper bound of activation of the lth rule of the fuzzy controller is computed as a possibility measure of A with respect to X_l,

$$\lambda_l = \Pi\left(X | X_l\right)$$

Definition 6.2

The lower bound of activation of the lth rule of the fuzzy controller is computed as a necessity measure of X taken with respect to X_l,

$$\mu_l = N\left(X | X_l\right)$$

Note that by virtue of the fundamental properties of these measures the inequality $\lambda_l \geq \mu_l$ is always satisfied. The difference between the values of λ_l and μ_l depends upon the relationships existing between X and X_l. Usually, for a fixed granularity of X_l the coarser granularity of X yields higher values in the differences $\lambda_l - \mu_l$.

Let us introduce the notation

$$\lambda_{kl} = \Pi\left(X_k | X_l\right) \qquad \text{and} \qquad \mu_{kl} = N\left(X_k | X_l\right)$$

which will be useful in describing the interaction between the fuzzy relations found in the control rules. We first note that

- $\lambda_{kk} = 1$ for a normal fuzzy relation X_k,
- $\lambda_{kl} = \lambda_{lk}$.

The coefficient λ_{kl} expresses the extent to which X_k activates the lth rule. μ_{kl} tells us to what extent the complement of X does not activate the remaining fuzzy relations that differ from the lth one.

Combining the values of λ_{kl} into matrix form, $\lambda = [\lambda_{kl}]$ [5], we get a concise representation of the interaction between the control rules, mainly from the standpoint of input fuzzy relations. If X_l are fuzzy numbers in **R**, we may introduce an additional interpretation. Consider an ordering:

$$X_l \prec X_k, \text{ if } \arg\sup_{x \in \mathbf{R}} X_l(x) \le \arg\sup_{x \in \mathbf{R}} X_k(x)$$

Then the following holds:

$$\text{if } X_l \prec X_k \prec X_i, \text{ then } \lambda_{ik} \le \lambda_{li}$$

This means that the matrix contains information on how the activation of the concrete rule, say the ith one, spreads over all the remaining rules. Arranging the coefficients μ_{kl} in the form of a matrix $\mu = [\mu_{kl}]$, we describe the (λ,μ) evaluation of the controller.

Now we can study the inconsistency of the control rules more formally. First, we answer the question: what does 'a set of inconsistent rules' mean? We suppose that the ith and kth rules are consistent when a slight difference between X_i and X_k produces a slight difference between U_i and U_k. These differences may be expressed as the possibility measures of the respective fuzzy relations (sets). We now introduce:

Definition 6.3

An index of the inconsistency of a pair of the ith and kth rules, c_{ik}, is the absolute difference of the possibilities:

$$c_{ik} = \left| \Pi(X_i \mid X_k) - \Pi(U_i \mid U_k) \right|$$

If $X_i = X_k$ and $\Pi(U_i \mid U_k) = 0.0$, then c_{ik} attains its maximal value, $c_{ik} = 1$. For the same X_k and X_i, U_k and U_i, we have $c_{ik} = 0.0$, which indicates the lowest level of inconsistency between the rules (complete accordance). Carrying out summation over the second index, we get

$$c_i = \sum_{k=1}^{N} c_{ik}$$

which characterises the level of inconsistency of the ith rule and the remaining ones.

Now let us illustrate the proposed concepts with the help of the fuzzy controller studied by Mamdani [9], designed to control a combined steam engine and boiler. The system is treated as having two inputs and one output. The inputs are: the heat input to the boiler, and the throttle opening at the input of the engine cylinder. Steam pressure in the boiler and the speed of the engine are the outputs of the system. The linguistic control rules are summarised in the Table below.

	PE	CPE	SE	CSE	HC
1	NB	NB or NM	ANY	ANY	PB
2	NB or NM	NS	ANY	ANY	PM
3	NS	PS or NO	ANY	ANY	PM
4	NO	PB or PM	ANY	ANY	PM
5	NO	NB or NM	ANY	ANY	NM
6	PO or ZO	NO	ANY	ANY	NO
7	PO	NB or NM	ANY	ANY	PM
8	PO	PB or PM	ANY	ANY	NM
9	PS	PS or NO	ANY	ANY	NM
10	PB or PM	NS	ANY	ANY	NM
11	PB	NB or NM	ANY	ANY	NB
12	NO	PS	ANY	ANY	PS
13	NO	NS	ANY	ANY	NS
14	PO	NS	ANY	ANY	PS
15	PO	PS	ANY	ANY	NS

NB = Negative Big; NM = Negative Medium; NS = Negative Small; NO = Negative Zero; PO = Positive Zero; PS = Positive Small; PM = Positive Medium; PB = Positive Big.

From this Table, the format of the control rules can be deduced:

if (PE & CPE), then HC

The inputs of the controller relevant to the steam engine do not influence the Heat Change, HC.

We now discuss the control rules in terms of λ_{ik}, as visualised in the Table below.

(i,k)	1	2	3	4	5	6	7	8	9	10	11	12	13	14	15
1	1	0.9	0	0	0	0	0	0	0	0	0	0	0	0	0
2		1	0.5	0	0.1	0.1	0	0	0	0	0	0	0.1	0	0
3			1	0.6	0	0.6	0	0	0	0	0	0.6	0.5	0	0
4				1	0	0	0	0	0	0	0	0.7	0	0	0
5					1	0	0	0	0	0	0	0	0.7	0	0
6						1	0	0	0.6	0.1	0	0.5	0.5	0.5	0.5
7							1	0	0	0.1	0	0	0	0.7	0
8								1	0.6	0	0	0	0	0	0.7
9									1	0.5	0	0	0	0.5	0.6
10										1	0.7	0	0	0	0.1
12												1	0	0	0
13													1	0	0
14														1	0
15															1

The matrix of possibilities shows how the rules activate one another. Calculation of the c_i s gives:

Rule Number	c_i
1	2.4
2	3.4
3	4.2
4	3.8
5	4.2
6	1.8
7	4.5
8	3.5
9	4.0
10	3.9
11	1.7
12	3.3
13	4.1
14	3.7
15	3.3

The seventh rule gives the highest inconsistency in the entire collection, which suggests its omission in the creation of the fuzzy relation of the controller. In the c_is, we can distinguish a subset of rules with a higher inconsistency index. These rules should be omitted in computations of the fuzzy relation of the controller by imposing an appropriate threshold level. More formally, the ith rule is skipped in the calculation of the fuzzy relation if $c_i \geq \gamma$. The performance of the controller constructed with the aid of the specified subset of control rules is evaluated using the sum of the Hamming distances between the fuzzy sets of control equal to U_i and $X_i \bullet R$:

$$Q = \sum_{i=1}^{N} d(U_i, X_i \bullet R)$$

The fuzzy relation is now the union of the Cartesian product of X_i and U_i:

$$R = \bigcup_{i=1}^{N} (X_i \times U_i) \tag{6.4}$$

where the above union is taken over all 'i' for which c_i does not exceed the threshold level:

$$I = \{1 = 1, 2, \ldots, 15 \quad c_i < \gamma\}$$

For the subsets I:

Case	Omitted Control Rules	Threshold Level
a	none	4.5
b	{3,5,7,9,13}	4.0
c	{3,4,5,7,9,10,13,14}	3.7
d	{2,3,4,5,7,8,9,10,12,13,14,15}	3.0
e	{1,2,3,4,5,7,8,9,10,12,13,14,15}	2.0
f	all	1.7

we get:

Case	a	b	c	d	e	f
Q	40.6	35.8	34.3	34.1	37.4	39.2

The performance index Q is lowest when some rules are skipped (case 'd'); however, Q is not very sensitive to the number of rules omitted. The above can help in the evaluation of the control protocols, allowing us to resolve conflicts in the entire set of rules. If some rules have been created on at least two more or less competitive criteria, this is detected by the above indices. Roughly speaking, when performing a sup-min composition for the fuzzy relations, this fact is reflected in a lack of convexity of the fuzzy set of control. N.B. If the mean-of-maxima approach is used to choose the nonfuzzy value, it is unsatisfactory from both the first and the second criteria.

This deficiency of convexity of the fuzzy set of control can be expressed quantitatively by taking into account the difference between the energy of a convex hull of U and the fuzzy set U itself. The higher this difference, the more evident it is that conflicting performance indices have been used and the stronger their influence on the entire construction of the controller. If a threshold level is imposed, the strength of this conflict can be measured.

6.3 Fuzzy Controller as a Nonlinear Mapping

Referring to the modes of use of the fuzzy controller, in the first of them it is treated as a nonlinear mapping. Once the input information is pointwise and the output is again required to be provided in a numerical format, the rules along with the inference mechanism:

$$X_k \to U_k \quad , \quad X \to U$$

are represented as a static nonlinear mapping:

$$u = \Phi(\mathbf{x})$$

where **x** stands for a vector of input variables (i.e. composed of the error e and its change de). One should stress that this nonlinear mapping is not unique and significantly relies on the conversion process of the fuzzy sets of control into their pointwise representatives. As shown in Chapter 3, there are three of these transformations: MOM, COG, and modified COG. Before studying them in greater detail, we give this example, illustrating the character of the nonlinearities existing there.

Example 6.1
The fuzzy controller consists of the rules specified in the Table in Example 6.2 below. The fuzzy sets of the control protocol for E, DE, and U are defined in Fig. 6.1.

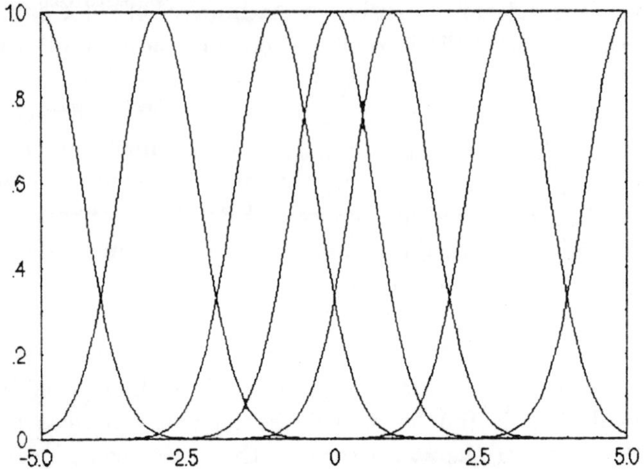

Fig. 6.1 Fuzzy sets E, DE, and U used in the control protocol

The characteristic u = Φ(e,de) are summarised in Figs. 6.2(a) and (b) (note that one of the input variables is fixed). All of them are highly nonlinear.

6.4 Output Interface of the Fuzzy Controller

We recall the main transformation methods applied to fuzzy sets of control. We make a clear distinction between continuous and discrete modes of processing:

MOM: in the mean-of-maxima method, we select the elements with the highest degrees of membership, say λ:

$$U = \{u_1, u_2, \dots, u_r\} \quad , \quad \lambda = \max_i U(u_i) \quad , \quad U(u_j) = \lambda \qquad \text{for all } u_j \in U$$

and compute:

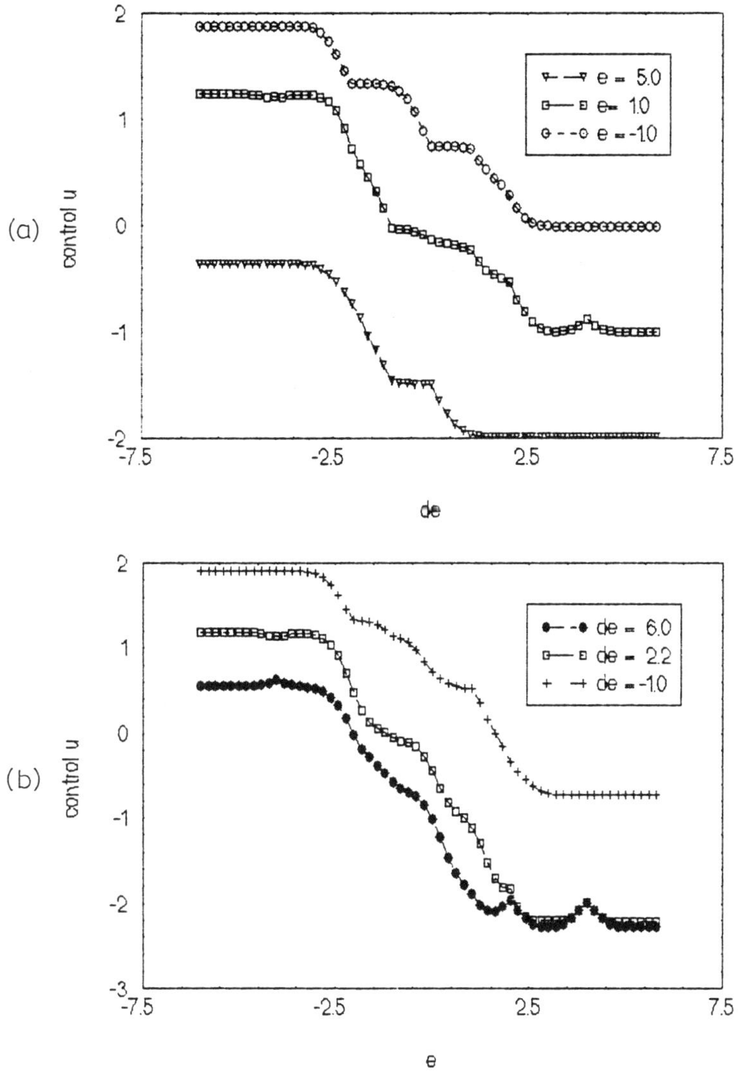

Fig. 6.2 Nonlinear characteristics of control u = Φ(e,de) for e fixed (a), de fixed (b)

$$\overline{u} = \sum \frac{u_j}{r}$$

(in the continuous case, we select $\overline{u}$ by taking the average value:

$$\bar{u} = \frac{\int\limits_{U(u)>\lambda} u\,du}{\int\limits_{U(u)>\lambda} du}$$

where the integrals are taken over all the disjoint intervals in which $U(u) > \lambda$ is satisfied).

COG: the centre of gravity accommodates contributions from all the grades of membership. In its discrete format, we have:

$$\bar{u} = \frac{\sum\limits_{j} U(u_j)\,u_j}{\sum\limits_{j} U(u_j)}$$

while the continuous version implies:

$$\bar{u} = \frac{\int U(u)\,u\,du}{\int U(u)\,du}$$

The modified COG applies to the continuous universes of discourse:

$$\bar{u} = \frac{\sum\limits_{k}^{N} \bar{u}_k\,\lambda_k}{\sum\limits_{k}^{N} \lambda_k}$$

where $\bar{u}_k$ is a pointwise representative of the fuzzy set of control standing in the kth control rule.

6.5 Relay Analogy of the Fuzzy Controller

The fuzzy controller equipped by the MOM method exhibits an interesting nonlinear-relay characteristic. We assume that U_k are normal fuzzy sets symmetric around their maxima. The proof is straightforward. The numerical input x_0 activates the X_ks to some degrees $\lambda_k = X_k(\mathbf{x})$, k = 1, 2, ..., N. We select the highest one, say k_0:

$$\lambda_{k_0} = \max \lambda_k \tag{6.5}$$

The index k_0 identifies a certain region of the inputs of the controller, say $X(k_0) = \{x \mid \text{condition (6.5) is satisfied}\}$. Overall, these regions form a

Boolean partition of the input space, namely:

$$\mathbf{X} = \bigcup_{k=1}^{N} X(k) \quad , \quad X(k_i) \cap X(k_j) = \varnothing \qquad \text{for } i \neq j$$

The index k_0 picks up the fuzzy set of control U_{k_0}, which, owing to the MOM method and the symmetry of the fuzzy sets of control, is converted into a single numerical quantity u_{k_0}. For the two-input fuzzy controller with fuzzy labels E_1, E_2, E_3, DE_1, DE_2, we obtain these relay-like characteristics:

	DE_1	DE_2
E_1	u_{11}	u_{12}
E_2	u_{21}	u_{22}
E_3	u_{31}	u_{32}

6.6 Robustness of the Fuzzy Controller

The discussion in Chapter 3 showed clearly that fuzzy sets forming the fuzzy partition for the input variables of the controller exhibit a certain noise immunity over its Boolean counterpart. In general, this fault-tolerance should also be visible for the fuzzy controller. Setting the experiment in the same way as in Chapter 3, the difference between the control value u obtained for exact input information (e and de) and that u_{error} generated by the controller for the noisy version of the input (the Gaussian noise is imposed on the exact input value) is viewed as a suitable indicator of fault-tolerance.

Figs. 6.3 and 6.4 summarise the values of $u-u_{error}$ for the fuzzy controller using the fuzzy partition and the induced Boolean one.

The results for the COG method have been used to determine a pointwise value of fuzzy control. The results show that the fuzzy controller with the fuzzy partition features significantly higher values of fault-tolerance (lower values of the absolute differences). A significant portion of the input noise is absorbed by the fuzzy partition and does not affect the output of the controller.

A more general approach is to include noise directly in terms of its probability functions. The level of activation of the ith control rule then becomes a random variable. We denote it by Λ_i. Its distribution function

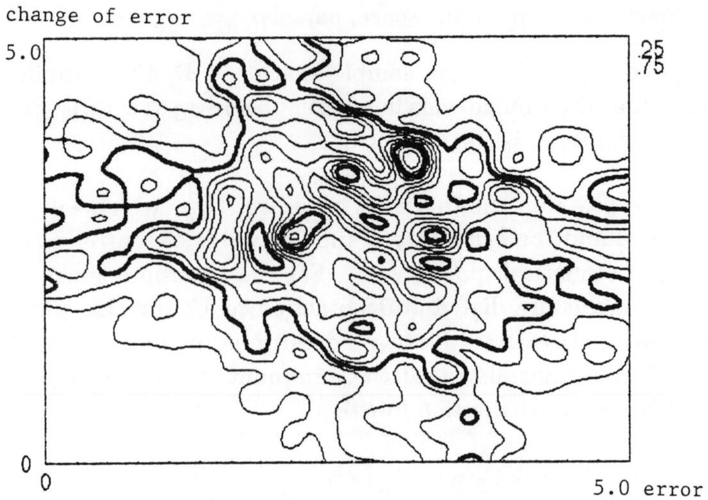

Fig. 6.3 u-u_{error} for the fuzzy controller with the fuzzy partition for input variables; standard deviation of noise σ = 1.0

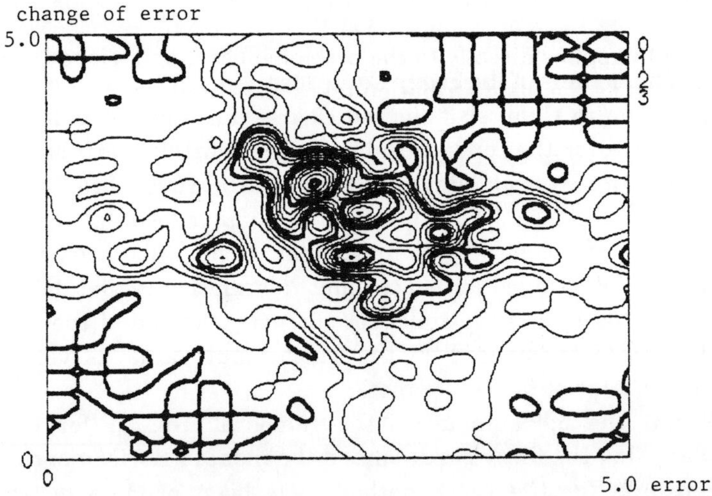

Fig. 6.4 u-u_{error} for the fuzzy controller with the Boolean partition for input variables; standard deviation of noise σ = 1.0

F_{Λ_i} comes from a basic formula of probability calculus:

$$F_{\Lambda_i}(w) = P\left\{\omega \mid \Lambda_i(w) < w\right\} = \int_{\{x:X_i(x)<w\}} dF_x(x)$$

$w \in [0,1]$

where ω is en element of the sample space $\omega \in \Omega$. P(.) stands for the probability that the random variable is kept below w, $w \in [0,1]$. Since Λ_i is a random variable, so is the control U:

$$U(u,\omega) = \bigvee_{i=1}^{N} \left[\Lambda_i(\omega) \wedge U_i(u)\right] \quad , \quad u \in \mathbf{U} \quad , \quad \omega \in \Omega$$

We now calculate the distribution function of U; for this purpose, we assume that $u \in \mathbf{U}$ is fixed. First, we determine the distribution function of Λ_i and U_i. We remember that the minimum of two random variables, say A and B, has a distribution function:

$$F_{min(A,B)}(w) = F_A(w) + F_B(w) - F_{AB}(w,w) \tag{6.6}$$

If A and B are independent, we obtain:

$$F_{min(A,B)}(w) = F_A(w) + F_B(w) - F_A(w)\,F_B(w) \tag{6.7}$$

This formula applies directly to the problem above, and, since U_i is not a random variable (i.e. its distribution takes on 0-1 values):

$$F_{U_i}(u) = \begin{cases} 0 \quad , \quad \text{if } w < U_i(u) \\ 1 \quad , \quad \text{otherwise} \end{cases}$$

(6.7) simplifies thus:

$$F_{min(\Lambda_i,U_i)}(w) = F_{\Lambda_i}(w) + F_{U_i}(u) - F_{\Lambda_i}(w)\,F_{U_i}(u)$$

Next, if $w < U_i(u)$, then

$$F_{min(\Lambda_i,U_i)}(w) = F_{\Lambda_i}(w)$$

otherwise

$$F_{min(\Lambda_i,U_i)}(w) = 1$$

In the next step, the distribution function of the union of $\Lambda_i \cap U_i$ (if the random variables are independent) is:

$$F_U(w) = \prod_{i=1}^{N} F_{\min(\Lambda_i, U_i)}(w)$$

Denoting $w_{min} = \min_i \; U_i(u)$:

$$F_U(u) = \begin{cases} \prod_{i=1}^{N} F_{\Lambda_i}(w) & , \text{ if } w < w_{min} \\ 1 & , \text{ otherwise} \end{cases}$$

This procedure is realised for every element of **U** that gives the required distribution function of the control.

The derived formula enables us to study the robustness of the fuzzy controller, finding how the disturbances (noises) influence the fuzzy control of the fuzzy controller. We can consider the ratio of the mean value and the standard deviation as a measure of the randomness of the variable. A plot of this ratio versus the same ratio of the input variable, namely:

$$\frac{\sigma_u}{m_u} = g\left(\frac{\sigma_x}{m_x}\right)$$

where

$$m_u = \int_0^1 w \, dF_U(w) \quad , \quad \sigma_u = \int_0^1 (w - m_u)^2 \, dF_U(w)$$

may serve as an indicator of the robustness of the fuzzy controller. If m_u/σ_u is fairly constant for increasing m_x/σ_x, the fuzzy controller is robust for the specified range of variance of input.

In comparison to the previous experimental approach, the latter one is useful in carrying out parametric studies on different membership functions and analysing their influence on the maximal noise suppression.

6.7 Analysis of the Dynamic Properties of the Fuzzy Controller

The dynamic properties of the fuzzy controller are analysed in a closed-loop control structure involving the fuzzy controller and the system under control. Some relevant facts are summarised below:

• fuzzy controllers are successfully applied to ill-defined processes where their models and/or performance indices (goals of control) are not defined explicitly. As emphasised before, the model is 'coded' directly into the control protocol of the fuzzy controller.

• fuzzy controllers are used in cases where control is left to the operator. Therefore, the fuzzy controller, in addition to its evident advantages, inherits some shortcomings arising in this type of control. In particular, it reaches the set point quickly and without significant overshoot, but may have some oscillations around the set point or a certain steady-state error. Thus, precision is not as good as with PID controllers, where an integral action (I-part) of the controller eliminates this type of discrepancy.

In studying the dynamic properties of the fuzzy controller, we must have a model of the process so that we can monitor the impact of the successive control actions.

Since a model of the system is not available (this is simply a cornerstone feature of the idea of fuzzy controllers), the dynamic properties of the closed-loop structure have to be derived experimentally. The dynamic properties of the controller can be adjusted by a series of carefully designed experiments. This implies a trial-and-error method. The dynamic analysis is completed at two conceptually distinct levels:

• at a numerical level. The fuzzy controller cooperating with the system is viewed as a certain nonlinear numerical transformation that, in conjunction with the system under control, gives rise to a certain phase portrait of the overall structure.

• at a linguistic level. At this level, the linguistic phase plane is exploited and the resulting trajectories generated by the system-controller structure are mapped there. This means that the overall analysis is performed at a completely different level.

One possible way of improving the dynamic properties is to adjust the scale factors used in the construction of the fuzzy controller. We recall (Section 4) that they scale up or down the entire universe of discourse. As verified by many experiments, they play a primordial role in the formation of the dynamics of the closed-loop structure leading to the desired response of the controlled system (Fig. 6.5).

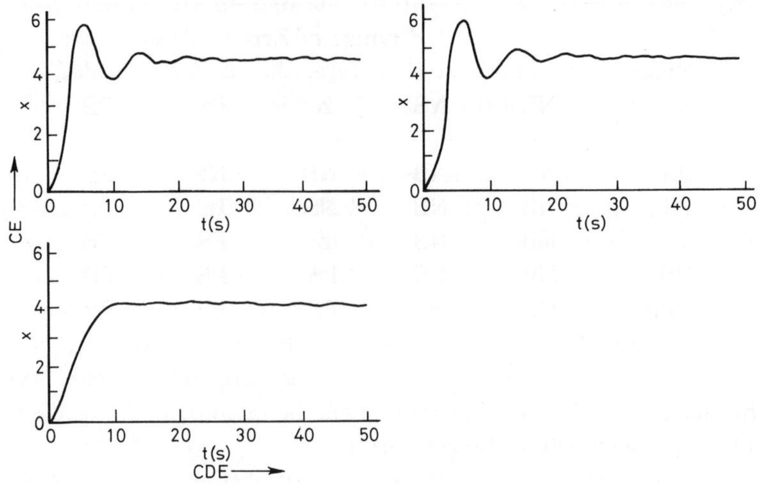

Fig. 6.5 Response of the system for different scale factors

For a controller with two inputs (error and change of error), the corresponding scale factors influence the dynamics of the system:

- if both CE and CCE increase, the control becomes more sensitive around the set point until oscillations are observed,

- where CE and CCE decrease, a tolerance band exists around the set point and a large steady-state error is quite common; so if the scale factors are too small, the system gives a poor response.

Example 6.2
Let the transfer function of the function be:

$$K(s) = \frac{K}{(s\,T_1 + 1)\,(s\,T_2 + 1)} = \frac{5}{(4s+1)\,(5s+1)}$$

The fuzzy controller consists of a set of rules driven by the linguistic values of the error and the change of error:

Error	Change of Error				
	NB	NS	Z	PS	PB
PB	NB	NB	NB	NS	PB
PS	NB	NB	NS	PS	PB
Z	NB	NS	Z	PS	PB
NS	NB	NS	PS	PB	PB
NB	NB	PS	PB	PB	PB

Furthermore, we add rules describing the behaviour of the controller in a small neighbourhood of the zero error,

Sum of Errors	Error is CB, and Change of Error is CB
NEG	CBN
POS	CBP

Thus, the fuzzy controller may be a three-input:single-output system. The corresponding membership functions of error, change of error and sum of errors are given in Fig. 6.6. Experiments with different scale factors give us these results: For increasing scale factors attached to the error (CE), the response of the system is faster, with larger overshoots. Significant oscillations are also reported. When changing the value of the scale factor responsible for the change of error (CCE), we observe that the rise-time and peak-time (the moment when the system response x(t) is maximum) remain much the same.

For another view of the performance of the controller, it is useful to describe the system response by some indices. However, we can also analyse the results produced by the controller by ‘inspection’ of its response. We deal with three indices:

(i) a squared error between the system output and a set point x_0:

$$Q_1 = \int_0^{\infty} e^2(t)\, dt$$

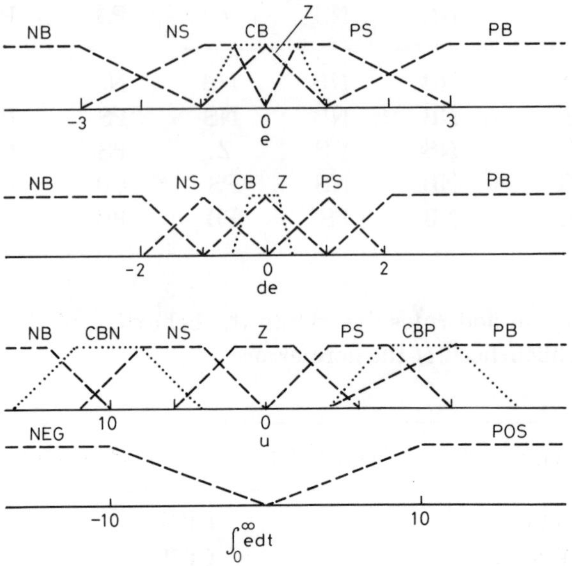

Fig. 6.6 Membership functions of fuzzy sets of error, change of error, and sum of errors

(ii) $Q_2 = \dfrac{\sup x(t) - x_0}{x_0}$

(iii) $Q_3 = \min\left\{t \,|\, x(t) = x_0\right\}$

Q_2 expresses a relative overshoot recorded in the response of the controlled system, while Q_3 describes the moment at which the response of the system is equal to the set point.

Before giving a detailed insight into plots visualising the relationships between them and the parameters of the fuzzy controller, we show that an 'optimal' set of parameters ensuring a proper system performance (obtained by 'inspection' only) is:

- for the error, CE = 0.3,
- for change of error, CCE = 0.45,

- for sum of errors, CSE = 0.15,
- for control, CU = 0.5,

while the set point is x_0 = 5. The resulting system response is shown in Fig. 6.7.

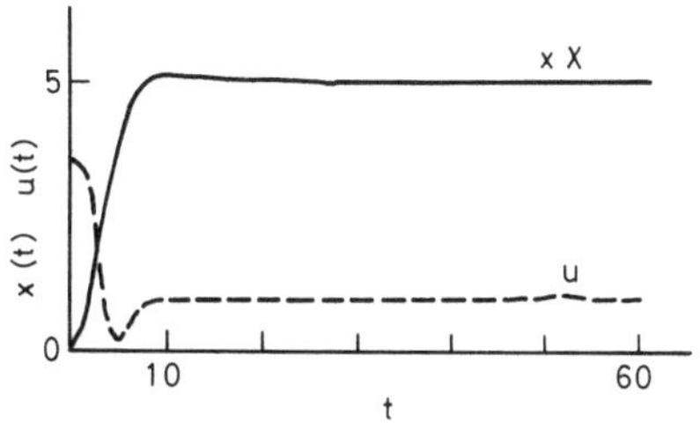

Fig. 6.7 System response under fuzzy control

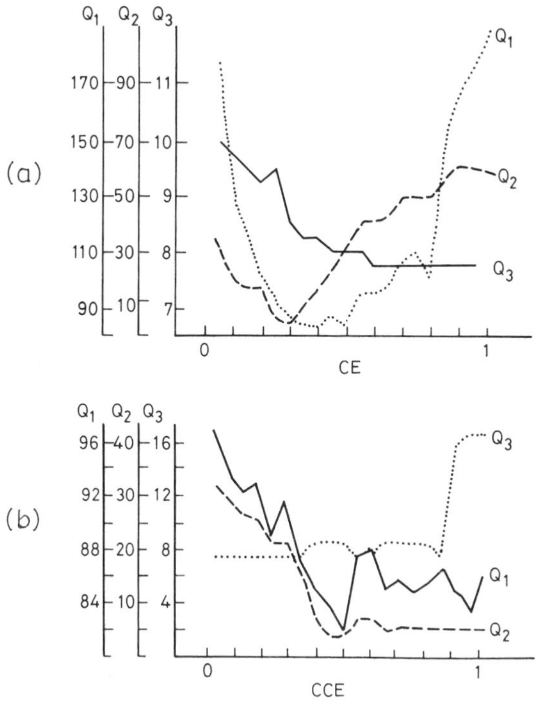

Figs. 6.8(a),(b) Q_1, Q_2 and Q_3 vs CE and CCE

Figs. 6.8(a) and (b) plot Q_1, Q_2, Q_3 versus CE and CCE for the remaining scale factors set to the values above. In both cases, Q_1 has several local minima; however, the changes in it are not as drastic as the changes in CE and CCE. In the remaining performance indices, say Q_2 and Q_3, the changes are different. Q_2 forms, in general, an increasing function of CE, while Q_3 does the opposite.

In changing CCE, a mirror behaviour is seen: Q_2 is a decreasing function of CCE, and Q_3 increases when CCE increases.

Results on the sensitivity of the fuzzy controller and the relevant PID controller after changes in the parameters of the system (given in % of nominal values at which the controllers are tuned), namely K and the greater constant T = max (T_1,T_2), are given in Figs. 6.9 and 6.10. For changes in K, both controllers act similarly except when K is significantly increased, say by more than 60%. In this region, the fuzzy controller is more robust. For T, a similar behaviour is observed at the same region of its changes, as discussed above. Also, the fuzzy controller is less sensitive to negative changes in the time constant.

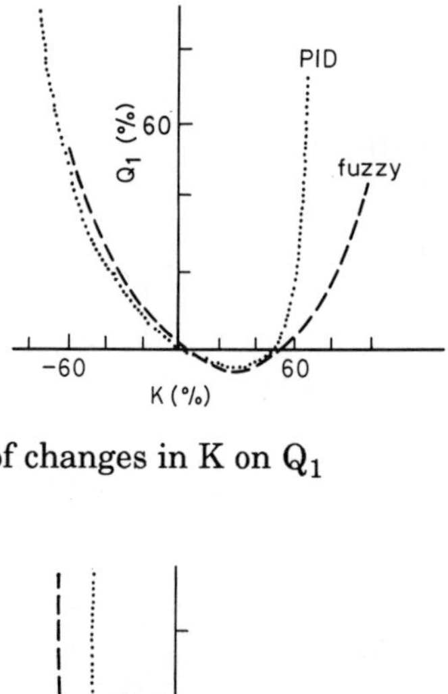

Fig. 6.9 Influence of changes in K on Q_1

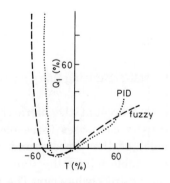

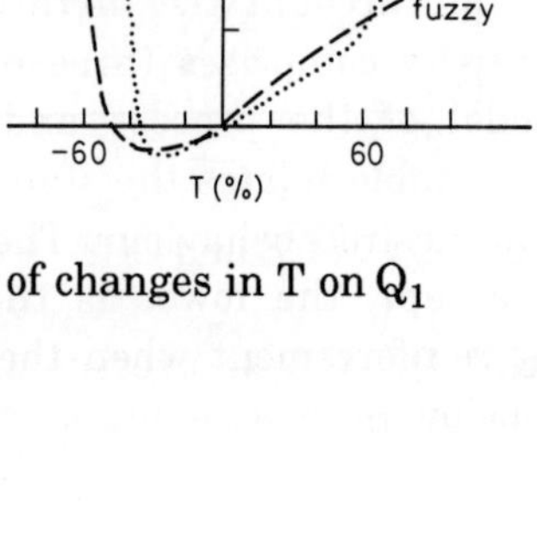

Fig. 6.10 Influence of changes in T on Q_1

6.8 Self-Organising Fuzzy Controller

An important generalisation is the addition of a self-organising structure to enable us to achieve a better performance of the controlled process, by on-line improvement of the properties of the fuzzy controller. This is especially useful when the set of control rules is inadequate for the dynamics of the controlled process. A glance at the fuzzy controller shows that many of its parameters must be optimised if it is to do its job better. They may be listed in these groups:

- the scale factors of the input and output variables,
- the set of control rules,
- the membership functions of the variables.

We have already noted that the scale factors have a strong influence on the dynamics of the closed-loop system (rise-time, oscillation, amplitude, overshoot etc.). Nevertheless, they cannot help us to remove the few unsuitable rules, since they influence the entire control protocol.

An effective way of tuning the control rules as the background of a self-organising mechanism is presented in [7] and [10]. The structure of the controller is shown in Fig. 6.11.

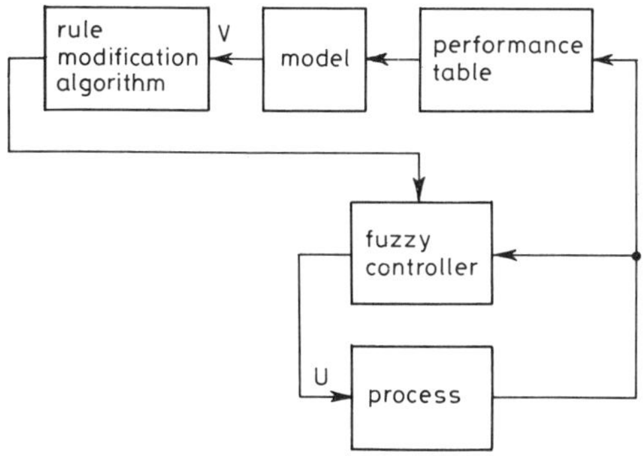

Fig. 6.11 Structure of a self-organising fuzzy controller

We now have to take into account the hierarchical structure. The higher layer of this hierarchy comprises three blocks: a performance table, a (simplified) model of the process and a rule-modification algorithm. The performance table relates the state of the process to the cost of a deviation from its desired behaviour. The larger the deviation that we are willing to accept, the lower is the associated cost. A performance table giving reinforcement when the state has error and change of error is shown below.

Error	Change of Error												
	−6	−5	−4	−3	−2	−1	0	1	2	3	4	5	6
−6	0	6	4	4	4	7	7	7	7	7	7	7	7
−5	0	6	4	4	5	5	6	6	6	6	6	6	6
−4	0	6	4	4	4	4	6	6	6	6	6	6	6
−3	0	5	4	4	4	4	4	4	4	4	4	5	5
−2	−4	−4	4	4	4	4	4	4	4	4	4	4	4
−1	−4	−4	−4	−3	−2	−2	−2	4	4	4	4	4	4
−0	−4	−4	−4	−3	−1	−1	0	1	1	3	4	4	4
+0	4	4	4	3	1	1	0	−1	−1	−3	−4	−4	−4
1	4	4	4	3	1	1	−4	−4	−4	−4	−4	−4	−4
2	4	4	−4	−4	−4	−4	−4	−4	−4	−4	−4	−4	−4
3	0	−6	−4	−4	−4	−4	−4	−4	−4	−4	−4	−6	−6
4	0	−6	−4	−4	−4	−4	−6	−6	−6	−6	−6	−6	−6
5	0	−6	−4	−4	−6	−6	−7	−7	−7	−7	−7	−7	−7
6	0	−4	−4	−4	−7	−7	−7	−7	−7	−7	−7	−7	−7

This Table [10] expresses the tolerable states of a process needing no correction of the actions of the fuzzy controller. It is quite general and does not relate to control of a specific process. If no reinforcement is needed, the respective entries of the Table are set to zero.

Any required reinforcement is transformed into a value of control by means of the model of the process. The model is a simplified linear incremental version of the process (construction of an overall model would be too difficult). Having the control computed via the model of the process contributes to the rule-modification algorithm. In essence, let V be nonzero, meaning that the control produced by the fuzzy controller has not been determined properly (assuming that the poor performance of the closed loop is a consequence of the control taken some sampling periods previously, say τ). We denote by R′ the fuzzy relation, contributing to the entire relation of the controller, that has caused its observed performance:

$$R' = E \times DE \times (U + V)$$

(as usual, '+' stands for the sum of fuzzy numbers). Then R′ must be replaced by R″. This operation can be written in logical fashion as:

$$R_{new} = R \text{ and } (\text{not } R') \text{ or } R''$$

where R_{new} is the new relation of the fuzzy controller and R is the old one. Using the respective membership functions, the above formula reads:

$$R_{new}(e, de, u) = \max\left[\min(R(e, de, u), 1 - R'(e, de, u)), R''(e, de, u)\right]$$

Simulation studies on a linear system of first, second and third order approved the algorithm for learning the effective rules of the fuzzy controller [10]. The structure satisfying the learning task is based on a simplified model of the process. However, it is difficult to verify how far the simplification and the approximate character of the model may preserve the property of self-organisation. In further studies, we show how the fuzzy model, not a linear deterministic one, may be applied and how the rules of the fuzzy controller may be computed explicitly by omitting the learning phase.

The above scheme of learning is based on a modification of the control rules: the old ones are replaced by new ones that reflect the new shape of the membership functions of error and change of error. The new shape is based on heuristics, making use of changes in such performance indices as averaged square error, maximum absolute error, and average error of the controlled variable.

Another mechanism in a self-organising fuzzy controller is described in [2].

Further schemes of modification are based on verification of the fuzzy controller with respect to numerical data that are treated as 'optimal' in a certain sense and should be followed by the controller. The numerical data are given as a series of the triples (e_l, de_l, u_l), $l = 1, 2, \ldots, N$. The basic idea is to test whether each numerical entry (e_l, de_l) with its associated control u_l activates an appropriate linguistic label of the control protocol. Consider the lth pair of data (e_l, de_l). It activates the rules to some degrees $\lambda_1, \lambda_2, \ldots, \lambda_N$. Denote its maximum by λ_t,

$$\lambda_t = \max_{l=1,2,\ldots,N} \lambda_l$$

Then the corresponding fuzzy set of control U_t coincides with the label activated by u_l, namely

$$U_t = U_t{}^*$$

where

$$U_t{}^*(u_l) = \max_{k=1,2,\ldots,N} U_k(u_l)$$

If this condition holds, we say that the control rule is consistent with the lth element of the data set. If, on the other hand, the rule is inconsistent, all the cases of inconsistency are summarised and the rules should be properly updated. One possible adaptation mechanism is to shift fuzzy sets of control along their universe of discourse. In particular, we take into account the modal value of each of the membership functions of the fuzzy sets of control U_k and shift them so that the mismatch situations are eliminated or their number is reduced. Let λ_t and λ_k denote the levels of activation of the tth and kth fuzzy sets of control. Since U_k should have been activated instead of U_t, the modal value of U_k, say u_k, should be updated. This adjustment is expressed as:

$$u_k(\text{new}) = u_k - \alpha\left(u_k - u_a\right)$$

where the adjustment component u_a is computed as a weighted average of the form:

$$u_a = \frac{\lambda_k\, u_k + \lambda_t\, u_t}{\lambda_k + \lambda_t}$$

and α stands for the learning rate, $\alpha \in [0,1]$.

The reinforcement expressed by the above formula depends on the level of activation of U_k and U_t. For instance,

- for $\lambda_k = 0$ and $\lambda_t = 1$, which portrays the most conflicting case, the adjustment is equal to $u_a = u_t$ and the new modal value of u_k becomes:

$$u_k(\text{new}) = u_k - \alpha\left(u_k - u_t\right) = u_k - \alpha\, u_k + \alpha\, u_t$$

so $u_k(\text{new})$ is moved towards u_t. In particular, if α equals 1 (or nearly equals 1), $u_k(\text{new})$ is replaced by u_t.

- for $\lambda_k = 1$ and $\lambda_t = 0$, the adjustment gives $u_a = u_k$, and $u_k(\text{new}) = u_k$; so no adjustment is necessary.

The adjustments are performed successively, covering all the data available in the learning process.

6.9 Linguistic Phase-Plane Analysis of the Fuzzy Controller

The fuzzy partition introduced for the input variables of the controller (i.e. the areas in which the linguistic labels have a dominant effect) determines the regions of attraction of the individual fuzzy sets. For the two-input case incorporating error E and its change DE, the resulting formulas describing a linguistic cell (i_0, j_0) are:

$$(i_0, j_0) = \left\{ e \in \mathbf{E}, de \in \mathbf{DE} \,\middle|\, E_{i_0}(e) = \max E_i(e), DE_{j_0}(de) = \max DE_j(de) \right\}$$

The idea of linguistic phase-plane analysis is to describe and study the behaviour of the closed-loop control structure in terms of the linguistic labels. We develop notations that are essential in the analysis of the controller: some well known concepts of stability (such as fixed point, limit cycle etc.) apply at the global linguistic (set-theoretic) level; some new phenomena call for additional definitions.

We start from a collection of data describing the behaviour of the closed-loop system with a fuzzy controller. By the grades of membership of E_i and DE_j, we uniquely derive their position in the linguistic phase plane. By connecting these data points, we build a linguistic trajectory (Fig. 6.12).

At a certain point of time, the system stays in the (i_0, j_0) sector of the plane. To determine the general tendency regarding possible linguistic transitions, we consider all the transitions starting at (i_0, j_0) and ending up at one of its adjacent cells, say (i_1, j_1), $i_1 = i_0$, $j_1 = j_0$. A simple count gives the number of paths originating from (i_0, j_0) and entering (i_1, j_1):

$$n(i_0 \Rightarrow i_1, j_0 \Rightarrow j_1)$$

By selecting the maximum over all the adjoining cells:

$$\max_A n(i_0 \Rightarrow i_1, j_0 \Rightarrow j_1)$$

where A denotes all the cells (i_1, j_1) adjacent to (i_0, j_0), we aggregate all those trajectories determined at the 'microscopic' (i.e. numerical) level and identify a part of the global (linguistic) trajectory, say:

$$E_{i_0} \Rightarrow E_{i_2} \quad , \quad DE_{i_0} \Rightarrow DE_{j_2}$$

where i_2 and j_2 yield a maximum taken over all the elements of A. A series of transitions completed over time constitutes the linguistic trajectory of

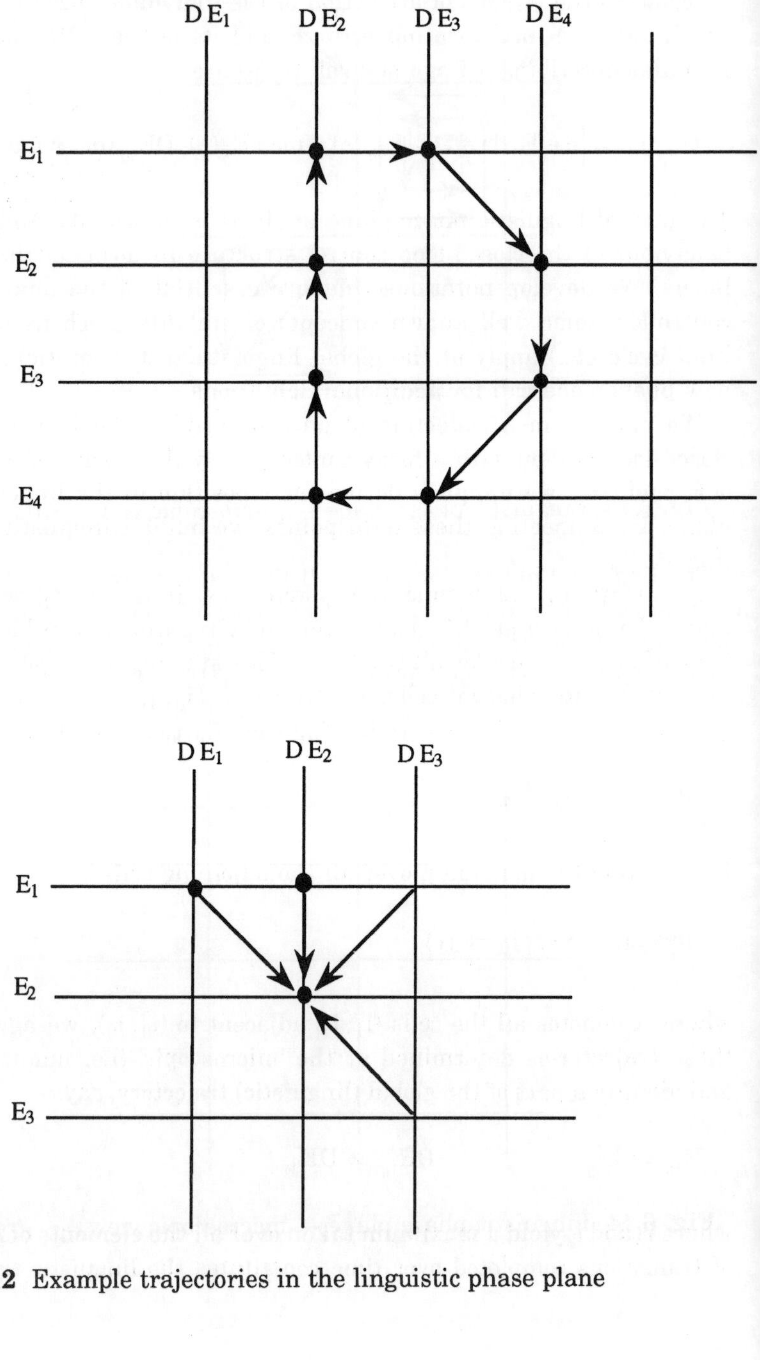

Fig. 6.12 Example trajectories in the linguistic phase plane

the closed-loop system. The standard patterns of stability (like fixed point, limit cycle) have the interpretation, with respect to the linguistic labels shown in Fig. 6.13.

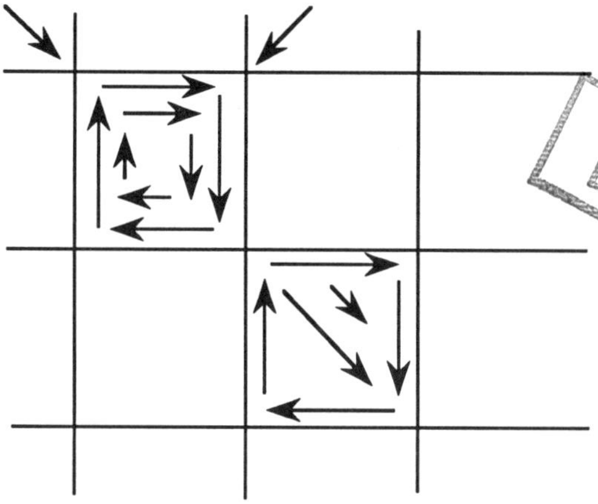

Fig. 6.13 Linguistic phase plane — microscopic view

The linguistic phase plane puts all the states as corresponding points of intersection of the grid. An interesting phenomenon occurs when some transitions start at (i_0, j_0) and do not leave the region, even though some different numerical values are obtained (Fig. 6.14).

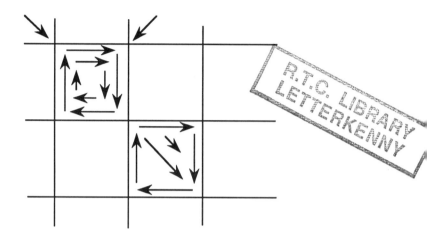

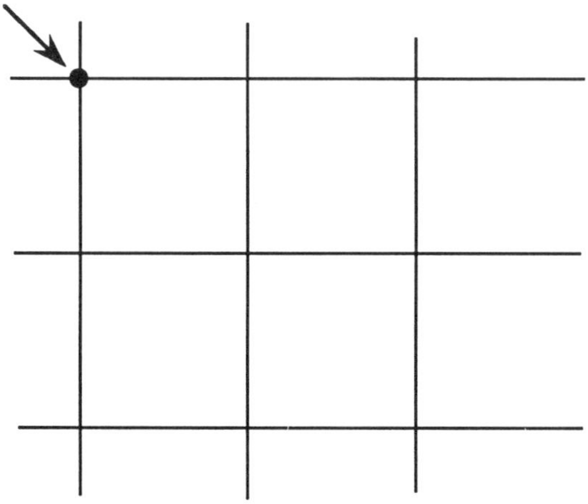

Fig. 6.14 Linguistic phase plane — macroscopic view

From the microscopic point of view, some low-level numerical dynamics are visible (Fig. 6.13). From the level of the linguistic phase, these are fluctuations and are completely ignored (absorbed by the linguistic label) — the trajectory ends up at a single point of the linguistic phase plane. We can envisage that the notion of stability and its definitions for fuzzy and numerical models may vary significantly: a system that is unstable at the microscopic level may be stable at the linguistic level.

6.10 Interpreted and Compiled Versions of the Fuzzy Controller

Subject to many technical details, the fuzzy controller can be utilised in its interpreted or compiled version. The interpreted version includes all the control rules, which, for any input information, are invoked in parallel and 'interpreted' with respect to the existing circumstances, yielding a final fuzzy control. This version applies to all the modes discussed in Chapter 4. The main disadvantage there is that the controller requires some specialised computing resources. In hardware implementation, it calls for a fairly specialised architecture; for details, refer to [1,3].

An evident advantage lies in the flexibility of the controller developed in this version: it can easily be modified by changing its components and/or its structure (e.g. rules, conditions, actions, scaling factors etc.)

The compiled version is mainly used to achieve a compact and simple realisation from conventional and inexpensive digital components. A popular technique is based on look-up tables. For the number-to-number type of mappings realised by the fuzzy controller (the first basic mode of operation; see Chapter 4), we determine the nonlinear numerical relationship of the fuzzy controller, say $u = \Phi(e,de)$, and summarise its discrete version in tabular form.

For instance, the look-up table for the rules in Example 6.2 and the MOM transformation method appears like that shown below. A look-up table obtained in this way can be invoked for given numerical input values (e and de), and the corresponding numerical value of the control determined immediately. The realisation is straightforward and does not require special hardware; standard memories can do very well here. An interesting architecture is discussed in [1]; for a general overview, the reader is referred to [3]. Some ideas in neurocomputation could be found useful in designing optimal look-up tables with the aid of the counterpropagation method; cf. [6].

The structure of the look-up table is rigid; any, even slight, changes necessitate a completely new table (i.e. the modified control knowledge has to be recompiled). Furthermore, in this basic form the structure can

Error	Change of Error												
	−6	−5	−4	−3	−2	−1	0	1	2	3	4	5	6
−6	0	0	−1	−2	−3	−4	−6	−6	−6	−6	−6	−6	−6
−5	0	0	0	−1	−2	−3	−4	−4	−4	−5	−5	−6	−6
−4	0	0	0	0	−1	−2	−3	−3	−4	−5	−5	−6	−6
−3	1	0	0	0	0	−1	−2	−2	−3	−4	−5	−5	−6
−2	2	1	0	0	0	0	−1	−1	−2	−3	−4	−5	−6
−1	3	2	1	0	0	0	−1	−1	−1	−2	−3	−4	−5
0	4	3	2	1	1	0	0	0	−1	−1	−2	−3	−4
1	5	4	3	2	1	1	1	0	0	0	−1	−2	−3
2	6	5	4	3	2	1	1	0	0	0	0	−1	−2
3	6	5	5	4	3	2	2	1	0	0	0	0	−1
4	6	6	5	5	4	3	3	2	1	0	0	0	0
5	6	6	6	5	5	4	4	3	2	1	0	0	0
6	6	6	6	6	6	6	6	4	3	2	1	0	0

accept only numerical quantities as input information. Genuine fuzzy-set information can be accepted by admitting some special classes of regular fuzzy set as inputs (e.g. triangular, trapezoidal or Gaussian-like fuzzy numbers) and storing more information in the table.

6.11 Concluding Remarks

The properties of the fuzzy controller viewed as a control algorithm have been examined. These properties are both static and dynamic. Static properties include the completeness and consistency of the control rules, and the robustness of the controller. The results reveal the most characteristic features of the fuzzy controller. However, the nonfuzzy representation of the fuzzy controller (such as the appropriate look-up table or relay-like characteristics) must be studied in a carefully chosen context (with a specified nonfuzzy method of control determination). Studies of the dynamics of the controller are encouraging. Owing to its nonlinearity, responses of the system under fuzzy control are faster than in the system under PID control. Some additional phenomena are due to the specificity of the control task solved by a human being (e.g. some small oscillations around a set point).

The design process, even when significantly scrutinised, still involves a significant experimentation component and calls for an intensive trial-

and-error design activity. This in turn can be conveniently worked out by using one of the standard design platforms (both software and hardware) easily available on the market.

6.12 References

[1] Arikawa, H., & Hirota, K. 1988. 'Fuzzy inference engine by address-look up and paging'. Proc. 1st Int. Workshop on Fuzzy System Applications, Iizuka, pp. 45–48

[2] Bartolini, G., Casolino, G., Daroli, F., & Mortem, M. 1985. 'Development of performance adaptive fuzzy controllers with application to continuous casting plants'. *In* 'Industrial Applications of Fuzzy Control' (Ed. M. Sugeno). North Holland, Amsterdam, pp. 73–86

[3] Bonissone, P.P. 1991. 'A compiler for fuzzy logic controller'. Proc. Int. Fuzzy Engineering Symp., Yokohama, Vol. 2, pp. 706–717

[4] Daley, S., & Gill, K.F. 1985. 'The fuzzy logic controller: an alternative design scheme'. *Comput. in Ind.*, **6**, pp. 3–14

[5] Gottwald, S., & Pedrycz, W. 1984. 'Analysis and synthesis of fuzzy controller'. *Probl. Contr. & Inf. Theory*, **1**, pp. 33–45

[6] Hecht-Nielsen, R. 1987. 'Nearest matched filter classification of spatiotemporal patterns'. *Appl. Opt.*, **26**, pp. 1892–1899

[7] Mamdani, E.H., & Baaklini, N. 1975. 'Prescriptive methods for deriving control policy in a fuzzy logic controller'. *Electron. Lett.*, **11**, pp. 625–626

[8] Mamdani, E.H. 1977. 'Application of fuzzy logic to approximate reasoning using linguistic systems'. *IEEE Trans. Comput.*, **26**, pp. 1182–1191

[6] Mamdani, E.H., & Assilian, S. 1981. 'An experiment in linguistic synthesis with a fuzzy logic controller'. *In* 'Fuzzy Reasoning and its Applications' (Eds. E.H. Mamdani & B.R. Gaines). Academic Press, New York, USA, pp. 311–323

[7] Procyk, T.J., & Mamdani, E.H. 1979. 'A linguistic self-organizing process controller'. *Automatica*, **15**, pp. 15–30

[8] Ray, K.S., Ghosh, A.M., & Dutta Majumder, D. 1984. 'L_2-stability and the related design concept for SISO system associated with fuzzy logic controller'. *IEEE Trans. Syst., Man & Cybern.*, **6**, pp. 932–939

[9] van der Veen, J.C.T. 1976. 'Fuzzy sets, theoretical reflections, application to ship steering'. M.Sc. Thesis, Dept. Elect. Eng., Delft Univ. Technol., Netherlands

CHAPTER 7

Theoretical and Conceptual Developments in the Construction of Fuzzy Controllers

7.1 Introduction

We turn now to some developments in the fuzzy controller involving more-advanced concepts and architectures. These embrace ideas derived from the principles of many-valued logic (e.g. Lukasiewicz logic) and their fuzzy-set-based extensions. Probabilistic sets studied in the development of fuzzy controllers allow us to construct metarules preventing the controller from absorbing biases coded into some standard rules. We exploit the theory of fuzzy-relational equations in designing mechanisms of knowledge acquisition and inference. The Chapter also includes studies on the hierarchical (multilevel) architectures of fuzzy controllers. In the course of presenting the underlying ideas, we include numerical examples that are useful for comparison.

7.2 Fuzzification of Lukasiewicz Many-Valued Logic

Research [16,10] offers an extended (fuzzified) version of Lukasiewicz logic. This method of constructing the fuzzy controller has an appealing logical background.

An original implication studied by Lukasiewicz in three-valued logic is:

$$r = (1 - p + q) \wedge 1 \tag{7.1}$$

where p, q denote truth values of propositions, namely antecedent P and consequent Q, respectively, while r is the value of the implication P implies Q, $P \to Q$. All the truth values are treated as pointwise values lying in the [0,1] interval. We now tackle the situation where the truth values of the proposition P and the implication R (= $P \to Q$) are given; we are interested in calculating the value of the conclusion Q. This is the well known scheme of *modus ponens*:

$$P$$

$$R = P \rightarrow Q$$

$$\overline{}$$

$$Q = ?$$

from which we infer the true value of proposition Q. From the computational point of view, we need to solve (7.1) for a given p and r. The solution may be unique, it may not exist, or it may be a subinterval of the [0,1] elements that satisfy (7.1). After calculation [16], we reach these conclusions:

(i) if $r = 1$, then $q \in [p,1]$,
(ii) if $0 \le p + r - 1 < 1$, and $r = 1$, then $q = p + r - 1$,
(iii) if $p + r < 1$, there is no solution.

The set of solutions of (7.1) can be given as a multivalued mapping, $[0,1] \times [0,1] \rightarrow 2^{[0,1]}$, so that:

$$q = F(p,r) = \begin{cases} \{p + r - 1\} & , \text{ if } 0 \le p + r - 1 < 1, r = 1 \\ [p,1] & , \text{ if } r = 1 \\ \varnothing & , \text{ otherwise} \end{cases}$$

We admit the extended version of Lukasiewicz logic, accepting that the truth values of propositions P, Q, R are no longer treated as single numerical quantities but as fuzzy sets defined in the unit interval, namely P, Q, R : $[0,1] \rightarrow [0,1]$. Thus Q is calculated by applying the extension principle to the multivalued mapping F(p,q):

$$Q(q) = \sup_{(p,q) \in [0,1] \times [0,1] : F(p,r) = q} [P(p) \wedge R(r)] \tag{7.2}$$

$$q \in [0,1]$$

In terms of α-cuts of P and R, Q reads as follows:

$$Q_\alpha(q) = \left\{ q \,|\, F^{-1}(q) \cap (P_\alpha \times [0,1]) \cap (Q_\alpha \times [0,1]) \right\}$$

$$\alpha \in [0,1]$$

Furthermore, if we restrict ourselves to P and R having α-cuts in the form of closed intervals (i.e. P and R are convex fuzzy sets):

$$P = [p_1, p_2] \qquad , \qquad R = [r_1, r_2]$$

then Q can be obtained in a more readable form, i.e. as a subinterval of [0,1] derived accordingly:

$$Q_\alpha = \begin{cases} [(p_1 + r_1 - 1) \vee 0, 1] & , \text{ if } r_2 = 1 \\ [(p_1 + r_1 - 1) \vee 0, (p_2 + r_2 - 1)] & , \text{ if } p_2 + r_2 > 1, r_2 = 1 \\ \varnothing & , \text{ otherwise} \end{cases} \tag{7.3}$$

Thus, from the linguistic values of propositions R and P, we know how the membership function of the truth value of Q can be derived. Note, however, that there may exist α-cuts for which Q is an empty set.

Making a further restriction and putting r_2 equal to 1 for all α, (7.3) reduces to:

$$Q_\alpha = [(p_1 + r_1 - 1) \vee 0, 1] \tag{7.4}$$

This α-cut of Q exists for all α.

The fuzzy set of control is inferred from the linguistic control protocol in this way: The key notion is linked to a so-called truth qualification and a converse truth qualification. The input of the fuzzy controller is X. For every rule, this fuzzy set X is matched with X_i to get a measure of fit that indicates how the ith rule contributes to the control generated. Here we do not use the measure of fit studied in the previous construction of the controller, but express it by a truth qualification. This enables us to preserve a direct logical interpretation:

We consider the following proposition describing the relationships between condition X_i and a new fact X. Given the proposition:

(input of the controller is X_i) is P_i = input of the controller is X

with P_i denoting the linguistic truth value, the truth qualification enables us to convert the proposition into a formal expression:

$$X(x) = P_i\,(X_i(x)) \tag{7.5}$$

$$x \in \mathbf{X}$$

where P_i is treated as the truth value of the proposition 'the input of the controller is X_i'. Thus one can say:

'the input of the controller is X_i' is

more or less true,
true,
false etc.

In other words, one can quantify the truth in linguistic terms such as *more or less true*, *very true*, *almost false* etc.

With the fuzzy sets X_i and X, the truth value P is computed accordingly:

$$P_i(v) = X_i^{-1}(v)$$

$$v \in [0,1]$$

assuming that X_i is one-to-one mapping. If this is not true, the above formula is replaced by:

$$P_i(v) = \sup_{x:X_i(x)=v} X(x)$$

which results from the extension principle. The above method of deriving the truth value is called the converse truth qualification. Given the fuzzy truth value of the ith rule, say $R_i : [0,1] \rightarrow [0,1]$, the truth value attached to the fuzzy set of control of this rule Q_i is computed via scheme (7.5) thus:

$$U_i'(u) = Q_i\left(U_i(u)\right)$$

$$u \in \mathbf{U}$$

The fuzzy control is encoded by the truth qualification applied to U_i. The aggregation of U_is, i = 1, 2, ..., N, is carried out by performing their intersection:

$$U' = \bigcap_{i=1}^{N} U_i' \tag{7.6}$$

If a nonfuzzy value of control, say u*, is desired, it can be derived by one of the methods already described.

The structure of the fuzzy controller using fuzzified Lukasiewicz logic is shown in Fig. 7.1. The reasoning scheme realised in the second block

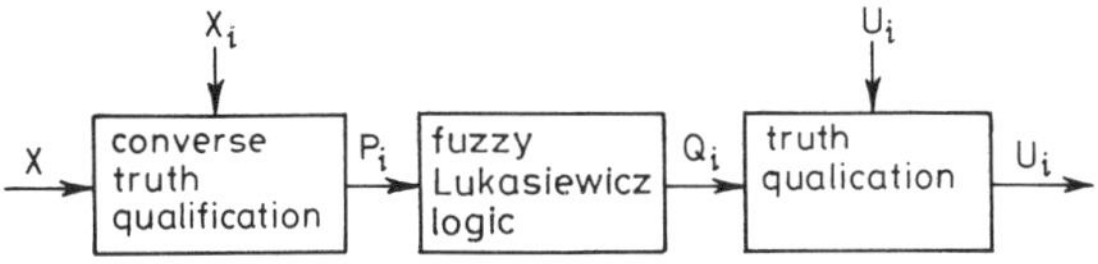

Fig. 7.1 Structure of the fuzzy controller using fuzzified Lukasiewicz logic

in this Figure is performed in the unit interval where fuzzy truth values have been defined rather than in the original physical spaces where X_i and U_i were defined.

Two features distinguish this approach from previous ones:

• First, a strong logical background of the method is observed: instead of the compositional (max-min) rule of inference that derives from set-relation calculus rather than from a logical setting, a fuzzification of a well established logical formula is performed. In this sense, it would be better to call a controller using the compositional rule of inference a fuzzy controller, reserving the name of fuzzy logic controller for one that actually uses fuzzification of a construction in many-valued logic.

• Secondly, the control action (the fuzzy set U′) is obtained as an intersection of its components coming from different rules, which enables us to handle the case where they are assembled on the basis of contradictory or competitive criteria. The rules can have different importance modelled by fuzzy relations. This stream of investigation can be extended to any multiple-valued implication such as a Gödelian one. By virtue of the extension principle, for the truth values P(p), Q(q), one has the relationship:

$$R(r) = \sup_{p,q \in [0,1]: r = p \to q} [P(p) \wedge Q(q)] \tag{7.7}$$

$r \in [0,1]$

which for the Gödelian implication '→' reads:

$$r = p \rightarrow q = \begin{cases} 1 \ , & \text{if } p \leq q \\ q \ , & \text{if } p > q \end{cases}$$

For '→' treated as φ-composition, (7.7) reads:

$$R(r) = \sup_{p,q \in [0,1]: r = p\varphi q} [P(p) \wedge Q(q)] \tag{7.8}$$

We can proceed to sketch how the fuzzy truth value Q can be worked out in a setting of fuzzy-relational equations. To use again the results in Chapter 5, we rewrite (7.8) so that they apply directly. We introduce the Boolean relation:

$$B(p,q,r) = \begin{cases} 1 \ , & \text{if } r = p \ \varphi \ q \\ 0 \ , & \text{otherwise} \end{cases}$$

$$p, q, r \in [0,1]$$

Inserting this relation B into (7.8), we get:

$$R(r) = \sup_{p,q \in [0,1]} [P(p) \wedge Q(q) \wedge B(p,q,r)]$$

$$r \in [0,1]$$

so in fact we have:

$$R = P \bullet Q \bullet B$$

Treating the above as an equation with Q unknown, the greatest solution, if it exists, is:

$$Q = (P \bullet B) \ \alpha \ R$$

7.3 Hierarchical Structures of Fuzzy Controllers

The fuzzy controller discussed so far is a single-level information-processing structure where reasoning is performed on the basis of input information through a single-step reasoning (e.g. through the max-min composition operation). In this Section, we introduce and study some classes of hierarchical structure in which additional conceptual levels are added or the controller itself is restructured into a net of hierarchies (formed by subcontrollers).

7.3.1 Hierarchy of control rules

The control rules of the fuzzy controller can be structured in such a way that the rules dealing with the more specific and detailed condition parts are invoked by more general ones. The concept of this hierarchy arises from the different levels of granularity of the fuzzy sets included in the rules. In this respect, the idea of zooming the control rules is one of the simplest solutions. The zooming effect takes place in rules operating at a 'standard' level of specificity. Some of them have a different format: their condition parts, instead of generating control actions, activate a collection of specialised rules describing the control policy in more detail. These rules indicate control actions.

The detailed rules can be nested even further, leading to the compound effect of zooming [14]. The zooming idea is illustrated in Fig. 7.2.

Note that the context of the fuzzy set named *zero* for both the error and the change of error triggers all the associated rules that express more-detailed control policy (the rules are called context-invoked). The granularity of the fuzzy sets standing there is higher than that of those present at the 'standard' level (Fig. 7.3).

So we produce more-specialised control rules. If necessary, the zooming effect can be repeated with respect to one of the context-invoked rules, again producing rules of a higher specialisation. Each zooming level is homogeneous with respect to the granularity of the condition part of the rules defined there.

The hierarchy of fuzzy controllers can be developed in order to avoid a combinatorial explosion in the number of rules occurring while increasing the number of subconditions in the condition part of each rule. For instance, considering 'n' conditions (fuzzy sets):

A and B and ... and W

and assuming that each of them takes on 'm' linguistic values, the total number of rules is equal to m^n. Thus the growth is exponential with respect to the number of variables.

Note that a complete specification of any Boolean function of 'n' variables calls for defining the values of the function at any one of the 2^n possible combinations of the variables.

In our case, the situation is even more computationally demanding, since usually $m > 2$. Even though some of the combinations are not relevant (hopefully a significant portion of them will be found physically not

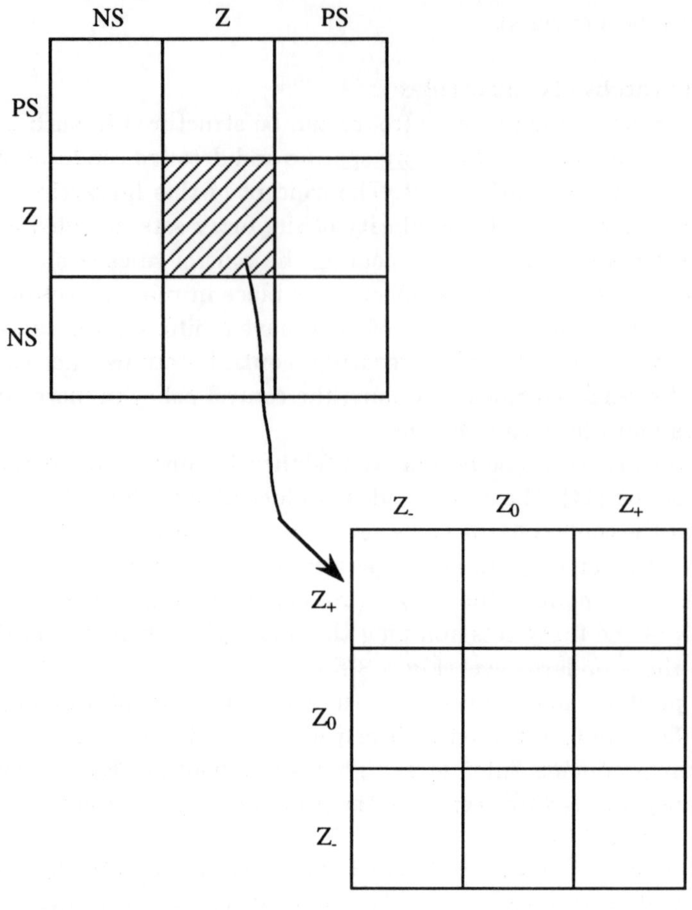

Fig. 7.2 Zooming effect in control rules

realisable), the formal criterion of rule completeness requires that all of them be fully inspected. A summary of rule number is provided below:

Number of Linguistic Terms Used by Each Variable	Number of Variables			
	2	3	4	5
m = 2	4	8	16	32
m = 3	9	27	108	324
m = 4	16	64	256	1024

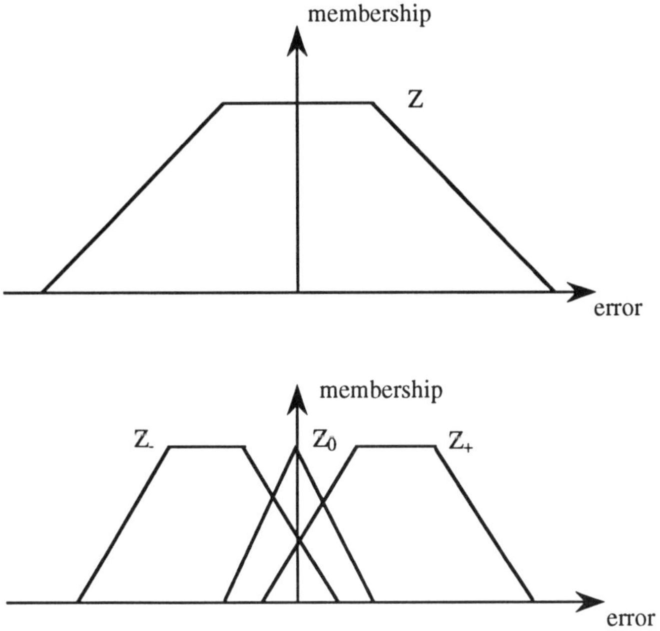

Fig. 7.3 Granularity of the fuzzy sets in the control rules involved in the zooming procedure

Bearing this in mind, the idea proposed in [13] is to design a series of fuzzy controllers with a smaller number of input variables (subset of the original variables). The same paper tackles the question of an optimal distribution of input variables per controller. To illustrate the concept, we consider a structure with n = 4 variables and compare the results with the hierarchical structure contained in Fig. 7.4.

At each level of the hierarchy, we discuss two input variables. This in total yields $m^2 + m^2 + m^2$ rules compared with the m^4 rules necessary in a single controller with four inputs. Thus, savings are obtained if $3m^2 < m^4$, i.e. $m^2 > 3$. This inequality holds for every natural 'm'. The savings increase for higher values of 'm' (Fig. 7.5).

While the computational efficiency of the distributed structure of the fuzzy controller is evident, some associated shortcomings have to be resolved. The individual controllers at successive levels of this hierarchy generate some intermediate variables, say Y, Z (Fig. 7.4). These are not fuzzy sets of control but coordinating variables describing 'aggregate'

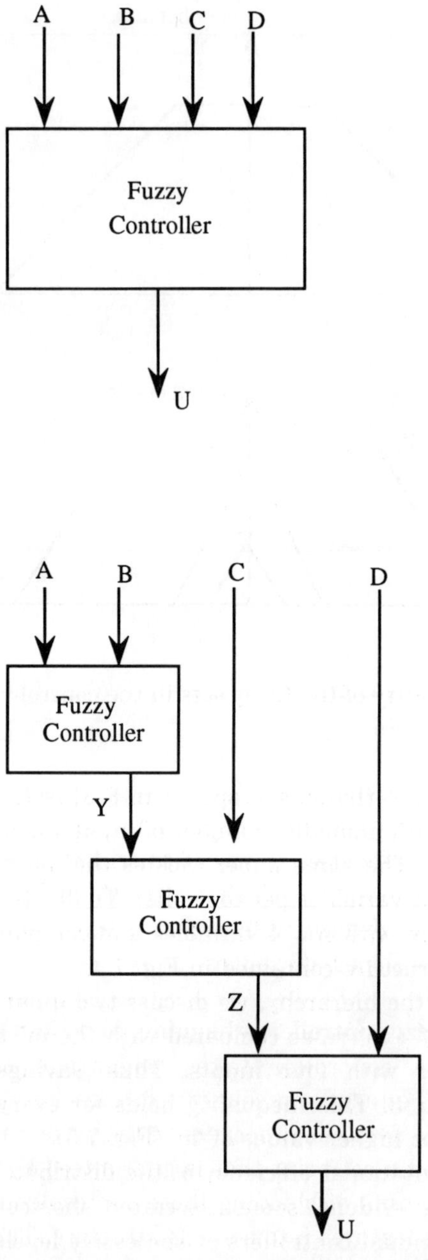

Fig. 7.4 Hierarchical structure of the fuzzy controller for four input variables

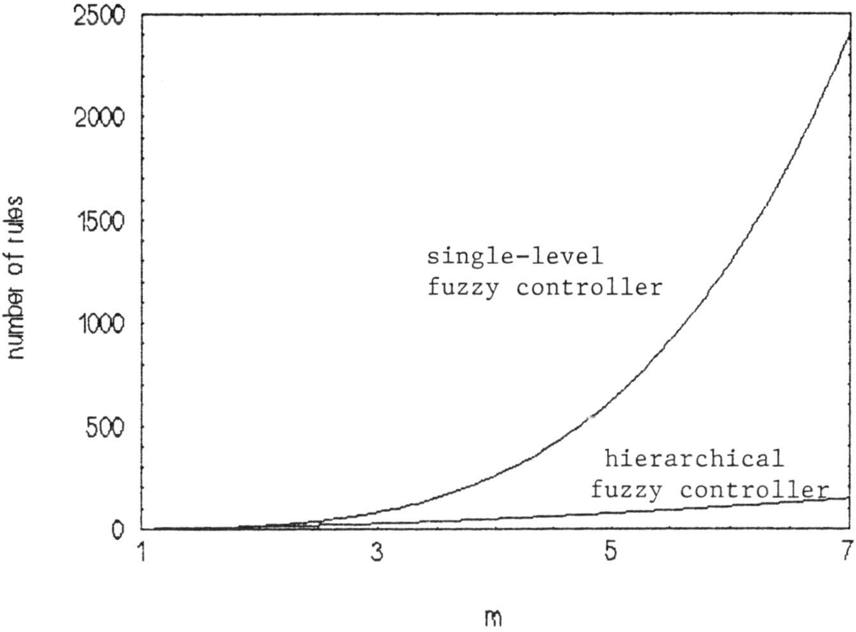

Fig. 7.5 Rule dimensionality in hierarchical and single-level fuzzy controllers

situations in the controller at the lowest level of hierarchy; e.g. A and B contribute to a certain generalised character of the state of the system (Y). The rules at consecutive levels of this hierarchy are expressed in terms of Y and the input variables. In the design procedure, we must ensure that the rules dealing with this structure (input and generalised variables) can be developed efficiently.

7.3.2 Hybrid fuzzy controller — blending fuzzy and PID controllers

The hybrid fuzzy controller takes advantage of the nonlinear characteristics of the fuzzy controller (which are particularly important in ensuring suitable dynamic properties of the system under control) and the accuracy around a set point that is guaranteed by the standard PID controller. We can think of blending these two controllers so as to bring their advantageous features into a single structure. The hybrid structure is shown in Fig. 7.6.

The essential part of this architecture is the switch S. Its role is to

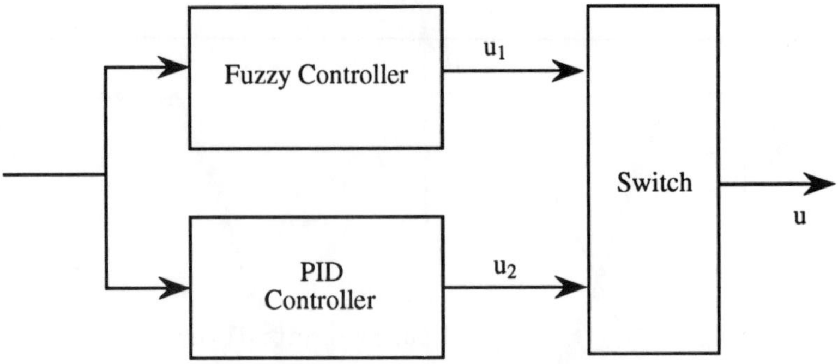

Fig. 7.6 The overall architecture of the hybrid fuzzy controller: close-by and faraway control modes

provide a control signal u that is a composite version of the signal produced by the fuzzy controller (u_1) and the one coming from the PID controller (u_2). The switch operates by providing discrimination between the Close-by and Faraway control modes of operation of the system. To characterise these modes, we introduce two fuzzy sets describing the degree of satisfaction of the goals (say error and change of error) pertaining to the notion of 'closeness'. The first one is:

$$close(\mathrm{e,de}) = close\text{-}zero \text{ error (e) AND } close\text{-}zero \text{ change of error (de)}$$

where the two fuzzy sets refer to the error and its change (Fig. 7.7).

Similarly, for the notion of control applied in the faraway mode, we define this fuzzy relation:

$$faraway(\mathrm{e,de}) = faraway \text{ error (e) AND } faraway \text{ change of error (de)}$$

Then the control value u is computed as a sort of weighted sum, where the contributions of u_1 and u_2 are affected by the above grades of membership:

$$\mathrm{u} = \frac{close(\mathrm{e,de})\, \mathrm{u}_2 + faraway(\mathrm{e,de})\, \mathrm{u}_1}{close(\mathrm{e,de}) + faraway(\mathrm{e,de})}$$

For a given control situation in which the membership function of *close-zero* attains 1 (and the *faraway*(e,de) approaches 0), the control value

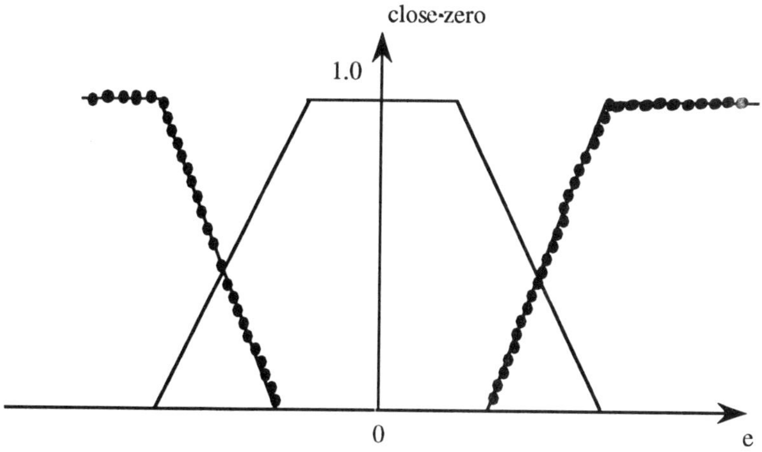

Fig. 7.7 Membership functions describing the notion of *closeness*

becomes almost exclusively dominated by the signal coming from the PID part of the controller. In the opposite situation, the structure exhibits the dominant features of the fuzzy controller.

7.3.3 Supervisory structures of the fuzzy controller

As opposed to the structure discussed in the previous Section in which both the controllers (fuzzy and PID) operated at the same level, we study the use of the fuzzy controller as a component performing a supervisory role with respect to a series of PID controllers. The basic concept is to decompose a complex control policy into a group of 'local' control algorithms. The local control algorithm applies in a fairly limited range of values of input variables (usually around a given set point). Usually, even for nonlinear and weakly nonstationary systems, these control policies, owing to their limited range of validity, can be linear. The associated design process can then be straightforward. A schematic of the complete structure is given in Fig. 7.8.

The rules of the fuzzy controller are more abstract (and less precise), as well as not being related to some control variables but rather invoking the local control algorithms. For example, the rules activating local and specialised PD controllers can read:

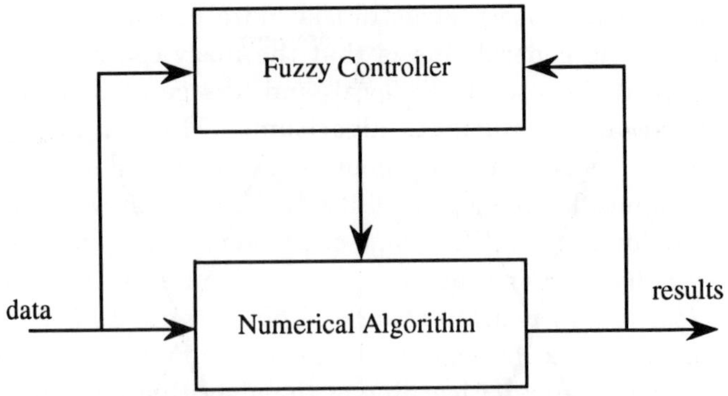

Fig. 7.8 Supervisory structure of the fuzzy controller

- if the error is E_1 and the change of error is DE_1:

(controller 1) $u(1) = a_{01} + a_{p1}\, e + a_{d1}\, de$ (rule 1)

- if the error is E_2 and the change of error is DE_2:

(controller 2) $u(2) = a_{02} + a_{p2}\, e + a_{d2}\, de$ (rule 2)

...

where a_{0i}, a_{1i}, a_{2i} are the parameters of the local controllers appearing as the right-had side of the rules, i = 1, 2, ..., N. The associated inference procedure is completed in two basic steps:

- first, a given situation (e,de) activates each of the above control rules, where the degree of activation is treated as a possibility of E_i and DE_i, i.e. $\mu_i = \min(E_i(e), DE_i(de))$,

- secondly, we determine the overall control action based on the contributions from the rules. In other words:

$$u = \frac{\sum_{i=1}^{N} \mu_i\, u(i)}{\sum_{i=1}^{N} \mu_i}$$

where u(i) denotes the control value of the ith PD controller.

This type of supervisory structure is quite common in many quite distinct areas. The general clue is that the fuzzy controller represents knowledge about the use of the 'local' and less general (more detailed and computationally efficient) algorithms. The aim of the fuzzy controller is to encapsulate the heuristics about problem-solving and 'delegate' computational responsibility on how to carry out individual actions to the lower level. The range of problems that can be structured in this way is broad. For example, [1] reports on an interesting task of improving learning in neural networks. The learning itself is performed using a standard Back-Propagation method. The fuzzy controller is used to adjust the parameters of learning of the algorithm (learning rate and momentum). The control rule here relates the values of the performance criterion (e.g. the mapping error) to the parameters of the algorithm.

The primary rôle of the fuzzy controller is to tune up the numerical algorithm by advising on the most relevant values of the parameters of learning (Fig. 7.9).

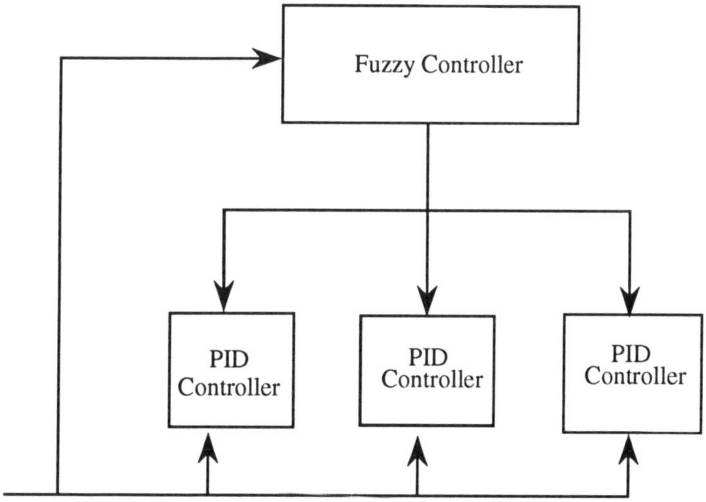

Fig. 7.9 Fuzzy controller in a supervisory structure of PID controllers

The control structure coordinated by the fuzzy controller can also involve a single PID (or any other) controller with adjustable parameters. For instance, for a PD controller:

- if the error is E_1 and the change of error is DE_1:

  a_p is P_1 and a_d is D_1

- if the error is E_2 and the change of error is DE_2:

  a_p is P_2 and a_d is D_2

  ...

where a_p and a_d are variables describing parameters (P and D respectively) of the PD controller, while $P_1, P_2, \ldots, P_N$ and $D_1, D_2, \ldots, D_N$ denote fuzzy sets used to quantify these parameters.

The parameters are adjusted with respect to the status of the system recognised by the fuzzy controller. The overall structure is compact, but requires the use of a more sophisticated controller with adjustable parameters. This, unfortunately, prevents us from using standard low-cost control modules.

7.4 Probabilistic Sets in Constructing the Fuzzy Controller

We have posed questions on the construction of the fuzzy controller, stemming from an inconsistent knowledge coded in linguistic protocol. This is especially so where several control protocols have to be taken into account, varying with respect to sets of control rules as well as to the membership functions of the linguistic terms. Thus we seek a rationale to implement them all, and try to resolve conflicting situations. Additionally, even though most of the control rules are similar, some differences may occur and the resulting form of the controller should reflect this.

Let us consider a collection of control rules:

'if X_i, then U_i' , $i = 1, 2, \ldots, M$

which consists of all the rules coming from different control protocols. So the number M is a sum of the control rules standing in each of them, namely:

$$M = N_1 + N_2 + \ldots + N_p$$

p being the number of the control protocols. Assume that the spaces in which X and U are expressed are of a finite dimensionality, namely card(**X**), card(**U**) < +∞. At first glance, we judge that the relation of the controller can be derived as a union of all R_js, j = 1, 2, ..., p. However, recalling the above remarks, we see that this approach is not appropriate because all the conflicting situations influence the inference method, leading to improper results. But if the rules are grouped (clustered) and some representative elements of each of the groups established are chosen as contributing to the fuzzy relation, we reach a plausible solution of the problem. Constructing the fuzzy controller with the aid of representative rules enables us to avoid inserting conflicting pairs of the rules [6].

As a byproduct of this construction, we obtain a measure of precision that is useful in expressing the probabilistic-fuzzy character of the inputs and outputs (state and control) of the controller. Equipped with a rough idea of how the rules can be combined, we describe an algorithm that is useful for this purpose. Its results are interpreted in a framework of probabilistic sets; we point out below how their two lowest monitors, namely membership and vagueness functions, can be estimated:

1. Consider fuzzy sets of input and output (control) standing in each rule as points in an r-dimensional space, namely $[0,1]^r$, where r = n + m, and the first 'n' coordinates are constituted by the membership functions of X_i and the rest are contributed from the fuzzy set of control U_i. Denote them by v_i, i = 1, 2, ..., N.

2. Detect the structure of this data set by means of a clustering technique: two such are described in Appendix 7.9. For the actual presentation, it is sufficient to note the form of the clustering results. Basically, the data set is partitioned into 'c' disjoint or overlapped subsets, each of them consisting of elements that are similar (where similarity is thought of in the sense of an optimised performance index). Usually, the number of clusters is not precisely given, but it is expected that the clustering algorithm will identify it. In general, the number of clusters varies from 1 to N, but we usually restrict ourselves to a range, say $[c_1, c_2]$. In our approach, we discuss a certain way of using the clustering technique. Since we have no idea how many clusters contribute to the structure searched, we wish to detect the number of classes (subsets) varying in a range. We compute the fuzzy relation of the controller, not using the entire data set but performing calculations on the basis of the representatives of the clusters (e.g. their prototypes). The idea is to take the subset of the original data set whose disturbances have been 'filtered' by the clustering performed. This avoids the situation in which data

significantly corrupted by noise contribute to the calculation of the fuzzy relation of the controller. As we showed in Chapter 6, the Hebbian-like way in which the fuzzy relation is formed does not prevent us from overriding it with some dominant noise components of the data.

Thus, let $U = [u_{ji}]$, $j = 1, 2, \ldots, c$, $i = 1, 2, \ldots, M$, be a partition matrix of the data set studied. Then the prototype of each cluster is calculated and used for the derivation of the controller relation. The prototype is viewed as the membership function of the probabilistic set (see Chapter 2). The partition matrix gives an estimate of the vagueness function. More formally, denoting X_j, U_j as probabilistic sets of state and control associated with the jth cluster, $j = 1, 2, \ldots, c$, the membership functions EX_j (EU_j) and the vagueness functions VX_j (VU_j) are calculated by these formulas:

$$EX_j(x) = \frac{\sum_{i=1}^{M} X_i(x)\, u_{ji}}{\sum_{i=1}^{M} u_{ji}}$$

$$VX_j(x) = \frac{\sum_{i=1}^{M} \left(X_i(x) - EX_j(x)\right)^2 u_{ji}}{\sum_{i=1}^{M} u_{ji}}$$

$j = 1, 2, \ldots, c$

$x \in \mathbf{X}$

The fuzzy relation of the controller has to satisfy this system of fuzzy-relational equations:

$$EX_j \bullet R = EU_j \quad , \quad j = 1, 2, \ldots, c$$

By virtue of Chapter 5, the calculations are completed in terms of the α-composition of EX_j and EU_j:

$$R = \bigcap_{j=1}^{c} \left(EX_j \,\alpha\, EU_j\right) \tag{7.9}$$

The choice of number of clusters is based on the performance index attached to the built controller. This performance index indicates how the fuzzy control coming from the equation of the controller, equal to $X \bullet R$, is close to the fuzzy control in the individual control rule, assuming that X is equal to some X_i. Then the number of clusters determining the prototypes of the fuzzy controller can be optimised by minimising the performance index considered as a function of 'c'. We denote this number by c_{opt}. The minimum of this index determines a family of subsets of the minimum rules that can be viewed in a certain sense as creating metarules of the controller. It is evident that c_{opt} is a compromise between an ability to filter the noise (for c tending to small values, this ability increases) and an ability to discover the structure of the controller (in this situation, the more extensive the set of error-free data, the better the determination of the fuzzy relation). At both extremes, i.e. for c = 1 and c = M, we cannot expect these requirements to be satisfied. A compromise can be reached for a value of c lying between 1 and M, depending on the structure of the data set and the characteristics of the disturbances existing in the data. Although we cannot determine c_{opt} analytically, we can roughly predict some trends.

The probabilistic sets used here derive naturally from the clustering of the original data from the control statements. Their advantages are twofold: First, the prototypes of clusters with derived membership functions allow us to determine the fuzzy relation of the controller. Secondly, the clusters detected in searching the fuzzy relation of the controller reveal the complexity of their metarules. So we can arrive at their membership and vagueness functions. Thus we get a deeper insight into how different fuzzy sets contribute to the same probabilistic set treated as the condition (action) of the metarule of the controller. The original control statements containing a significant number of rules now convert into a control protocol with a fairly small number of metarules:

$$\text{if } X_j, \text{ then } U_j \quad , \quad j = 1, 2, \ldots$$

7.5 The Fuzzy-Relational-Equations Approach to the Construction of the Fuzzy Controller

As already shown, the fuzzy controller can be implemented by applying different operators to compute its fuzzy relation [10]. Calculations of the fuzzy relation $R(R_i)$ vary from one case to another, but the sup-min composition as a dominant construct is preserved [8]. The question in implementing *modus ponens* is: what scheme of reasoning should be applied in a real-world application of the reasoning with fuzzy premises? The local properties of fuzzy logic [3] suggest this answer: all the implementation methods should be explored in order to find the one that best fits the given application. This, in turn, implies the problem of how to compare two different ways of implementing the reasoning method. We propose to consider a criterion of consistency by which we analyse how well the original state conditions included in the control protocol are mapped onto the corresponding control values. We discuss a certain implementation of the reasoning scheme, say method I. By it, we perform computations of the fuzzy relations R_i, i = 1, 2, ..., N, produce an overall relation R, and accept the inference mechanism. We apply method I in order to infer U for $X = X_i$, i = 1, 2, ..., N. We expect the inferred control set U to become equal to U_i. Usually, this is not the case. Some inconsistencies in the data set (i.e. the rules themselves), as well as some implementation discrepancies in the inference mechanism itself, mean that the above equality is not satisfied. The sum of distances between the fuzzy sets $U = X \bullet R$ and U_i:

$$D = \sum_{i=1}^{N} d(X_i \bullet R, U_i) \tag{7.10}$$

may be viewed as a performance index of the implementation method. For two methods, say I and II, we say that I is preferred to II if the distances associated with it fulfil the inequality D(I) < D(II).

The set of control rules may be written:

$$X \to \{\text{if } X_i \text{ then } U_i,\ i = 1, 2, \ldots, N\} \to U$$

which reads:

U and X are *related* by the relation R coming from the set of antecedents: if X_i then U_i. The fuzzy relation R is chosen to satisfy a set of equations:

$$X_i \bullet R = U_i$$

Then, if a set of solutions

$$\bigcap_{i=1}^{N} R_i$$

is nonempty (remember than R_i stands for a set of solutions of the ith equation, $R_i = \{R \mid X_i \bullet R = U_i\}$), fuzzy relation R is derived directly from the theory of fuzzy-relational equations:

$$R = \bigcap_{i=1}^{N} (X_i \,\alpha\, U_i) \tag{7.11}$$

If we compute R as a union of the Cartesian product using this formula:

$$R = \bigcup_{i=1}^{N} (X_i \times U_i) \tag{7.12}$$

i.e.

$$R(x,u) = \max_{1 \le i \le N} \left[\min\left(X_i(x), U_i(u)\right)\right]$$

then the mechanism compatible with the operator applied to R, by which the inference was performed, is:

$$U = X \,\alpha\, R$$

(this is just an adjoint fuzzy-relational equation). The crosstalk that results from the disjunctive way of combining the fuzzy relation of the rules and the max-min composition operator usually produces modified versions of the U_i s. To visualise these two ways of constructing the fuzzy relation of the controller, we look at the three control rules $A_1 \rightarrow A_3$, $A_2 \rightarrow A_2$, $A_3 \rightarrow A_1$, where the membership functions of the A_i s are shown in Fig. 7.10.

The two-dimensional contours of the fuzzy relation R(x,y), obtained as:

(a) a union of Cartesian products of the rules,
(b) a fuzzy relation constructed using (7.11),

are summarised in Figs. 7.11(a) and (b).

The difference is remarkable: for (b), we get a bunch of smooth contours of grades of membership following a diagonal of the square [0,5] × [0,5]. For (a), we get three families of characteristic nested 'boxes', strongly indicating the 'disjunctive' effect of this way of combining the rules.

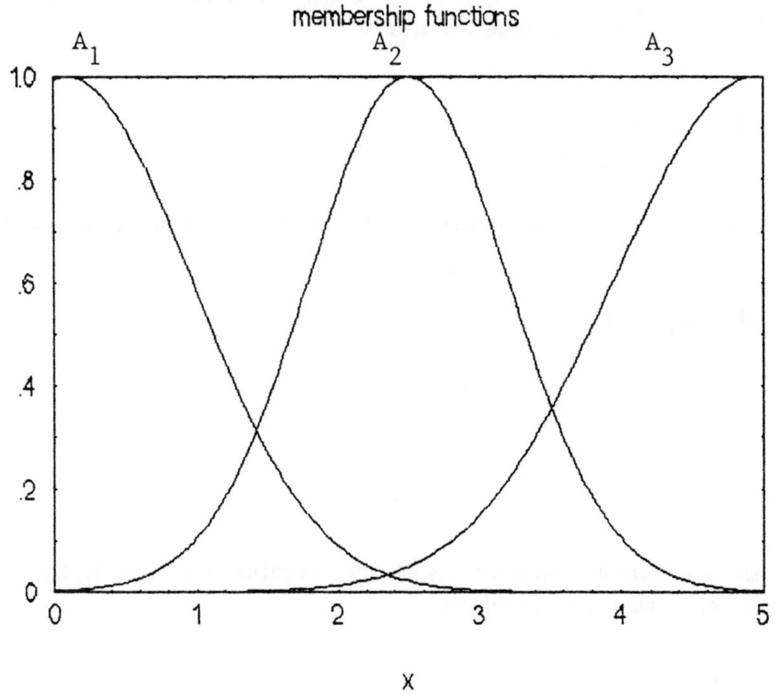

Fig. 7.10 Membership functions of A_1, A_2, and A_3

In practice, the assumptions about nonempty sets of solutions may be too restrictive. To overcome this difficulty, some methods of estimating the fuzzy relation have been proposed [11]. The fuzzy relation so obtained is the 'best' in the sense of the performance index formulated in the optimisation task. We propose to evaluate the antecedents in the reasoning scheme, introducing a certainty factor for each rule. To get its value, we calculate the fuzzy relation by formula (7.11) or (7.12). Afterwards, we determine the distances between the fuzzy sets:

U_i and $X_i \bullet R$ (or $X_i \;\alpha\; R$), namely $d_i = d(X_i \bullet R, U_i)$

After normalisation:

$$CF_i = \frac{(d_i - d')}{(d'' - d')}$$

where

$$d' = \min_{1 \le i \le N} d_i \quad , \quad d'' = \max_{1 \le i \le N} d_i$$

$i = 1, 2, \ldots, N$

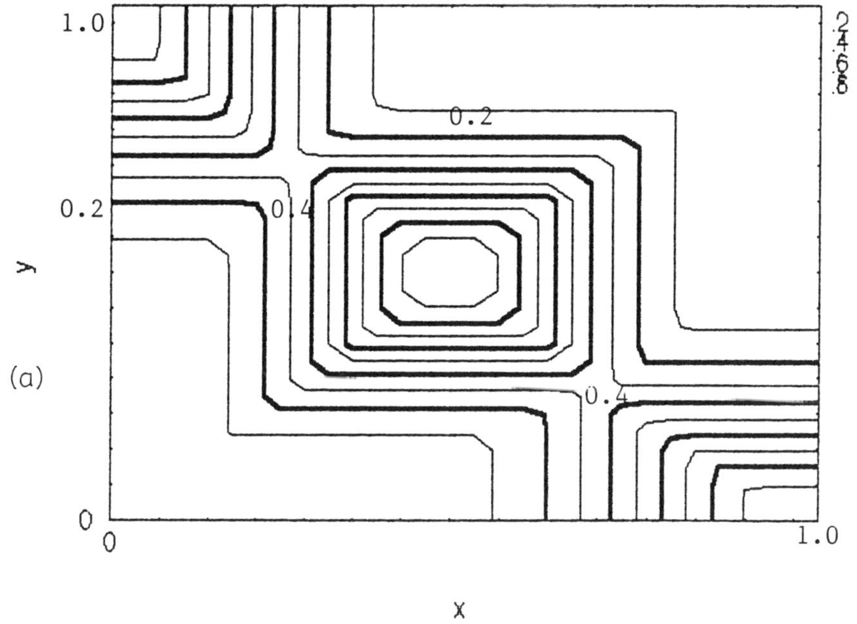

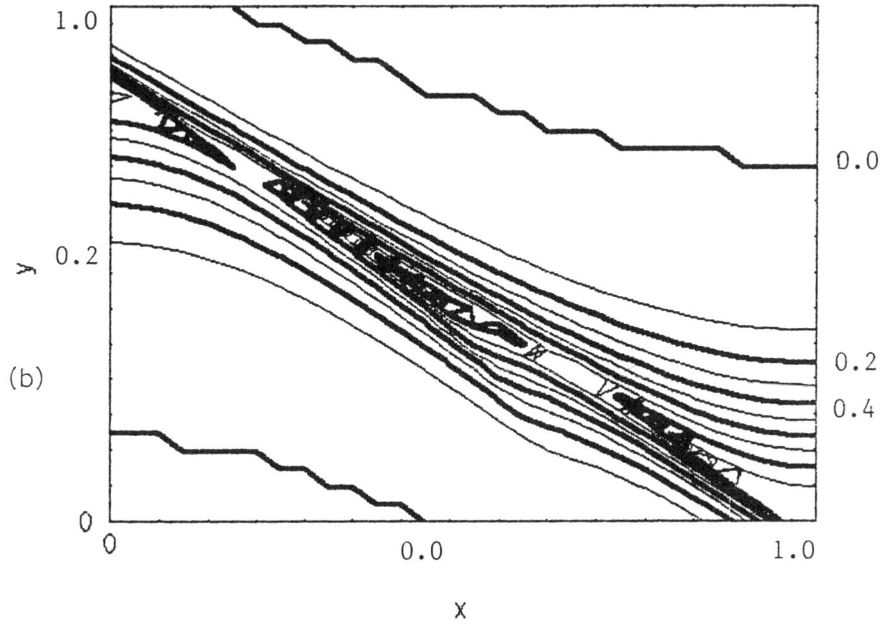

Figs. 7.11(a) and (b) Two-dimensional plots of the fuzzy relations

which gives us the certainty factor $CF_i \in [0,1]$ associated with the ith rule. The control rules are rewritten to include their certainty factors. Thus, certainty factors CF_i are included in the general:

- if X_i, then U_i with a certainty (CF_i)

Thus, rules based on a knowledge of the level of certainty should be included in the formulas expressing the fuzzy relation of the controller. They are added to the basic formulas (7.11) and (7.12) so as to express the influence of each control rule in a global equation of the controller. Thus, for the sup-min fuzzy relational equation, the modified formula for the fuzzy relation is:

$$R = \bigcap_{i=1}^{N} (\nu R_i) \tag{7.13}$$

where νR_i denotes modification of the fuzzy relation R_i so that:

$$(\nu R_i)(x,u) = [R_i(x,u)]^{CF_i} = [X_i(x)\ \alpha\ U_i(u)]^{CF_i} \tag{7.14}$$

If CF_i is equal or almost equal to zero, there is no contribution of the ith rule to the entire relation, yielding νR_i with a membership function identical to 1.0. For $CF_i = 1$, this computation ensures that νR_i equals R_i. For the fuzzy-relational equation (7.12), inclusion of the certainty factor proceeds thus:

$$R = \bigcup_{i=1}^{N} (\nu R_i) \tag{7.15}$$

i.e.

$$(\nu R_i)(x,u) = [R_i(x,u)]^{2-CF_i} = [X_i(x) \wedge U_i(u)]^{2-CF_i} \tag{7.16}$$

which is based on the same rationale as in the modification of R.

The foregoing numerical examples illustrate the methods investigated. To get a comparison, two fuzzy controllers are studied.

The first is designed for the control of a combined steam engine and boiler [9]. The structure of the antecedents of the control algorithm permits us to write it in this format: if X_i, then U_i. The control algorithm is proposed separately for the boiler and the steam engine. For the steam engine, the number of rules is 15, while for the boiler it is 9. Details of the membership functions and the universes of discourse are given in [9]. For the steam engine, the use of methods utilising fuzzy-relational equations gives these results (index D is computed using (7.10), with the distance specified as the Hamming one):

- for the sup-min fuzzy-relational equation, D = 10.4
- for the inverse composition operator, D = 1.7

For comparison, the results obtained by the previous method are characterised by D = 40.6 (case 3). The fuzzy-relational equations give superior implementation of the generalised *modus ponens*. Fig. 7.12 gives the values of the distance function 'd' for various control rules. The values of the certainty factor assigned to them are also visualised. The values of the certainty factor clearly indicate the superiority of the use of fuzzy-relational equations (cases 1 and 2) over the mechanism in case 3. We now look at the influence of the different inputs of the controller, i.e. error E and change of error CE of the state variable of the process under control, on the fuzzy control. The influence of a given input of the controller is expressed by the sum of the distances between the fuzzy sets of control resulting from the equation of the controller when this input is treated as 'unknown' (i.e. modelled by a fuzzy set with a membership function identical to 1.0) and the original fuzzy sets of control are found in the protocol. Here are the results:

	Case 1	Case 2	Case 3
D (E_i = ANY)	27.0	15.8	40.4
D (CE_i = ANY)	28.0	17.0	24.2

i = 1, 2, …, 15. This ordering is induced:

E } **CE** for case 3 ; **E** { **CE** for cases 1 and 2

This means that, in case 3, error is more important than change of error, while the converse holds for cases 1 and 2. Performing the same synthesis of the fuzzy controller for the second part of the combination under control, we get:

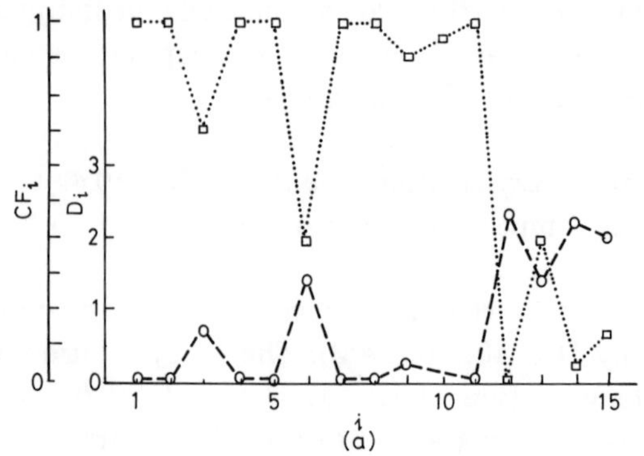

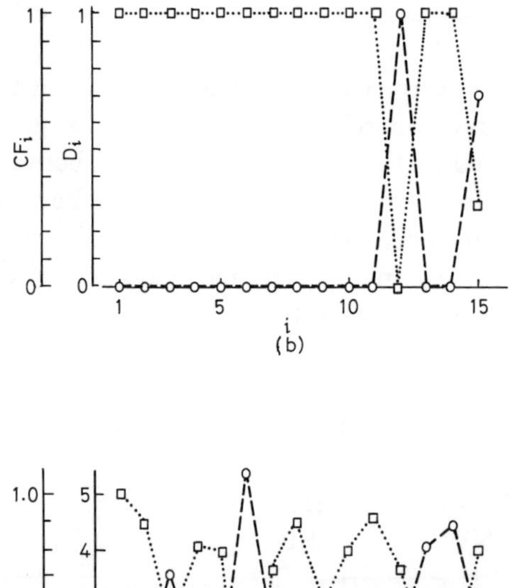

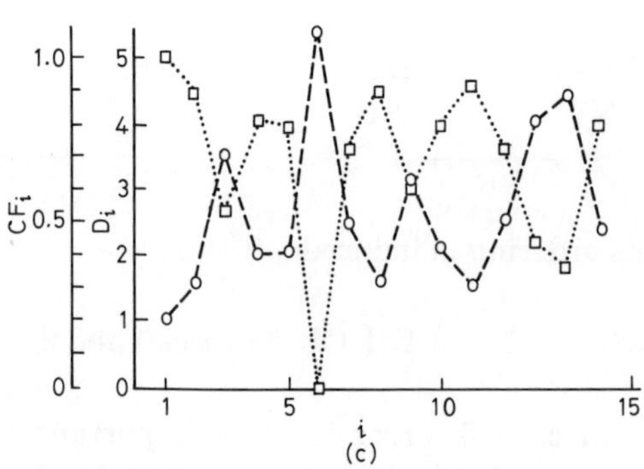

Fig. 7.12 Values of the distance function 'd' for different control rules for the three fuzzy controllers, (a), (b), (c)

	Case 1	Case 2	Case 3
D	0.6	0.0	4.8

So again the fuzzy-relational equation describes the control protocol better than the fuzzy controller implementing the Hebbian-like scheme of learning and the compositional rule of inference used in other papers.

The second numerical example comes from [5] and deals with the analysis of a fuzzy controller for a nonlinear process (engine) described by the differential equation:

$$\dot{x} = f(x)\,(u - g(x))$$

where f(x) and g(x) are nonlinear functions representing the overflow and steady- running characteristics of the engine. The model controlled is a nonlinear one, and this may cause significant obstacles in applying any algorithm derived from the control theory of linear systems. The entire structure of the system under control and of the controller itself is shown in Fig. 7.13. Notice that the fuzzy controller used now possesses two

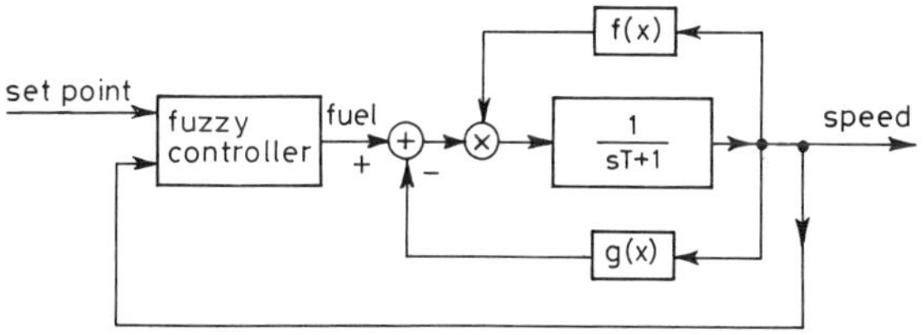

Fig. 7.13 The structure of the nonlinear system under control

inputs and one output. The fuzzy control refers to the level of fuel supplied, which is related to the speed of the engine and the set point, both expressed linguistically. The control rules are summarised in this Table:

Set Point	Speed						
	ST	VS	SL	SC	FC	FA	VF
ST	MN	MN	MN	MN	VL	MN	VL
VS	BA	NM	NM	VL	VL	MN	VL
SL	BA	LO	LO	VL	VL	MN	VL
SC	BA	BA	LO	VL	VL	MN	VL
FC	BA	HI	BA	BA	LO	BA	BA
FA	BA	HI	AA	AA	BA	AA	AA
VF	BA	HI	AA	HI	HI	HI	MX

where the abbreviations used are:

ST = stall; VS = very slow; SL = slow; SC = slow cruise; FC = fast cruise; FA = fast; VF = very fast; MN = minimum; VL = very low; BA = below average; NM = negative medium; LO = low; HI = high; AA = above average; MX = maximum.

The Table above contains 49 control rules. For the membership functions of these variables, refer to [5]. The results derived using the reasoning methods are:

	The Value of D
Case 1	63.0
Case 2	12.9
Case 3	283.3

The distance D is viewed, as before, as the Hamming one. Evidently, the methods using fuzzy-relational equations produce significantly better results than those from the max-min composition approach.

To gain a deeper insight into the performance of the controller, we construct a look-up table for the relation in case 2 (since this model appears the best among the controllers analysed). A nonfuzzy value of control is calculated:

$$u^* = \frac{\sum_{\mathbf{U}_\alpha} u_i \, U(u_i)}{\sum_{\mathbf{U}_\alpha} U(u_i)}$$

$\alpha \in [0,1]$, and $\mathbf{U}_\alpha$ is the α-cut of the fuzzy set U. The universe of discourse of control **U** consists of 13 elements:

$$\mathbf{U} = \{u_1, u_2, \ldots, u_{13}\}$$

Plots of u* versus the values of the set point for a fixed parameter α are shown in Fig. 7.14.

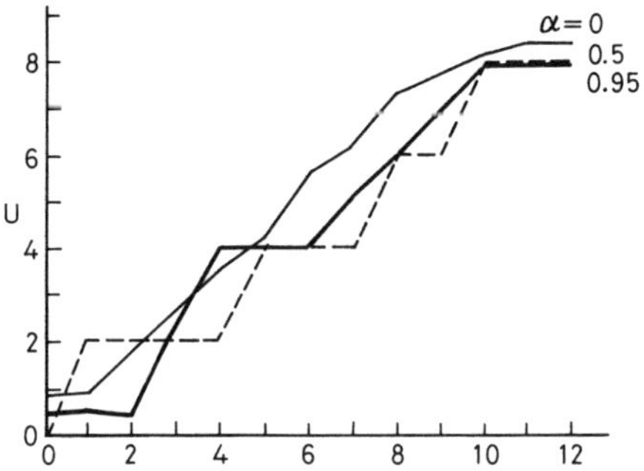

Fig. 7.14 Plots of u* versus set point for constant α

Summarising, fuzzy-relational equations are a convenient, flexible and easy tool for constructing the fuzzy controller. They allow us to avoid some implementation problems that were evident for the basic type of the fuzzy controller. The main advantages can be listed accordingly:

- there is no hesitation about choosing the correct fuzzy relation: the set of data contained in the control rules immediately implies the form of the equation. Also, one can easily verify how well the implementation scheme used fits the control rules.

The illustrative examples clearly show the advantages of this approach. A similar conclusion can be reached about the results [7], where different implication operators are studied for fuzzy models of the d.c. motor. The Gödelian implication (α-composition operator) yielded one of the best results in comparison with other forms of implication.

- There is no problem of interaction.
- The control rules can be evaluated; the certainty factor describes the importance (relevance) of a given control rule.

7.6 Numerical Examples

To illustrate the approaches used in the construction of the fuzzy controller, it is instructive to consider some numerical examples. They give a better insight into the methods discussed.

Consider a control protocol containing three rules, with condition parts covering fuzzy sets X_1, X_2 and X_3 and action parts with fuzzy sets U_1, U_2, U_3 defined in discrete spaces **X** and **U**, respectively. Their finite character is assumed only for computational purposes (Fig. 7.15). They

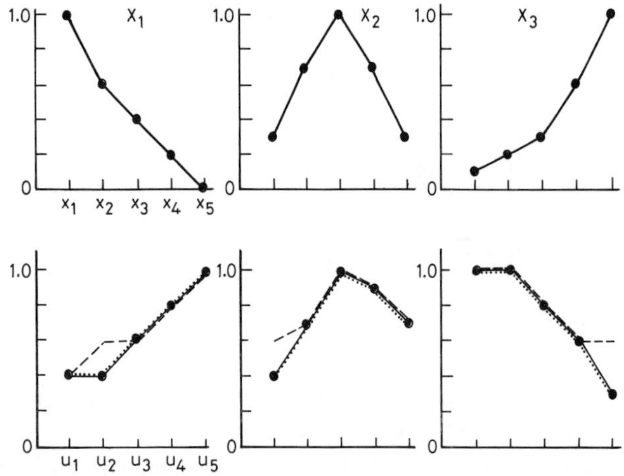

Fig. 7.15 Fuzzy sets of state and control contained in the control protocol and obtained using fuzzy-relational equations

can model linguistic notions such as *small*, *medium*, and *large*. The rules are of this format: if X_i, then U_i. Recalling the basic method and the approach using fuzzy-relational equations, the resulting fuzzy sets of control are also summarised in Fig. 7.15. In the first case, the fuzzy sets of control are a bit 'fuzzier' than the original ones, while in the second case there is a complete fit of the corresponding fuzzy sets.

Deriving the fuzzy control using fuzzy Lukasiewicz logic, the steps described in Section 7.2 have to be performed. We discuss a fuzzy set, X equal to X_1. Moreover, we consider $R_i(r) = r$, $r \in [0,1]$; so all the rules are viewed as *true*. The fuzzy truth values P_1, P_2 and P_3 obtained from comparing X and X_1, X_2 and X_3 are summarised in Fig. 7.16. Comparing

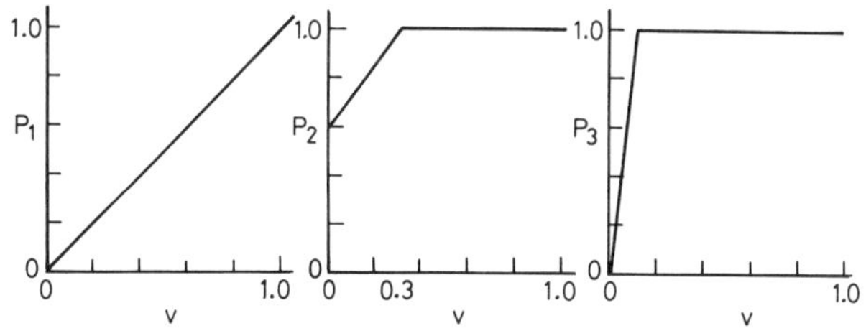

Fig. 7.16 Fuzzy truth values for the control rules

the fuzzy sets of truth, it is evident that P_3 is closer to 1.0 than, for instance, P_2; this underlines the fact that the third rule is almost irrelevant when the input of the controller is equal to X_1. Thus the so-called truth-qualification $U_i'(u) = Q_i(U_i(u))$ yields fuzzy sets of control equal to:

$$U_1' = [0.4 \quad 0.4 \quad 0.6 \quad 0.8 \quad 1.0]$$

$$U_2' = [1.0 \quad 1.0 \quad 1.0 \quad 1.0 \quad 1.0]$$

$$U_3' = [1.0 \quad 1.0 \quad 1.0 \quad 1.0 \quad 1.0]$$

so, taking their intersection, we come up with an ideal reconstruction yielding a genuine fuzzy set U_1. To illustrate this phenomenon, we study two rules: $X_1 \rightarrow U_1$, $X_2 \rightarrow U_2$, such that

$$X_1 = X_2 = [1.0 \quad 0.6 \quad 0.4 \quad 0.2 \quad 0.0]$$

and

$$U_1 = [0.4 \quad 0.4 \quad 0.6 \quad 0.8 \quad 1.0]$$

$$U_2 = [1.0 \quad 1.0 \quad 0.8 \quad 0.6 \quad 0.3]$$

where U_1 and U_2 contain two quite different fuzzy sets (e.g. *small* and *large*). Applying the methods previously discussed, the fuzzy control for $X = X_1$ for the basic fuzzy controller is equal to:

- for the method of Chapter 4:

$$U = [1.0 \quad 1.0 \quad 0.8 \quad 0.8 \quad 1.0]$$

- for the fuzzy-relational-equations approach:

$$U = [0.4 \quad 0.4 \quad 0.6 \quad 0.6 \quad 0.3]$$

- for the method of fuzzy Lukasiewicz logic:

$$U = [0.7 \quad 0.7 \quad 0.8 \quad 0.8 \quad 0.65]$$

For the first method, the fuzzy set of control has a multimodal membership function. This shape underlines the contradictions existing in the control rules. Both the remaining methods produce subnormal fuzzy sets of control, and their maximal membership functions occur at intermediate elements of the universe of discourse. In these methods, subnormality shows that a contradiction is present in the control protocol. The preference given to intermediate control values is well understood.

7.7 Conclusions

This Chapter has introduced new and more advanced concepts and architectures of fuzzy controllers. The underlying idea was to look at the knowledge representation of control rules from a general perspective. We showed how to cope with probabilistic uncertainty (including probabilistic sets) and produce controllers less susceptible to noises present in the control rules and implementation imperfections. This implied the concept of metarules, augmenting the controller by a higher level of generalisation and increased robustness. We have studied hierarchical structures where the knowledge becomes distributed around several well distinguished conceptual levels.

7.8 References

[1] Arabshahi, P., Choi, J.J., Marks II, R.J., & Caudell, T.P. 1992. 'Fuzzy control of backpropagation'. IEEE Int. Conf. on Fuzzy Systems, San Diego, pp. 967–972

[2] Ball, G., & Hall, A. 1967. 'A clustering technique for summarizing multivariate data'. *Behav. Sci.*, **12**, pp. 153–155

[3] Bellman, R.E., & Zadeh, L.A. 1977. 'Local and fuzzy logics'. *In* 'Modern Uses of Multiple Valued Logic' (Eds. J.M. Dunn & D. Epstein). Reidel, Dordrecht, pp. 103–165

[4] Bezdek, J.C. 1981. 'Pattern Recognition with Fuzzy Objective Function Algorithms'. Plenum Press, New York, USA

[5] Braae, M., & Rutherford, D.A. 1979. 'Theoretical and linguistic aspects of the fuzzy logic controller'. *Automatica*, **15**, pp. 553–577

[6] Hirota, K., & Pedrycz, W. 1982. 'Fuzzy system identification via probabilistic sets'. *Inf. Sci. (USA)*, **28**, pp. 21–43

[7] Kiszka, J.B., Kochanska, M., & Sliwinska, D. 1985. 'The influence of some fuzzy implication operators on the accuracy of a fuzzy model — II'. *Fuzzy Sets & Syst.*, **15**, pp. 223–240

[8] Mizumoto, M., & Zimmermann, H.J. 1982. 'Comparison of fuzzy reasoning methods'. *Fuzzy Sets & Syst.*, **8**, pp. 253–283

[9] Mamdani, E.H. 1977. 'Application of fuzzy logic to approximate reasoning using linguistic systems'. *IEEE Trans. Comput.*, **26**, pp. 1182–1191

[10] Pedrycz, W. 1980. 'On the use of fuzzy Lukasiewicz logic for fuzzy control'. *Arch. Autom.*, **3**, pp. 301–313

[11] Pedrycz, W. 1983. 'Numerical and applicational aspects of fuzzy relational equations'. *Fuzzy Sets & Syst.*, **11**, pp. 1–18

[12] Pedrycz, W. 1985. 'Applications of fuzzy relational equations for methods of reasoning in presence of fuzzy data'. *Fuzzy Sets & Syst.*, **16**, pp. 163–175

[13] Raju, G.V., Zhou, J., & Kisner, R.A. 1991. 'Hierarchical fuzzy control'. *Int. J. Contr.*, **54**, pp. 1201–1216

[14] Sugeno, M. (Ed.). 1985. 'Industrial Applications of Fuzzy Control'. North Holland, Amsterdam

[15] Sugeno, M., & Murakami, K. 1985. 'An experimental study on fuzzy parking control using a model car'. *In* 'Industrial Applications of Fuzzy Control' (Ed. M. Sugeno). North Holland, Amsterdam, pp. 125–132

[16] Tsukamoto, Y. 1979. 'Fuzzy logic based on Lukasiewicz logic and its application to diagnosis and control'. Ph.D. Thesis, Tokyo Institute of Technology, Tokyo, Japan

7.9 Appendix

We summarise two common iterative methods of clustering, FUZZY ISODATA and ISODATA [2,4].

Consider M vectors $x_1, x_2, \ldots, x_M$ viewed as elements in r-dimensional Euclidean space R^r equipped with a distance between any two elements denoted by d(. , .). The basic notion in the clustering method is the fuzzy partition matrix F consisting of 'c' rows and 'M' columns and possessing these properties:

$$0 \le f_{ik} \le 1$$

$$\sum_{i=1}^{c} f_{ik} = 1 \quad , \quad \forall_{1 \le k \le M} \tag{A.0}$$

$$0 < \sum_{k=1}^{M} f_{ik} < M \quad , \quad \forall_{1 \le i \le c}$$

The ith row of F denotes a discrete membership function of the ith cluster. f_{ik} denotes the grade of membership of the kth element, x_k, to the ith cluster. The value of this grade varies between 0 and 1, where 0 completely excludes x_k from the cluster and 1 totally confirms its belonging to the cluster. All elements (k) for which $f_{ik} > 0$ are more or less committed to this cluster. The second condition states that each cluster is nonempty and does not include all the data.

The minimised performance index, having an evident variance-like favour, reads:

$$Q = \sum_{k=1}^{M} \sum_{i=1}^{c} f_{ik}^2 \, d(x_k, v_i) \tag{A.1}$$

where v_i is a centroid (prototype) of the ith cluster. (A.1) is the sum of the dispersion of the elements around the prototypes. More formally, the optimisation task within which the clusters are generated is specified accordingly:

$$\min Q$$
$$\text{subject to constraints (A.0)}$$

Usually, through a series of iterations, we can determine the local minimum of Q; there is no guarantee that the global minimum of Q could be obtained in this manner.

The algorithm FUZZY ISODATA gives a local solution of the above problem. For its concise presentation, we denote by **F** a family of all the fuzzy matrices fulfilling (A.0). It comprises a sequence of steps:

1. Fix an initial partition matrix $F^{(0)} \in \mathbf{F}$.

2. Calculate centroids (prototypes) $v_i^{(0)}$ of the clusters by means of $F^{(0)}$:

$$v_i^{(0)} = \frac{\sum_{k=1}^{M} f_{ik}^{(0)2} x_k}{\sum_{k=1}^{M} f_{ik}^2} \tag{A.2}$$

3. Update the partition matrix, obtaining $F^{(1)} \in \mathbf{F}$ so that:

$$\left(f_{ik}^{(1)}\right)^{-1} = \sum_{j=1}^{c} \frac{d(x_k, v_i)^2}{d(x_k, v_j)^2} \tag{A.3}$$

4. Compare $F^{(0)}$ with $F^{(1)}$; if they are sufficiently alike, i.e. satisfy an assumed stopping criterion, then stop; otherwise, repeat 2, updating centroids (replacing the partition matrix $F^{(0)}$ by $F^{(1)}$).

If the constraints (A.0) are more restrictive, i.e. $f_{ik} \in \{0,1\}$, the corresponding clustering method is ISODATA. This optimisation differs from the one described in its way of updating the partition matrix. (A.3) is replaced by this expression:

$$f_{ik} = \begin{cases} 1 \ , & \text{if } d(x_k, v_i^{(0)}) = \min\limits_{1<j<c} d(x_k, v_j^{(0)}) \\ 0 \ , & \text{otherwise} \end{cases}$$

CHAPTER 8

Relational Neural Networks

8.1 Preliminaries: Logic Structures in Neural Networks

This Chapter is devoted to the interesting and general class of neural networks based on the fundamental operations and ideas of fuzzy-relational structures. As such, they convey an important logical structure and imply a variety of logic-oriented processing properties. We recall that neural networks are treated as a certain category of parallel and distributed computational structures with a significant level of learning capability, cf. [1,4,7]. They thus operate on an exclusively numerical level (usually with the input, internal (hidden) and output signals limited to a certain interval, such as [–1,1] or [0,1]). The basic computational components (computational neurons) are quite simple processing units. In their standard form, they realise a weighted sum of input signals $x_1, x_2, \ldots, x_n$, say $\sum_{i=1}^{n} w_i \, x_i + \upsilon$, where υ denotes a so-called bias, followed by a nonlinear static transformation. A common type of transformation used here is completed by a two-parameter sigmoid function:

$$f(z) = \frac{1}{1 + \exp\left(-(z - z_0)\,T\right)}$$

where z_0 denotes the translation of the function and T is used to control the shape of the function.

The neurons are arranged into a collection of layers. The number of these layers, as well as their dimensions, specify the representational capabilities of the resulting network. In contrast to the structures of Artificial Intelligence (AI), where the knowledge is expressed explicitly (as in expert systems), neural networks 'accommodate' knowledge by distributing ('coding') it among the connections of the network. Thus, the resulting structure is fairly difficult to interpret. Neural networks have no logical structure behind them; therefore, we cannot convert them into a series of readable 'if-then' statements or any other scheme of knowledge representation of an explicit character.

In this Chapter, we introduce logic-based neurons and develop neural networks produced by combining several categories of them. We highlight the evident correspondence between logic neurons and fuzzy-relational equations. First, bearing in mind basic logic operations on fuzzy sets, we study two standard categories of aggregative neurons, namely AND and OR computational nodes. These in turn give rise to several types of reference neuron. All are discussed in Section 8.3. Then we study the multilevel structures of neural networks. The aim of this presentation is to highlight their origin and the role they play as approximators of multivalued (fuzzy) logic functions. We present their basic properties. An interesting feature here is that these networks are evidently heterogeneous constructions, requiring several classes of logic neurons arranged into a structure directly reflecting the logical foundations of the networks. An important class of these networks is referred to as logic processors.

Throughout this Chapter, we treat fuzzy sets as points in the unit hypercubes, such as $[0,1]^n$, $[0,1]^m$ etc. Hence, $\mathbf{x} \in [0,1]^n$ denotes a fuzzy set $\mathbf{x}$ defined in an n-dimensional discrete space.

8.2 Aggregative Neurons: AND and OR Logical Computing Nodes

In this Section, we investigate two basic types of logical neuron performing aggregation of the AND and OR types (aggregative neurons). The first type, the AND computing node, is an n-input single-output computing structure $\mathbf{x} \to y$ performing the AND operation on input signals $x_1, x_2, \ldots, x_n$, say

$$y = \mathrm{AND}\,(x_1, x_2, \ldots, x_n)$$

The use of any t-norm in modelling AND connections implies this expression:

$$y = \mathop{\mathrm{T}}_{i=1}^{n} x_i$$

Usually, different x_is have a different impact on the output of the neuron. This can be accomplished by including connections $w_1, w_2, \ldots, w_n \in [0.1]$, associated with the corresponding inputs x_i. The aggregation is then carried out as follows:

$$y = (x_1 \text{ OR } w_1) \text{ AND } (x_2 \text{ OR } w_2) \text{ AND } \ldots \text{ AND } (x_n \text{ OR } w_n)$$

where OR-ing is realised by taking any s-norm. Vector notation applied to the above expression yields:

$$y = \text{AND}\,(\mathbf{x},\mathbf{w})$$

where $\mathbf{x}$ summarises all the inputs and vector $\mathbf{w}$ includes all the connections. In other words, the explicit formula is based on the t-s composition of fuzzy sets $\mathbf{x}$ and $\mathbf{w}$:

$$y = \operatorname*{T}_{i=1}^{n} (w_i \text{ s } x_i) \tag{8.1}$$

The connections w_i have a profound effect on the influence of the input x_i on the output value of the neuron. By virtue of the two boundary conditions of the s-norms, we read:

• if $w_i = 0$, then $w_i \text{ s } x_i = x_i$, and this input affects the output of the neuron in a straightforward fashion.

• On the other hand, if $w_i = 1$, it eliminates the impact of x_i on the output. We get $1 \text{ s } x_i = 1$ despite the values x_i takes on. To achieve a complete analogy with the standard model of the neuron studied in neurocomputations, the basic logic structures can be enhanced by these modifications:

• augmenting bias. The bias term is added as an additional term in (8.1) that is driven by a constant input signal always equal to 0, say $0 \text{ s } w_0$, where w_0 denotes the connection associated with this input. Thus, the AND neuron incorporating the bias is given by:

$$y = \operatorname*{T}_{i=0}^{n} (w_i \text{ s } x_i)$$

where, by convention, we put $x_0 = 0$. In the course of our discussion, we shall be using the same notation for the input signals. Obviously, by augmenting $\mathbf{x}$ by the zero coordinate of the bias, we make $\mathbf{x}$ to now lie in an $(n+1)$ hypercube. From a computational point of view, the bias shifts all the output values of the neuron.

• Some other ways in which the bias term can be introduced into this structure are discussed later.

The second generic computing node performs OR logical computations, namely:

$$y = \mathrm{OR}\,(\mathbf{w},\mathbf{x})$$

i.e.

$$y = \mathop{S}_{i=0}^{n} (w_i \; t \; x_i) \qquad (8.2)$$

where again w_0 stands for the connection of the bias term. Note, however, that x_0 is now kept as a constant value always equal to 1 (so that $w_0 \, t \, 1 = w_0$).

Owing to the monotonicity of the triangular norms, both the AND and OR nodes given above are exclusively excitatory in their behaviour. This means that higher values of x_i imply higher values of y:

$$\text{if } \mathbf{x} < \mathbf{x}', \text{ then (i) AND } (\mathbf{w},\mathbf{x}) < \text{AND } (\mathbf{w},\mathbf{x}'), \text{ and}$$
$$\text{(ii) OR } (\mathbf{w},\mathbf{x}) < \text{OR } (\mathbf{w},\mathbf{x}')$$

In contrast to that, depending on the signs of the connections, the basic neuron used in the literature [4] exhibits either an excitatory or an inhibitory performance. To add this feature to our construct (and still maintain the standard [0,1] range of the grades of membership), we extend the original hypercube of the inputs by including complementary values of x_i, say $1 - x_i$. Now the entire vector **x** is augmented by 'n' new coordinates and contains these entries:

$$\mathbf{x} = \left[x_1 \; x_2 \; \ldots \; x_n \;\vdots\; \bar{x}_1 \; \bar{x}_2 \; \ldots \; \bar{x}_n\right]$$

The AND and OR neuron with this extended vector of inputs makes it possible to admit both the inhibitory and excitatory characters of its behaviour, depending on the numerical values of the connections.

Example 8.1

Consider a single-input AND and OR neuron involving both a direct and a complementary form of the input signal x. The plots obtained of the characteristics:

$$\text{AND node: } y = \text{AND}\left([x \;\; \bar{x}],[w \;\; v]\right) = (x \text{ OR } w) \text{ AND } (\bar{x} \text{ OR } v)$$
$$\text{OR node: } y = \text{OR}\left([x \;\; \bar{x}],[w \;\; v]\right) = (x \text{ AND } w) \text{ OR } (\bar{x} \text{ AND } v)$$

for the minimum and the maximum are shown in Figs. 8.1(a), (b), (c) and (d) (several combinations of w and v are included here). Observe that

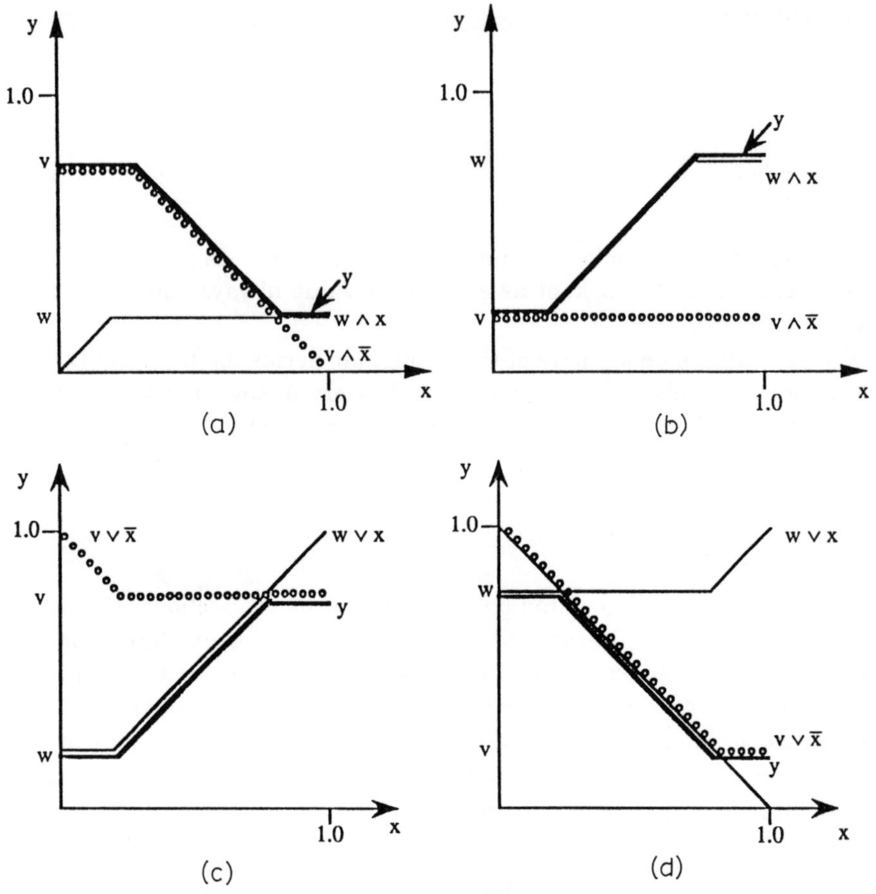

Figs. 8.1(a), (b), (c) and (d) Characteristics of AND and OR neurons for different combinations of connections w and v. (a),(b) OR neuron; (c),(d) AND neuron

the values of the connections have a primordial effect on the qualitative character of the response of the neuron. Depending on their values, the node exhibits either an excitatory or an inhibitory form of the characteristics. For all the combinations in which $w > v$, the OR node shows an inhibitory character, while the opposite phenomenon holds for the AND node. The characteristics of the AND and OR nodes are complementary. A special case occurs when both the connections are equal, w = v. This gives rise to a mixture of the inhibitory and excitatory characteristics, resulting in an output of the neuron that is kept constant, despite changes at its inputs.

It is worth noting that the AND and OR nodes generate a somewhat limited range of values of the output signal, which usually does not cover the entire unit interval. We can state that this static mapping completed by the neuron is 'into' rather than 'onto'. Consider now the case where all the connections w_i s are fixed. For the boundary conditions (namely $\mathbf{x} = 1$ and $\mathbf{x} = 0$), we derive:

$$\text{AND neuron: } y \in \left[0, \mathop{\mathrm{T}}_{i=1}^{n} w_i\right]$$

$$\text{OR neuron: } y \in \left[\mathop{\mathrm{S}}_{i=1}^{n} w_i, 1\right]$$

In order to cope with this issue and expand the range of possible output signals up to the entire unit interval, we can introduce a nonlinear sigmoid element that follows the neuron and is given by:

$$z = \frac{1}{1 + \exp\left(-(y - y_0)\,\delta\right)}$$

where y_0 translates the basic sigmoid function along the unit interval, and the second parameter, δ, is used to adjust the shape of the function (Fig. 8.2).

The graphical symbols used to identify the nodes are shown in Figs. 8.3(a) and (b).

The logic neurons have two interesting interpretations. The first one comes from the theory of fuzzy-relational equations; one can easily recognise that (8.2) and (8.1) are just s-t and t-s composition operators applied to **x** and **w** (note that y is a scalar quantity).

The second interpretation stems from the theory of Boolean functions and is of particular interest also for fuzzy functions [6]. The two fundamental notions in the theory of Boolean functions are minterms and maxterms [3] viewed as basic components forming this class of functions. When we restrict ourselves to the binary (0-1) values of the connections as well as consider the 0-1 values of the inputs, the AND node is used to represent any minterm (i.e. a product of some variables and their complements) while the OR node is applied to implement a maxterm (a sum of the variables and their complements). With

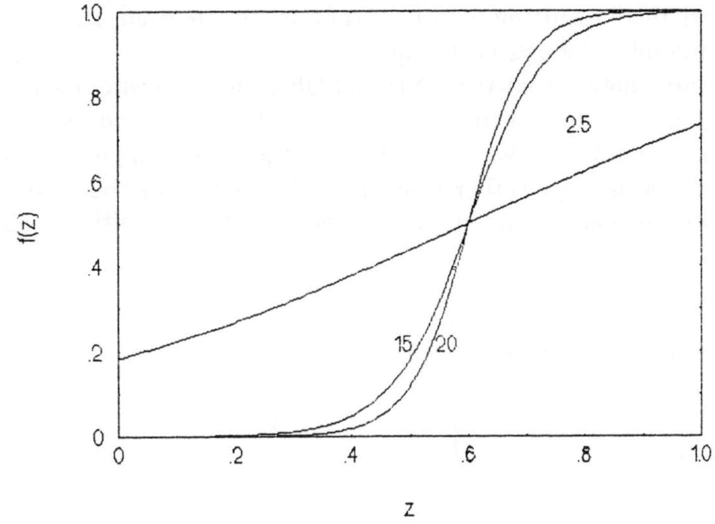

Fig. 8.2 Sigmoid functions as an example of 'onto' mapping ($y_0 = 0.6$, several values of δ, $\delta = 2.5, 15, 20$)

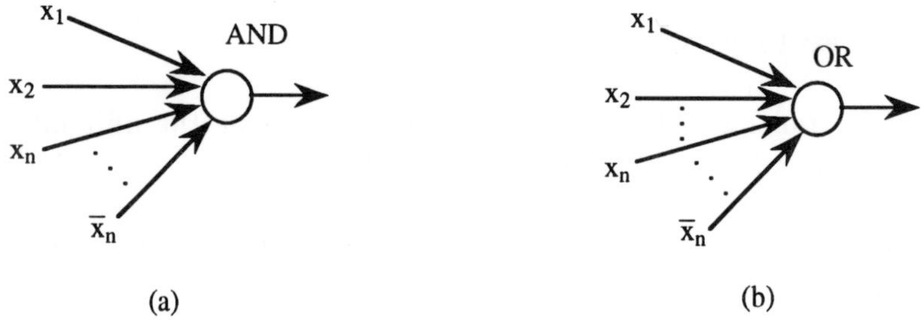

Figs. 8.3(a) and (b) Graphical notation of (a) AND and (b) OR neurons

intermediate values of the inputs and the connections, we talk about generalised minterms and maxterms.

8.3 Reference Neurons

The neurons we will be discussing are of a referential nature. This means that the inputs x_ks are first processed with respect to another fuzzy set viewed as a reference point and the grades of membership obtained are aggregated using AND or OR neurons. We distinguish several classes of neuron as of particular interest:

• equality neuron. Its basic formula reads:

$$y = \mathop{S}_{i=1}^{n} \left[(x_i = r_i) \, t \, w_i \right] \tag{8.3}$$

or more concisely:

$$y = \mathrm{OR}\,(\mathbf{x} = \mathbf{r}\,,\, w)$$

where $\mathbf{r} = [r_1 \;\; r_2 \;\ldots\; r_n]$ is a reference point in $[0,1]^n$. The result of matching, $\mathbf{x} = \mathbf{r}$, is aggregated by using the connections (w_i s) in the OR type of aggregation.

More descriptively, we can write (8.3) as follows:

$$y = [(x_1 \text{ EQUAL } r_1) \text{ AND } w_1] \text{ OR } [(x_2 \text{ EQUAL } r_2) \text{ AND } w_2] \text{ OR } \ldots$$
$$\text{OR } [(x_n \text{ EQUAL } r_n) \text{ AND } w_n]$$

where the predicate EQUAL corresponds to the equality operation used in the neuron.

• difference referential neuron. The dual neuron includes the difference operator $\dashv$, defined pointwise as a complement of the equality operation, say:

$$a \dashv b = 1 - (a = b)$$

The the basic formula for this neuron reads:

$$y = \mathop{S}_{i=1}^{n} \left[(x_i \dashv r_i) \, t \, w_i \right] \tag{8.4}$$

where $\mathbf{r}$ denotes a reference point with respect to which the difference operation is performed. The connections w_i reflect the impact that each of the inputs (x_i) has on the output of the neuron; the higher the value of w_i, the more evident the influence of this variable on the output of the neuron.

As before, we write this expression more descriptively:

$$y = [(x_1 \text{ DIFFER } r_1) \text{ AND } w_1] \text{ OR } [(x_2 \text{ DIFFER } r_2) \text{ AND } w_2] \text{ OR } \ldots$$
$$\text{OR } [(x_n \text{ DIFFER } r_n) \text{ AND } w_n]$$

The two remaining neurons apply when the reference operation is used to capture an inclusion or dominance relationship with respect to that reference point.

We have:

- neuron with an inclusion relationship:

$$y = \underset{i=1}{\overset{n}{S}}\left[(x_i \,\varphi\, r_i)\, t\, w_i\right]$$

where, as before, **r** is viewed as a reference point. Obviously, the property of inclusion is satisfied to the highest degree when $x_i \leq r_i$. The above formula can be written:

$$y = [(x_1 \text{ INCLUDED-IN } r_1) \text{ AND } w_1]$$
$$\text{OR } [(x_2 \text{ INCLUDED-IN } r_2) \text{ AND } w_2] \text{ OR } \ldots$$
$$\text{OR } [(x_n \text{ INCLUDED-IN } r_n) \text{ AND } w_n]$$

- reference neuron with a dominance relationship. This neuron reads:

$$y = \underset{i=1}{\overset{n}{S}}\left[(r_i \,\varphi\, x_i)\, t\, w_i\right]$$

or, more concisely:

$$y = [(r_1 \text{ DOM } x_1) \text{ AND } w_1] \text{ OR } [(r_2 \text{ DOM } x_2) \text{ AND } w_2] \text{ OR } \ldots$$
$$\text{OR } [(r_n \text{ DOM } x_n) \text{ AND } w_n]$$

The relationship $r_i \,\varphi\, x_i$ specifies a degree of dominance of r_i by x_i. As opposed to the previous neuron, the higher the dominance of **x** over **r**, the higher the value of y.

8.4 Multilevel Neural Networks

The neurons discussed in the previous Section can be put together to construct more general computational structures with significantly enhanced representational capabilities. In contrast to the well known multilevel neural networks known in the literature and applied in many practical cases, the networks composed with the aid of logic-based neurons are heterogeneous. This means that they include several neurons of different computational characteristics.

The neurons are organised into layers. Three-layer structures are studied first. We highlight their generality and draw some evident links between them and the canonical-representation forms of the two-valued functions, recalling the fundamental representation theorem formulated by Shannon [3]. Next, in a series of examples, we reveal how other topologies of the networks can easily be obtained by transforming the problem at hand into a collection of nested logical statements and finally converted into the topology of the network. Detailed learning schemes are investigated.

8.4.1 Logic processors — approximation of logic-oriented multidimensional relationships

The three-layer neural structure introduced and studied in [5] constitutes a suitable topology for coping with approximation-like problems when we are interested in eliciting logical and usually nonlinear relationships between elements of the input and the output hypercube. The neural network is built with the aid of the AND and OR neurons. It is composed of the three layers, each of them constructed of neurons of the same logical type (i.e. AND or OR only). An additional hidden layer enhances the representational capabilities of the entire structure. There are two types of the network, also called logic processors (LP) (Figs. 8.4(a) and (b)).

The first class of them is composed of AND neurons situated in the hidden layer, while the output layer consists of a single OR neuron. The second category has OR neurons in its hidden layer and a single AND neuron in the output layer. The formal expressions for the logic processor are provided in this form:

For the first class of the network (OR-AND):

- an input layer consists of 2n nodes including both the direct and the complementary versions of the x_is, say $x_1, x_2, \ldots, x_n, \bar{x}_1, \bar{x}_2, \ldots, \bar{x}_n$. The exclusive role of the nodes is to distribute the signals to all the nodes of the hidden layer,

- a hidden layer is composed of 'h' AND nodes. The intermediate signals z_l produced there are described as:

$$z_l = \mathrm{AND}\,(\mathbf{w}_l, \mathbf{x})$$

$$l = 1, 2, \ldots, h$$

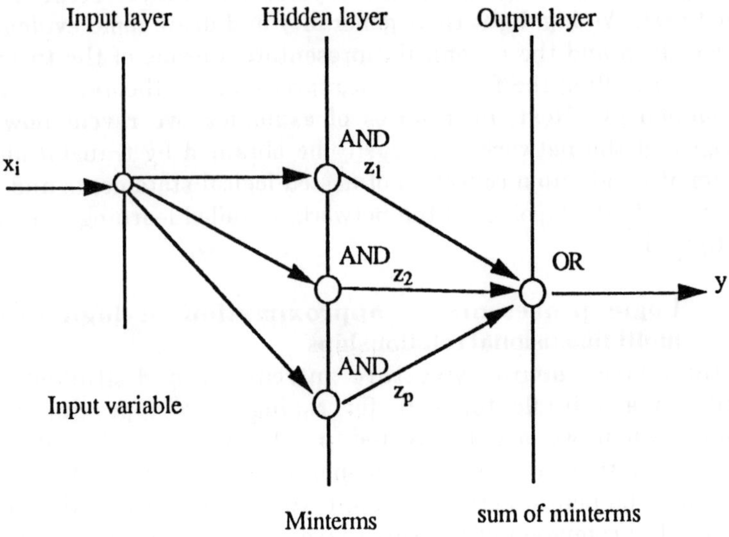

(a)

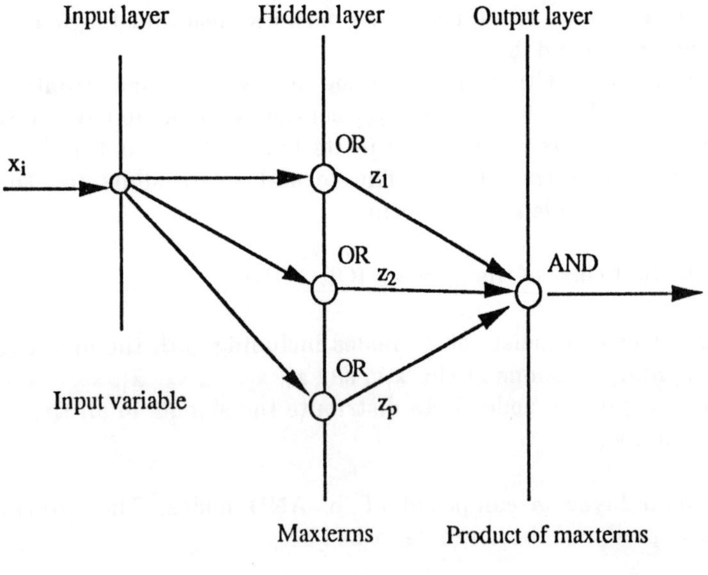

(b)

Figs. 8.4 Two types of logic processors (a) sum of minterms (SOM), (b) product of maxterms (POM)

The vector of connections, $\mathbf{w}_i$, summarises all the connections between the lth node of the hidden layer and the input nodes. In other words, we obtain this relationship:

$$z_l = \underset{i=1}{\overset{n}{T}} (w_{li} \text{ s } x_i) \text{ t } \underset{i=1}{\overset{n}{T}} \left(w_{l(n+i)} \text{ s } \overline{x}_i\right)$$

l = 1, 2, ..., h. The output layer consisting of a single OR node carried out an aggregation of z_ls:

$$y = \underset{l=1}{\overset{h}{S}} (v_l \text{ t } z_l)$$

Before we proceed with a more detailed description, it is instructive to restrict ourselves to the two-valued case, assuming that all the connections as well as the input signals are Boolean.

The first type of the network implements any two-valued function as a sum of minterms. Each node in the hidden layer realises a minterm. The dual structure becomes a product of maxterms. Now the hidden layer consists of OR neurons. The signals resulting there are summarised using the single AND node. Both of them are equivalent; i.e. they represent the two-valued function in two different ways. Of course, the number of nodes in the hidden layer may vary from one topology to the other.

In contrast to the two-valued case, the general multivalued (fuzzy) function can be approximated to by the two above structures. This approximation, though, is significantly legitimised. We refer to the first type of the network as a sum of minterms (SOM). The second architecture develops a product of maxterms (POM).

The next Section discusses the learning processes in logic neurons, and the logic processor in particular.

8.4.2 Learning in logic processors

The learning processes in logic processors are of a parametric nature. Their essence is to adjust the connections of the network so that a certain performance index is optimised (maximised or minimised). The learning procedure requires that the following components be clearly defined:

• a collection of learning data, i.e. pairs of input data $\mathbf{x}_k$ and the corresponding output values t_k, $k = 1, 2, \ldots, N$

• a performance index that expresses how well the logic processor maps inputs $\mathbf{x}$ into the corresponding values of the output t_k.

Our target is to maintain the closest resemblance of t_k and the corresponding output of the logic processor. We denote it by $LP(\mathbf{x}_k)$. The way in which we measure this matching can be expressed in two conceptually distinct manners. The first one uses any distance function between $LP(\mathbf{x}_k)$ and t_k. The distance function can employ well known distances like the Hamming, Euclidean etc. The Euclidean distance function is most frequently used. The performance index summarises all the learning data by putting the sum as follows:

$$Q = \sum_{k=1}^{N} (LP(\mathbf{x}_k) - t_k)^2$$

In this situation, our aim is to minimise the above performance index Q.

Another class of performance indices is constructed as the sum of the equality indices computed for $LP(\mathbf{x}_k)$ and t_ks:

$$Q = \sum_{k=1}^{N} (LP(\mathbf{x}_k) = t_k)$$

Obviously, our intention is to maximise Q, thus making the logic processor follow the data. These two performance indices are related. In some cases, we can show interesting relationships between them. Take, for instance, the equality index implemented by using the Lukasiewicz implication. The piecewise linear function can be rewritten:

$$LP(\mathbf{x}_k) = t_k = \begin{cases} 1 - LP(\mathbf{x}_k) + t_k \ , & \text{if } LP(\mathbf{x}_k) > t_k \\ 1 - t_k + LP(\mathbf{x}_k) \ , & \text{otherwise} \end{cases} = 1 - |LP(\mathbf{x}_k) - t_k|$$

It is clear that the minimisation of the Hamming distance is equivalent to the maximisation of the equality index generated by the Lukasiewicz implication.

The optimisation includes all the connections of the logic processor between the input layer and the hidden layer as well as the hidden layer and the output layer. The simplest update scheme is that in which the modifications are driven by a gradient of the performance index taken with regard to these connections. In a shorthand notation, we express the learning formula as follows:

$$\text{connections-new} = \text{connections} - \alpha \frac{\partial Q}{\partial \text{connections}}$$

where α denotes a learning factor, $\alpha \in (0,1)$. The detailed computations can be performed once the performance index and the detailed parametric description of the network have been defined. An example of a detailed learning scheme is described below.

Example 8.2

We discuss the neural network with the s-norm defined as the probabilistic sum while the product is specified for the t-norm. Furthermore, we assume that the performance index in the form of a sum of squared errors. Since we are studying an on-line learning version (where each pair $(\mathbf{x}_k, t_k)$ immediately affects the connections of the network), the index 'k' denoting the element in the training set can be dropped. In the off-line type of learning, we first summarise the changes coming from each element of the data set and then modify the connections based on this accumulated modification.

A single pair of data is written down as $\mathbf{x}$ and t. The detailed formulas of the network now read:

- hidden layer:

$$z_k = \overset{n}{\underset{i=1}{\mathrm{T}}} (w_{hi} \, s \, x_i) \, t \, \overset{n}{\underset{i=1}{\mathrm{T}}} \left(w_{h(n+i)} \, s \, \bar{x}_i\right)$$

- output layer:

$$y = \overset{H}{\underset{h=1}{\mathrm{S}}} (v_h \, s \, z_h)$$

where H denotes the dimension of the hidden layer. The adjustment of the connections is expressed accordingly:

$$\frac{\partial Q}{\partial w_{h_1 j}} = \frac{2\,[LP(\mathbf{x}) - t]\,\partial y}{\partial w_{h_1 j}}$$

$h_1 = 1, 2, \ldots, H, j = 1, 2, \ldots, 2n$

and

$$\frac{\partial Q}{\partial v_l} = \frac{2\,[LP(\mathbf{x}) - t]\,\partial y}{\partial v_l}$$

$l = 1, 2, \ldots, H$

Subsequently,

$$\frac{\partial y}{\partial v_l} = \frac{\partial}{\partial v_l}\left[\underset{h=1}{\overset{H}{S}} (v_h \text{ s } z_h)\right] = \frac{\partial}{\partial v_l}\left[A \text{ s } (v_l \text{ s } z_l)\right]$$

where

$$A = \underset{h \neq 1}{\overset{H}{S}} (v_h \text{ s } z_h)$$

By virtue of the triangular norms specified here, we derive:

$$\frac{\partial}{\partial v_l}\left[A + v_l\, z_l - A\, v_l\, z_l\right] = z_l\,(1 - A)$$

Then

$$\frac{\partial y}{\partial w_{h_1 j}} = \sum_{h=1}^{H} \frac{\partial y}{\partial z_h} \frac{\partial z_h}{\partial w_{h_1 j}}$$

Note, however, that the above sum reduces to a single component, since $\partial z_h / \partial w_{h_1 j} = 0$ for all $h \neq h_1$. In this way,

$$\frac{\partial y}{\partial w_{h_1 j}} = \frac{\partial y}{\partial z_{h_1}} \frac{\partial z_{h_1}}{\partial w_{h_1 j}}$$

And finally:

$$\frac{\partial y}{\partial z_{h_1}} = v_l(1-B) \qquad , \qquad B = \underset{h=h_1}{\overset{H}{S}}(v_h \; s \; z_h)$$

$$\frac{\partial z_{h_1}}{\partial w_{h_1 j}} = \frac{\partial}{\partial w_{h_1 j}}\left[\prod_{i=1}^{n}\left(w_{h_1 i} + x_i - w_{h_1 i}\, x_i\right)\prod_{i=1}^{n}\left(w_{h_1(n+i)} + \overline{x}_i - w_{h_1(n+i)}\, \overline{x}_i\right)\right]$$

Then we obtain:

$$\frac{\partial z_{h_1}}{\partial w_{h_1 j}} = \begin{cases} C_1\left(1-x_j\right) & , \text{ if } j \le n \\ C_2\left(1-\overline{x}_j\right) & , \text{ if } j > n \end{cases}$$

where C_1 and C_2 abbreviate the product terms not including x_j or its complement.

Regarding the internal derivatives, we should be aware that, for nondifferentiable triangular norms such as maximum and minimum, the derivatives should be defined with a certain caution since they can severely affect the learning algorithm [8]. Take, for instance, the minimum operation, min(x,w), and calculate its derivative with respect to w. Obviously:

$$\min(x, w) = \begin{cases} w & , \text{ if } x \geqslant w \\ x & , \text{ if } x < w \end{cases}$$

yielding this definition of the derivative with zero-one values:

$$\frac{\partial \min(x, w)}{\partial w} = \begin{cases} 1 & , \text{ if } x \ge w \\ 0 & , \text{ if } x < w \end{cases}$$

An optimisation problem driven by this type of ‘on-off’ updates can easily be affected by traps caused by some configurations of the connections and the data during the course of learning. A prudent inspection suggests a slight modification of this basic idea. The essence of this modification is to replace the above two-valued predicate by its multivalued counterpart. Thus, the two-valued predicate (relation) ‘less than’, taking on two values:

True(1) , if satisfied

and

False(0) , otherwise

is relaxed by admitting a multivalued predicate of partial inclusion ('included in') that is satisfied to a certain degree. A typical definition is the grade of inclusion $w \varphi x$. This expression takes on values within the entire unit interval. For more details, refer to [5,8].

Yet another solution to the problem of nondifferentiability of the maximum and minimum operations would be to approximate them by some smooth although very similar functions. Examples of this type of approximation are:

for minimum: $\frac{1}{2}\left[(x+w)-\sqrt{(x-w)^2+\delta^2}+\delta\right]$

for maximum: $\frac{1}{2}\left[(x+w)+\sqrt{(x-w)^2+\delta^2}-\delta\right]$

where δ is a small positive constant, say 0.05.

8.4.3 Other types of heterogeneous multilevel neural network

The referential type of neural network is developed in such a way that it realises some preliminary operations with respect to several reference points in the n-dimensional unit hypercube of the input signals. These operations pertain, in particular, to matching, difference, inclusion and covering (dominance). Below, we introduce them all:

- matching referential network [9]. First, the input signals $x_1, x_2, \ldots, x_n$ match the prototypes $\Omega_1, \Omega_2, \ldots, \Omega_H$ and the results of the matching are aggregated (Fig. 8.5).

The first layer performs coordinatewise matching as given in verbal form below:

$$z_h = [(x_1 \text{ EQUAL } \omega_{h1}) \text{ OR } w_{h1}] \text{ AND } [(x_2 \text{ EQUAL } \omega_{h2}) \text{ OR } w_{h2}] \text{ AND } \ldots$$
$$\text{AND } [(x_n \text{ EQUAL } \omega_{hn}) \text{ OR } w_{hn}]$$

$$h = 1, 2, \ldots, H$$

where $\Omega_h = [\omega_{h1}\ \omega_{h2}\ \ldots\ \omega_{hn}]$ denotes the grades of membership of the hth prototype and operator EQUAL is modelled as the equality index (cf.

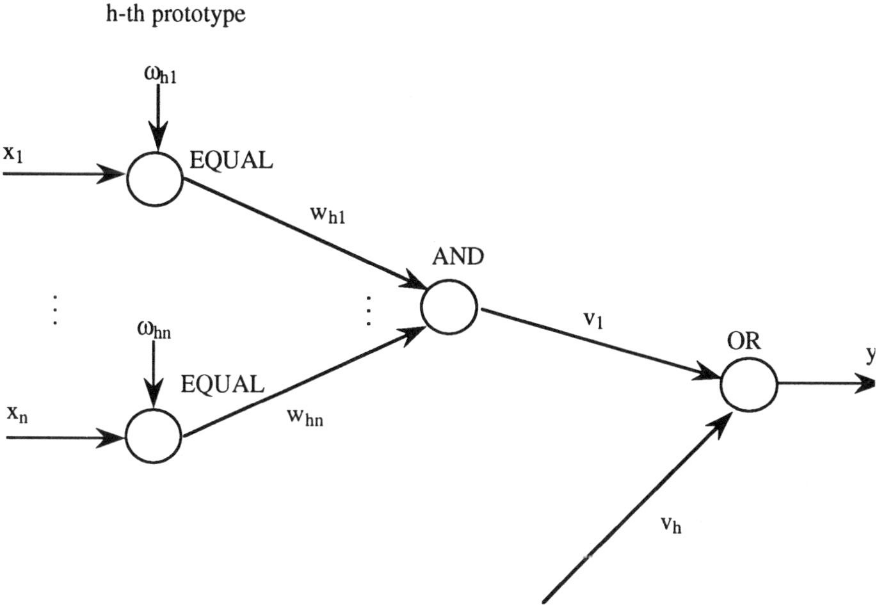

Fig. 8.5 Matching referential neural networks

Section 2.3). The second layer is implemented by OR-ing the signals from the hidden layer:

$$y = (z_1 \text{ AND } v_1) \text{ OR } (z_2 \text{ AND } v_2) \text{ OR } \dots \text{ OR } (z_H \text{ AND } v_H)$$

This type of neural network can be useful in problems of machine learning. For example, we can envisage positive instances as distributed around several disjoint regions in the feature space and centred at Ω_1, Ω_2, ..., Ω_H, while the values of the output y denote a level of class membership of the positive instances (assuming that they belong to this class to a certain nonzero degree). This can also be translated into this disjunctive description:

the pattern **x** *is a positive instance if it is similar to* Ω_I *or is similar to* Ω_2 *or ... or is similar to* Ω_H

Each subcondition is carried out by the corresponding node in the hidden layer, while the final aggregation is of a disjunctive format.

• difference (exception) reference neural network. The structure is driven by differences between x_is and some landmarks (acting as exception regions) ψ_1, ψ_2, ..., ψ_H. The corresponding formulas are:

$$z_h = [(x_1 \text{ DIFFER } \psi_{h1}) \text{ OR } w_{h1}] \text{ AND } [(x_2 \text{ DIFFER } \psi_{h2}) \text{ OR } w_{h2}] \text{ AND } \dots$$
$$\text{AND } [(x_n \text{ DIFFER } \psi_{hn}) \text{ OR } w_{hn}]$$

$$h = 1, 2, \dots, H$$

$$y = (z_1 \text{ OR } v_1) \text{ AND } (z_2 \text{ OR } v_2) \text{ AND } \dots \text{ AND } (z_H \text{ OR } v_H)$$

The difference operator can be implemented in many different ways. One can, for instance, use the complement of the equality index, a DIFFER b = $1-(a=b)$, $a, b \in [0,1]$.

The results of the hidden layer are described in the same manner as in the equality network. The difference referential structure can easily be interpreted in machine learning. The description becomes that of a conjunctive format highlighting the regions where positive instances are almost nonexistent (the vectors ψ_h represent exceptions — negative instances):

the pattern is viewed as a positive instance if it is excluded from ψ_I *and is excluded from* ψ_2 *and ... and is excluded from* Ω_H

The two remaining neural networks utilise inclusion and dominance neurons, respectively. The inclusion-driven neural network describes the inclusion relationship between a current input of the network and its reference points. For 'H' reference points, we obtain:

$$z_h = [(x_1 \text{ INCLUDED-IN } r_1) \text{ OR } w_{h1}]$$
$$\text{AND } [(x_2 \text{ INCLUDED-IN } r_2) \text{ OR } w_{h2}] \text{ AND } \dots$$
$$\text{AND } [(x_n \text{ INCLUDED-IN } r_n) \text{ OR } w_{hn}]$$

$$h = 1, 2, \dots, H$$

The aggregation can be specified either in its conjunctive form:

$$y = (z_1 \text{ OR } v_1) \text{ AND } (z_2 \text{ OR } v_2) \text{ AND } \dots \text{ AND } (z_H \text{ OR } v_H)$$

or in the disjunctive form:

$$y = (z_1 \text{ AND } v_1) \text{ OR } (z_2 \text{ AND } v_2) \text{ OR } \dots \text{ OR } (z_H \text{ AND } v_H)$$

When the inclusion predicate is replaced by the dominance relationship, we get the dominance-driven neural network expressing the constraints of a dominance character:

$$z_h = [(r_1 \text{ DOM } x_1) \text{ OR } w_{h1}] \text{ AND } [(r_2 \text{ DOM } x_2) \text{ OR } w_{h2}] \text{ AND} \ldots$$
$$\text{AND } [(r_n \text{ DOM } x_n) \text{ OR } w_{hn}]$$

Again, the aggregation occurring at the output layer may be either disjunctive or conjunctive.

The inclusion relationship (INCLUDED-IN) suggests some other ways of handling the biases in the logic neurons. So far, we have treated it as an additional constant input equal to 0 or 1 (depending on the character of the neuron). Here, we explore another avenue, using the inclusion relationship. Let y be the output of the neuron. The threshold T imposed on it generates a new output y′ according to this relationship:

$$y' = T \text{ INCLUDED-IN } y$$

which is just the degree of inclusion of T in y. Then, if T < y (the output exceeds the threshold level T), the neuron generates 1; otherwise, the values of y become slightly elevated (Fig. 8.6) for the inclusion relationship modelled with the aid of the Lukasiewicz implication:

$$y' = \begin{cases} 1 - T + y, & \text{if } y < T \\ 1, & \text{otherwise} \end{cases}$$

A second approach is to implement the inclusion relationship using the β operator (cf. Section 5.2.2). This yields:

$$y' = T \text{ INCLUDED-IN* } y$$

and then, for the β-operator implied by the Lukasiewicz OR conjunction, we obtain:

$$y' = \begin{cases} 0, & \text{if } T > y \\ y - T, & \text{otherwise} \end{cases}$$

see again Fig. 8.6.

8.5 Approximations of Logic Processors

An interesting issue arises with respect to some types of approximation of logic processors. The aim of these approximations is to come up with a

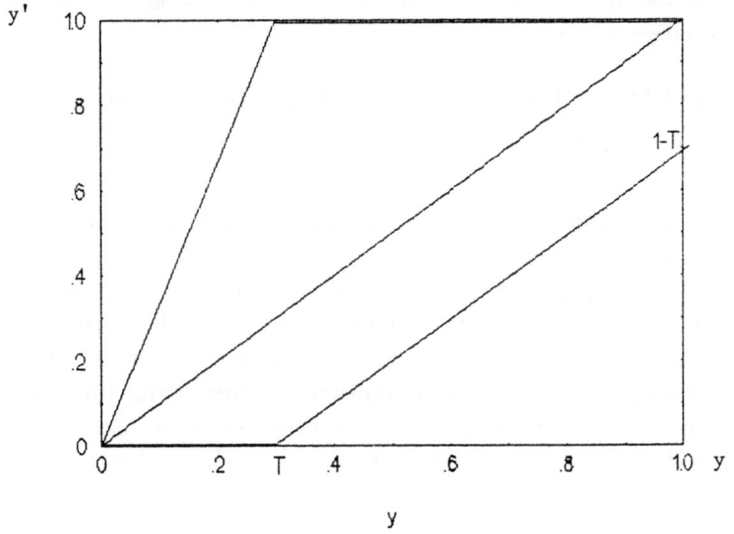

Fig. 8.6 Augmenting computing nodes by bias implemented using inclusion relations

qualitative or semiqualitative interpretation of the logical relationships captured by the neural network.

We distinguish two important classes of approximation. The first one prunes the 'weak' (meaningless) connections and so produces the core structure of the network. The second one converts the connections to either 0 or 1, producing a Boolean-logic processor (BLP).

8.5.1 Core structure of the logic processor

Some of the connections of the logic processor can be meaningless and so may be eliminated from it. This permits reduction of the logic processor (in the sense of its topology), and makes its interpretation in terms of the functions of the structure as well as its function more transparent.

The reduction is completed on the basis of the values of the connections between the nodes. The connections of any AND node are treated differently from those of the OR neuron. We recall (see (8.1)) that in the AND node the connections with values close to 1 eliminate the impact of the corresponding input on the output of the neuron. The value $w_i = 1 - \varepsilon$, where $\varepsilon \to 0$ implies that $\partial y / \partial x_i \to 0$. In particular, for the product and the probabilistic sum, we obtain $y = (x_i \text{ OR } w_i) \text{ AND } c$. Then:

$$\frac{\partial y}{\partial x_i} = c \frac{\partial}{\partial x_i}(x_i \text{ OR } w_i)$$

Subsequently:

$$\frac{\partial}{\partial x_i}(x_i + w_i - x_i w_i) = 1 - w_i = 1 - (1-\varepsilon) = \varepsilon$$

The higher the value of the connection, the lower the impact of its input variable on the output. So all the links with the high values are potential candidates to be eliminated from the node. Consequently, the node with all the meaningless connections is eliminated. The opposite effect holds for the OR node: its connections with the low values are likely to be dropped. Formally speaking, we introduce the two thresholding operations T_λ and T_μ so as to complete this pruning:

- for the AND node:

$$T_\lambda(a) = \begin{cases} a, & \text{if } a < \lambda \\ 1, & \text{otherwise} \end{cases}$$

- for the OR node:

$$T_\mu(a) = \begin{cases} 0, & \text{if } a < \mu \\ a, & \text{otherwise} \end{cases}$$

$a, \lambda, \mu \in [0,1]$

The threshold values λ and μ are selected to eliminate some of the connections. The operations T_λ and T_μ applied to the elements of the network convert it into a structure that is easier to interpret. Formally, we denote it by $y = \text{Core}(LP(\mathbf{x}),\lambda,\mu)$. The choice of the threshold levels is completed either on the basis of an analysis of the resulting network or through the optimisation of a relevant performance index capturing the relationships between the LP and its core. The obvious relationship is:

- if $\lambda_1 < \lambda_2$ and $\mu_2 < \mu_1$, then

$$\text{Core}(LP(\mathbf{x}), \lambda_1, \mu_1) < \text{Core}(LP(\mathbf{x}), \lambda_2, \mu_2)$$

In an experimental setting, we can select the thresholds λ and μ independently, so that the structure can be efficiently visualised. Too 'radical' values (i.e. too low values of λ and too high values of μ) may reduce the network to an empty structure. The choice of the values of the threshold should be mainly interpretation-oriented.

Yet another way of handling the selection procedure is to approximate the logic processor to its core structure. We require that:

$$\mathrm{LP}(\mathbf{x}) = \mathrm{Core}(\mathrm{LP}(\mathbf{x}), \lambda, \mu) \tag{8.5}$$

be satisfied for all $\mathbf{x}$s from a certain family of inputs X, $\mathbf{x} \in X$. The family of these inputs may be equivalent to the training set on which the original logic processor was trained (hence we are looking for the best approximation of the LP over the training set). The approximation requirement can also be formed with respect to any other family of inputs X.

Requirement (8.5) can be converted into a certain optimisation problem in which the distance function between the LP and its core structure is minimised with respect to λ and μ, say:

$$\min_{\lambda,\mu} \sum (\mathrm{LP}(\mathbf{x}) - \mathrm{Core}(\mathrm{LP}(\mathbf{x}), \lambda, \mu))^2 \tag{8.6}$$

where the above sum is taken over all $\mathbf{x} \in X$. Since this usually returns the value $\lambda = 1$ and $\mu = 0$ as an optimal solution to the problem, we can relax the original problem a bit by requiring that the distance function in (8.6) be less than a certain level. The resulting region of (λ,μ) summarises all the possibilities contributing to the formation of the core structure.

A straightforward enumeration of the values of the thresholds may be helpful in revealing the relationships between these values and the quality of the approximation process.

Example 8.3

We consider the logic processor with two inputs, three nodes in the hidden layer, and these connections:

$$\mathbf{w} = [0.65 \quad 0.44 \quad 0.87]$$

$$\mathbf{v} = \begin{array}{c} \begin{matrix} x_1 & x_2 & \bar{x}_1 & \bar{x}_2 \end{matrix} \\ \begin{bmatrix} 0.1 & 0.6 & 0.8 & 0.25 \\ 1.0 & 0.05 & 0.32 & 0.10 \\ 0.88 & 0.02 & 0.65 & 0.51 \end{bmatrix} \end{array}$$

(t-norm: product; s-norm: probabilistic sum).

The elements of X are selected by taking $x_1 = x_2$ distributed uniformly over the unit interval (N = 8). The resulting sum of the squared errors, specified as a two-dimensional plot with respect to λ and μ, is given in Fig. 8.7.

Note that a certain region of values of λ and μ induces a zero approximation error.

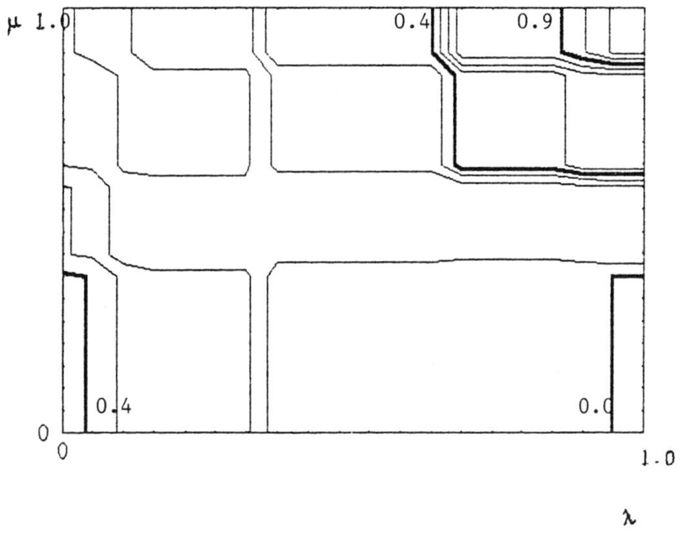

Fig. 8.7 Approximation of the LP to its core structure

8.5.2 Boolean-logic processors

Boolean-logic processors (BLP) constitute approximations to the logic processors in which the connections are two-valued (0-1) variables. These processors are used to represent multivalued functions in terms of their two-valued components:

$$\mathrm{BLP} : [0,1]^n \rightarrow [0,1]$$

so that:

$$\mathrm{BLP} : \{0,1\}^n \rightarrow \{0,1\}$$

The basic approach is to apply the thresholding operations to the original logic processor:

$$B_\lambda : [0,1] \to \{0,1\} \quad , \quad B_\mu : [0,1] \to \{0,1\}$$

so that:

$$B_\lambda(a) = \begin{cases} 0 \ , & \text{if } a < \lambda \\ 1 \ , & \text{otherwise} \end{cases}$$

and

$$B_\mu(a) = \begin{cases} 1 \ , & \text{if } a > \mu \\ 0 \ , & \text{otherwise} \end{cases}$$

The resulting matrix of connections consists of zeros and ones. The choice of the threshold values λ and μ is completed in the same fashion as for the core structure of the neural network. Again, this approximation is worked out with the aid of a given set of data X.

Example 8.4
For the network of Example 8.3, the characteristics obtained for the approximation to the different values of λ and μ are plotted in Fig. 8.8.

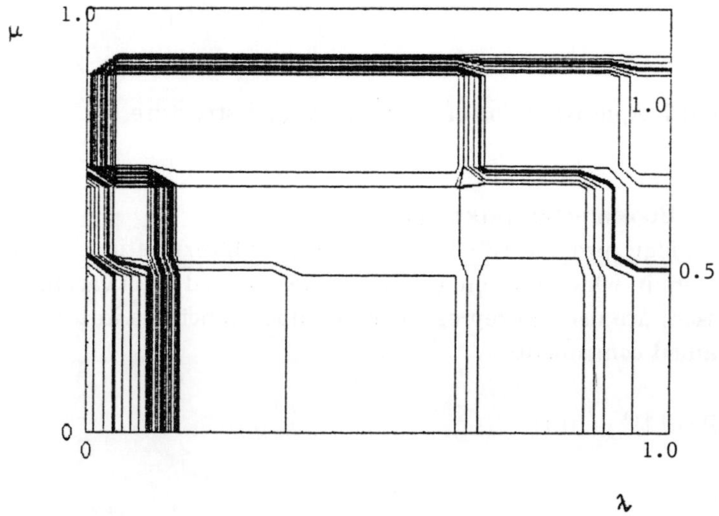

Fig. 8.8 Boolean approximation of the LP

First of all, the approximation is worse than for the core of the network. This is not surprising, since the current approximation is less detailed than the previous one. Secondly, the region of the minimal values of the approximation is centred on values of λ close to 1.0 and of μ around 0.5.

We can construct the Boolean network [2] through a regular learning procedure, as discussed for the logic processor. Nevertheless, some modification of the basic algorithm is indispensable. First, the connections are restricted to the values 0 and 1, and so the update (learning) procedure should keep these values within the two-element set. We consider a training set $(\mathbf{x}_1, t_1)$:

$$(\mathbf{x}_2, t_2), \ldots, (\mathbf{x}_N, t_N)$$

In this case, the off-line version of learning is preferred as leading towards a more stable scheme of learning. We recall that the increment of the connections is expressed:

$$\Delta \text{ connections} = -\frac{Q}{\text{connections}}$$

where

$$Q = \sum_{k=1}^{N} [\text{BLP}(\mathbf{x}_k) - t_k]^2$$

Depending on the sign of the above derivative (denoted by δ for a certain connection), the corresponding connection is modified according to the rule:

- if $\delta > 0$ and the connection is 0, convert it to 1,
- if $\delta < 0$ and the connection is 1, replace it by 0; otherwise, leave the value of the connection unchanged.

The starting-point of the learning procedure is selected randomly by assigning 0s and 1s randomly to the connections of the network. The above learning scheme, because of the extremal character of the connections (0-1), may easily be affected by the local minima.

8.6 Conclusions

Fuzzy neural networks combine the learning and the knowledge-representational capabilities of neural networks and fuzzy sets. The two general classes of neurons, i.e. the aggregative and the referential computing nodes, have been discussed at length and their properties carefully investigated. These neurons are then used in building fuzzy neural networks. Owing to the functional variety of the nodes introduced, the networks are usually formed as heterogeneous structures, reflecting in this way the logical nature of the problem in hand. We shall exploit this aspect in developing the networks for the specific identification of fuzzy control problems, including those of fuzzy controllers.

Fuzzy neural networks can easily be interpreted by a straightforward translation of the layers and the individual nodes into a series of 'if-then' statements. The core version of the network makes this interpretation more transparent, focusing our attention on the essential components of the description (approximation). Finally, the class of induced Boolean (two-valued) logical networks is capable of generating qualitative descriptions of the mapping extracted from the experimental data.

8.7 References

[1] Abu-Mostafa, Y.S. 1986. 'Neural network for computing'. *In* 'Proc. Neural Networks for Computing, Snowbird, UT' (Ed. J.S. Denker). American Institute of Physics, New York, pp. 3–8

[2] Carnavali, P., & Patarnello, S. 1987. 'Exhaustive thermodynamical analysis of Boolean learning networks'. *Europhys. Lett.*, **4**, p. 1199

[3] Harrison, M.A. 1965. 'Introduction to Switching and Automata Theory'. McGraw-Hill, New York

[4] Hecht-Nielsen, R. 1990. 'Neurocomputing'. Addison-Wesley, Reading, Mass.

[5] Hirota, K., & Pedrycz, W. 1991. 'Fuzzy logic neural networks: design and computations'. Int. Conference on Neural Networks, Singapore, Nov. 18-21

[6] Kandel, A., & Lee, S.C. 1979. 'Fuzzy Switching and Automata: Theory and Applications'. Crane, Russak, New York

[7] Keller, J.M., Yager, R.R., & Tahani, H. 1992. 'Neural network implementation of fuzzy logic'. *Fuzzy Sets & Syst.*, **45**, pp. 1–12

[8] Pedrycz, W. 1991. 'Neurocomputations in relational systems'. *IEEE Trans. Pattern Anal. & Mach. Intell.*, **13**, pp. 289–296

[9] Pedrycz, W. 1992. 'Fuzzy neural networks with reference neurons as pattern classifiers'. *IEEE Trans. Neural Networks*, **3**, pp. 770–775

CHAPTER 9

Developments of Fuzzy Controllers — Fuzzy-Neural-Network Approach

9.1 Preliminaries

In this Chapter, we study the different architectures of fuzzy controllers developed with the aid of the relational neural networks discussed in Chapter 8. In general, fuzzy sets and neural networks deal efficiently with the two very distinct areas of information processing. As we learned in Chapter 2, fuzzy sets are good at various aspects of uncertain-knowledge representation, while neural networks are efficient structures capable of learning from example.

Neither technique is free from some weaknesses. They become complementary, in the sense that some features that are dominant in one approach are significantly lacking in the other. In this regard, we can encounter fuzzy sets that are not powerful enough to carry out learning and adaptation to a changing environment. They also require extra numerical computation to process the grades of membership.

On the other hand, neural networks are purely numerical constructs. Their learning is accomplished at the numerical level, and therefore it does not contribute essentially towards developing new schemes of knowledge representation. The role of neural networks in representing and handling uncertainty is also fairly limited, and usually calls for some extra mechanisms that must be incorporated into the relevant algorithms of learning. It is not surprising that there is an evident interest in combining both fuzzy sets and neural networks in such areas of application as fuzzy control, pattern recognition, and identification.

The ultimate goal pursued here is to design new architectures in which enhanced learning and improved knowledge-representational capabilities can go hand in hand. We start this Chapter by identifying a general scheme in which this blending can be accomplished (Section 9.2). Particular emphasis is laid on understanding the role of fuzzy sets and neurocomputations as a unified design platform for fuzzy controllers. We study the role of this aggregation in developing input and output interfaces for the fuzzy controller.

Then, a discussion of the generic architecture of the fuzzy controller viewed as a collection of logic processors clarifies the further links existing between fuzzy sets and neural networks. We consider two detailed architectures, along with the pertaining learning schemes. The first includes reinforcement learning, giving rise to the so-called Cerebellar Model Articulation Controller (CMAC). The second develops the structure of a fuzzy controller incorporating the principles of learning by analogy. The Chapter utilises some notions of neural networks; the interested reader may refer to [3,6], which are good and concise references in this respect.

9.2 Neural Networks and Fuzzy Sets in Fuzzy Controllers — Knowledge Representation and Learning

Let us take another look at the standard form of the fuzzy controller. The three basic phases — matching, inference, and final transformation of the fuzzy set of control into a single numerical value of control — can also be treated as three operations occurring between the control environment and the logical processes of inference carried out for the space of state variables and the space of control variables. The input interface (physical-logical level) is realised through the matching process. The output interface (logical-physical level) completes the transformation of the grades of activation of the fuzzy sets of control into a single numerical value. The internal part (including the control rules and the inference mechanism) operates on logical quantities rather than on the physical ones that become visible at the input and output interfaces.

All these three functional blocks can be implemented using techniques of fuzzy sets and neural networks. Different approaches have been vigorously pursued in the existing literature, cf. [5,9]. The general tendency is to concentrate on the use of fuzzy sets in building the input and output interfaces. The role of fuzzy sets at this level of the structure is fully legitimate. During the underlying design procedures, we focus on representing the input variables in terms of a given frame of cognition. A similar role can be witnessed for the output interface, where fuzzy sets are used to retransform the logical quantities into numerical ones. The main advantages of the use of fuzzy sets in developing these two interfaces are discussed in great detail.

The inference scheme (which operates on logical variables) can sometimes be quite computationally demanding and may therefore require a significant level of learning capability. We have already seen several disadvantages and computational weaknesses in the Hebbian-like learning (aggregation) of the control rules. Here, it is useful to discuss some well known architectures, stemming from the class of feedforward neural networks, that can be utilised to implement this logical mapping. The use of fuzzy sets and neural networks, even localised as sketched above, can freely permeate across different functional blocks [11] (Fig. 9.1).

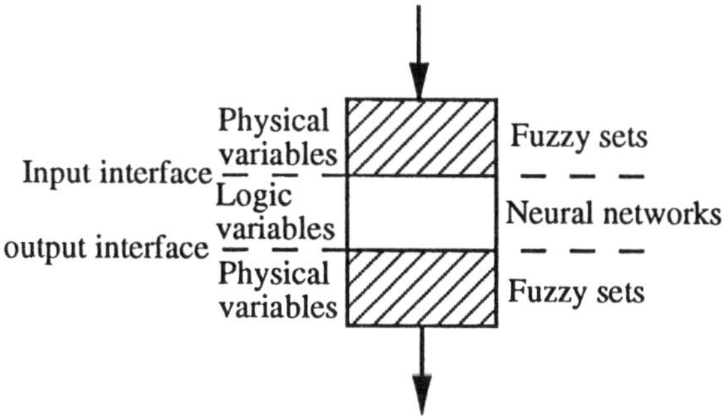

Fig. 9.1 Fuzzy sets and neural networks in developing fuzzy controllers

In our studies, we keep the structure at a high level of uniformity, discussing the logic-based neural networks (logic processors) introduced in the previous Chapter. They are characterised by the same learning capabilities as the standard neural networks, but are superior to them with respect to their clear interpretation. This interpretation allows us to convert the network into a series of 'if-then' rules.

The discussion on the input interface is first concentrated on a single input variable. In a neural network, this variable can be represented as a single input node in the input layer [3]. Its simplicity is definitely a strong argument for this representation scheme. Noting that there may be different input variables with different ranges of values (such as 10^{-3} and 10^5), we should apply normalisation, thus distributing all the input variables in a certain interval (for instance, [–1,1] or [0,1]). Note that this normalisation makes the normalised variables dimensionless.

Another way of representing the input variable is to carry out quantisation. In this form, we divide the entire range of the values of the variable into equal intervals and create input nodes associated with each of the intervals. This type of coding increases the dimensionality of the input layer of the network and affects the efficiency of learning. As we can easily see, this gives rise to the Boolean partition of the universe of discourse of the variable. By analogy, the frame of cognition can be viewed as a fuzzy quantisation of the space furnishing the input nodes of the network with the levels of activation of the linguistic labels in the frame. The fuzzy partition can easily capture uncertainty existing in the data. With an interval-like type of input information, several corresponding nodes in the input layer become activated (Figs. 9.2(a), (b) and (c)).

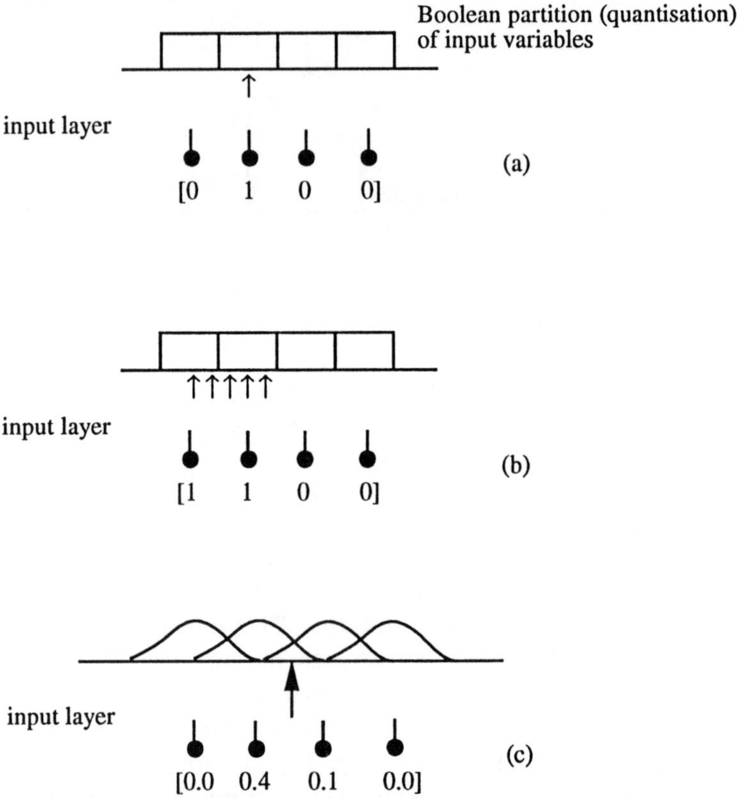

Figs. 9.2(a), (b) and (c) Activation of nodes in the input layer of the neural network for Boolean quantisation and a pointwise input datum (a), interval-valued information (b), and fuzzy quantisation and interval-valued information (c)

If nothing is known about the variable (total ignorance), all the input nodes for this variable are activated to the highest possible degree (for normal fuzzy sets, we get a vector of the input signals equal identically to **1**). Since the input signals are used by the neural network (e.g. in its training), the choice of the fuzzy discretisation may have a primordial effect on its learning. Once the regions of interest are properly defined by the frame of cognition, the learning can be significantly improved. For a single input neuron associated with the variable, we cannot represent its uncertainty (the single numerical value that we are forced to introduce as the input of the neuron does not leave any room for this type of representation).

In the next Section, we discuss the structure of the logic neural network that is used directly in implementing the fuzzy controller.

9.3 Fuzzy Controller as a Collection of Logic Processors

We assume that the control rules include the condition parts composed of two subconditions (say error and change of error). The logical structure of the control can be deduced from the rules interpreted as follows (see also the basic format of the control rules studied in Chapter 4):

the fuzzy control U_j occurs if:

condition$_{j1}$ (subconditions A_{j1} and B_{j2}) is satisfied
or
condition$_{j2}$ (subconditions A_{j2} and B_{j2}) is satisfied
or ...
condition$_{jh}$ (subconditions A_{jh} and B_{jh}) is satisfied,

where all the fuzzy sets A_{j1}, A_{j2}, ..., A_{jh} and B_{j1}, B_{j2}, ..., B_{jh} stem from a finite repertoire of the linguistic labels defined for the fuzzy controller.

We show that the control rule can be structured into a collection of AND and OR logic neurons. First, note that each condition can be formed by a single AND neuron:

$$\text{condition}_{jl} = (A_{jl} \text{ OR } w_{j1}) \text{ AND } (B_{jl} \text{ OR } w_{j2})$$

where the connections w_{j1} and w_{j2} modulate the contribution of the subconditions in the entire condition. The number of the conditions (h) determines the size of the hidden layer composed of these AND neurons. The fuzzy control U_j is derived by OR-ing the successive conditions standing in the rule, i.e.

$$U_j = (\text{condition}_{j1} \text{ AND } v_{j1}) \text{ OR } (\text{condition}_{j2} \text{ AND } v_{j2}) \text{ OR } \ldots \\ \text{OR } (\text{condition}_{jh} \text{ OR } v_{jh}) \quad (9.1)$$

We can recognise that the rule is constructed as a single logic processor; the number of neurons in its hidden layer depends on the structure of the rule. Technically, we can treat the conditions as degrees of activation of the corresponding fuzzy sets; while the left-hand side of expression (9.1) can be interpreted as a degree of activation of the jth fuzzy set of control. The logic processors are designed separately for each fuzzy set of control. Thus, 'm' fuzzy sets of control in the control protocol require a collection of 'm' logic processors (Fig. 9.3).

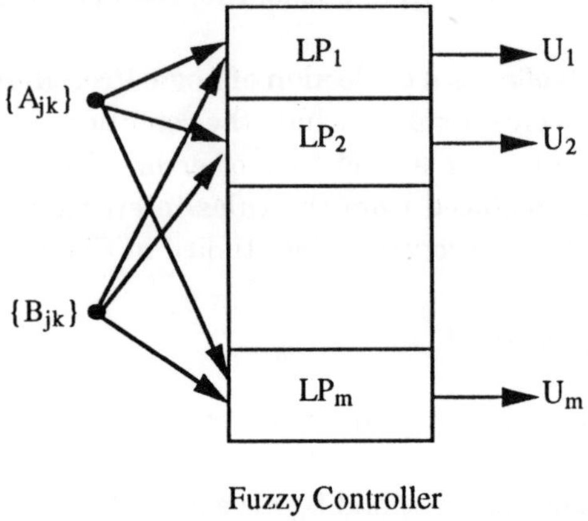

Fig. 9.3 Fuzzy controller as a collection of logic processors

Example 9.1
In this Example, we convert control rules specified in their 'if-then' format into relevant logic processors. The processors are constructed with the aid of available associations between the fuzzy sets of conditions and the fuzzy sets of control. The condition part has two subconditions. The first one takes on three linguistic variables A_1, A_2, A_3; the second contains B_1 and B_2 (Fig. 9.4).

The levels of their activation associated with the corresponding levels of the activation of the fuzzy set of control are given in this Table:

A_1	A_2	A_3	B_1	B_2	Activation of the Fuzzy Set of Control
0.9	0.1	0.0	0.7	0.1	0.1
0.8	0.2	0.1	0.2	0.9	0.75
0.1	1.0	0.05	0.9	0.05	0.92
0.2	0.8	0.1	0.0	1.0	0.05
0.0	0.0	1.0	0.95	0.05	0.0
0.0	0.01	0.95	0.1	1.0	1.0

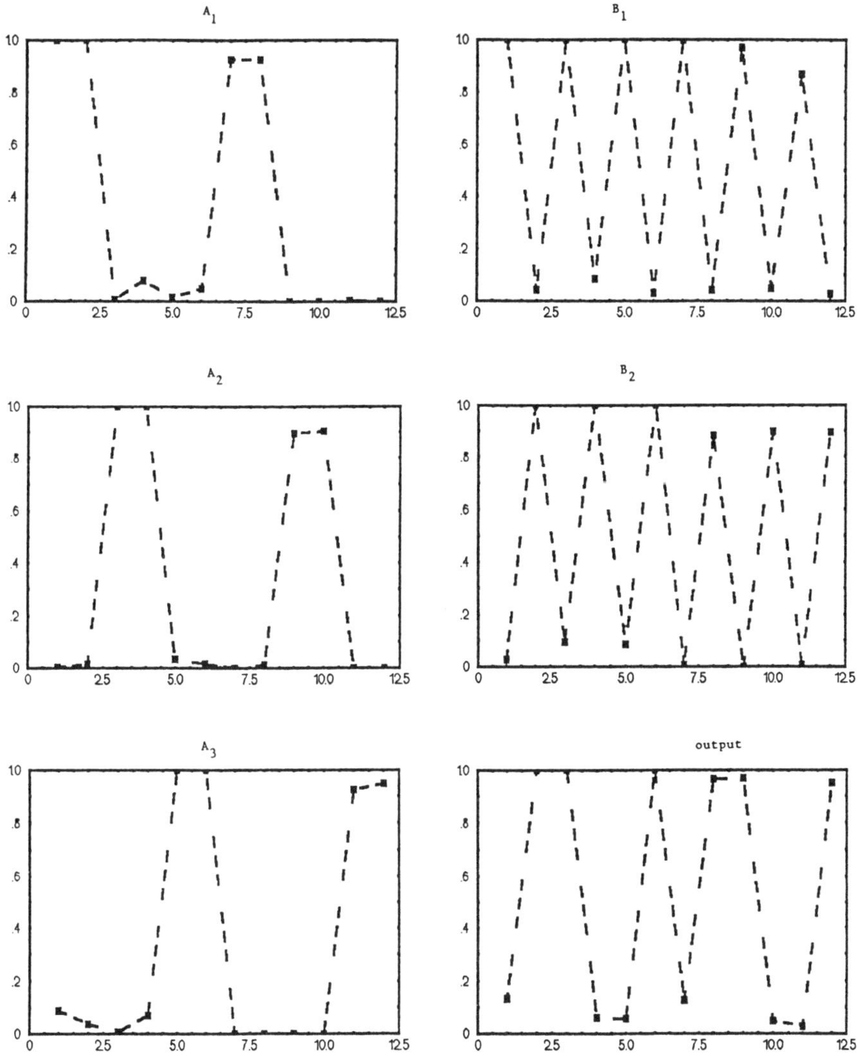

Fig. 9.4 Membership functions of A_1, A_2, A_3 and B_1, B_2

They are used as a training set in the learning scheme of the logic processors. Each logic processor has (3+2) + (3+2) = 10 input variables corresponding with the A_is and B_is and their complements. The dimension of the hidden layer is not available in advance and should be determined by optimising the results of learning obtained. In the current experiment, 'h' varies from 1 to 3. The results of learning (evaluated in terms of the sum of the squared errors between the target values in the learning set and those produced by the logic processor) are summarised in Fig. 9.5 (the learning rate was set to 0.1).

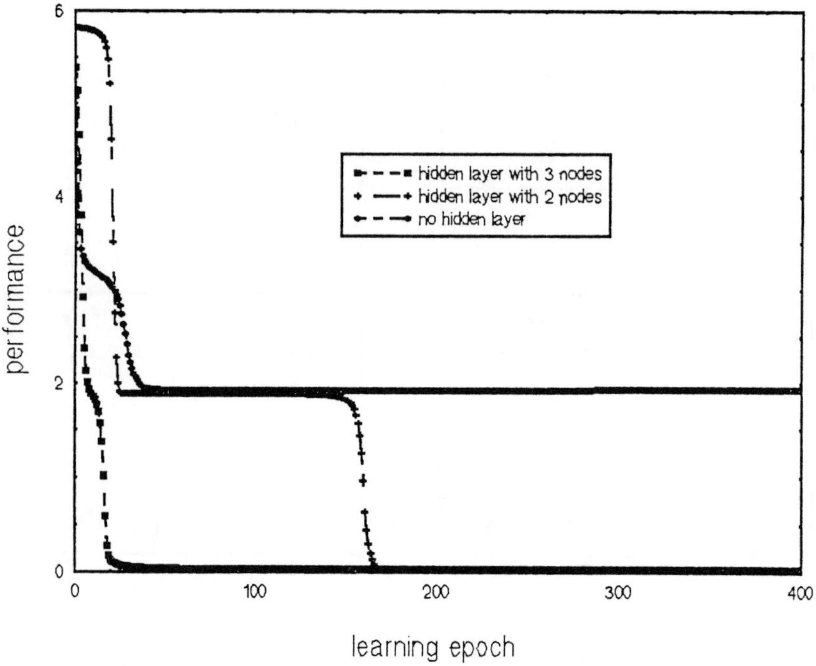

Fig. 9.5 Learning logic processors — different number of nodes in the hidden layer (t-norm: product; s-norm: probabilistic sum)

The hidden layer with 2 or 3 nodes revealed that the learning process had been completed without residual error. For h = 1 (which in fact led to the elimination of the hidden layer), the learning was unsuccessful — the error was constantly kept at a nonzero level. The connections of the network obtained (for h = 3) are:

hidden output layer:

1.0 1.0 0.2

hidden input layer:

1.0	1.0	1.0	1.0	0.55	1.0	0.0	1.0	0.0	1.0
1.0	0.34	1.0	0.0	1.0	0.36	1.0	0.0	1.0	0.66
0.49	0.28	0.16	0.61	0.74	0.75	0.08	0.07	0.06	0.12

The thresholding operation ignores some meaningless terms, and the logic processors can be converted into a collection of control rules of this form:

$$U = (\overline{A}_2 \text{ AND } \overline{B}_1) \text{ OR } (B_1 \text{ AND } \overline{A}_1 \text{ AND } \overline{A}_3)$$

Repeating the same pruning procedure in the structure with two nodes in the hidden layer, we derive:

$$U = (B_2 \text{ AND } \overline{A}_2) \text{ OR } (B_1 \text{ AND } \overline{A}_1 \text{ AND } \overline{A}_3)$$

Note that the neural network with h = 3 shows a certain structural redundancy (the third node in the hidden layer is quite inactive because of the associated connection). The core structure of the network exhibits the same collection of the subconditions in the rules (Fig. 9.6).

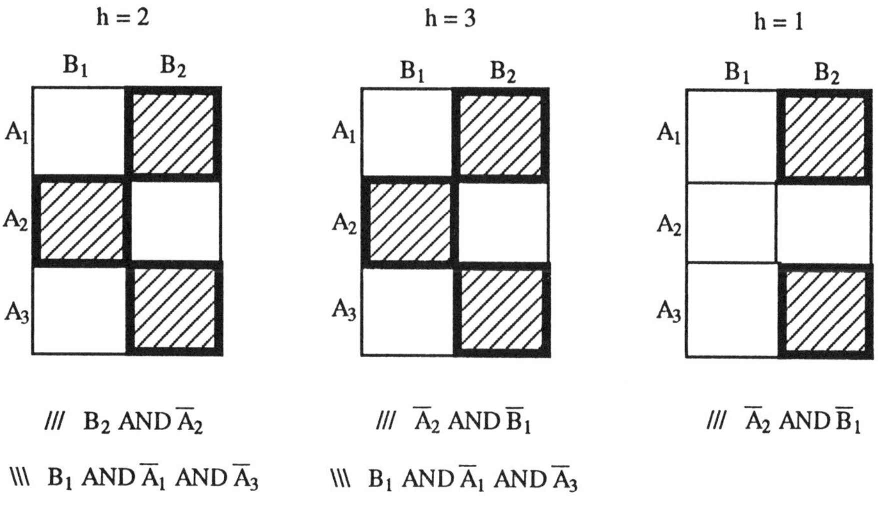

Fig. 9.6 Rules identified by logic processors

Interestingly enough, the learning for h = 3 is significantly faster than in the previous topology including the two elements in the hidden layer. We can take it as a suggestion to work with a slightly overdimensioned structure of the network to speed up the learning and hope that all the unnecessary connections can be trimmed down, producing a structure of a manageable size once the training has been completed. For h = 1, the structure of the network becomes less complicated, but there is no improvement in learning. For the threshold value $\lambda = 0.5$, the induced control rule reads:

$$U = \overline{A}_2 \text{ AND } \overline{B}_1$$

which, in fact, forms a subset of the rules determined by the previous structures. The derived approximation is quite unacceptable.

It is worth noting that, in general, fuzzy controllers should be modelled in terms of three-layer structures (logic processors). Attempts to restrict this representation to the two-layer topology are usually unjustified and in many cases misleading (however, this simplified approach has been attempted in the existing literature).

9.4 Fuzzy Controller as a CMAC Structure

The fuzzy controller, including the output interface, can be interpreted as a general structure of the Cerebellar Model Articulation Controller (CMAC), introduced in [1]. The general structure of the CMAC, as visualised in Fig. 9.7, reveals some striking similarities between it and

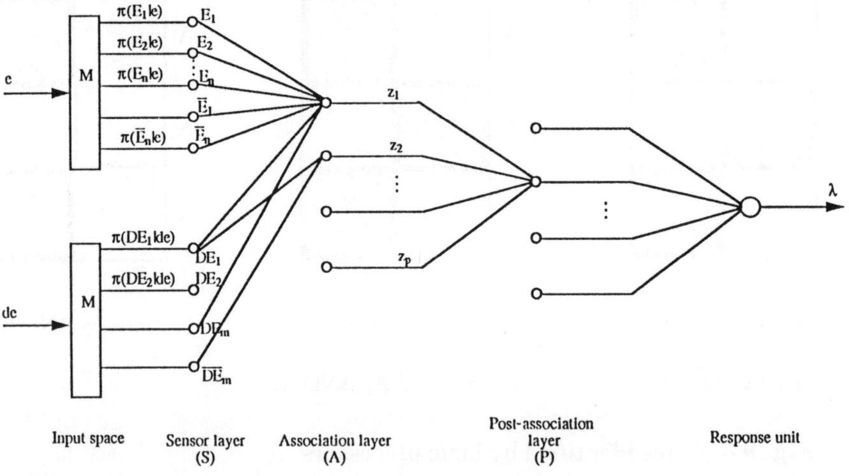

Fig. 9.7 General structure of the Cerebellar Model Articulation Controller

the fuzzy controller discussed in the previous Section. The functional characteristics of the elements (AND and OR nodes) can be instantly identified in the CMAC. The layers of the CMAC are:

• sensor layer (S). It is provided with the results of the matching obtained between E_i and DE_i (these are the fuzzy sets of error and change of error) and the input datum. The mapping between the sensor layer and the input space is essentially reflected by the matching procedure and proceeds from the physical measurement space into the logical space of the degrees of matching.

• association layer (A), consisting of AND nodes. Its role is to aggregate the pieces of evidence reported at the sensor layer (in our logic notions, these are the minterms of the signals originating at the sensor layer).

• post-association layer (P), consisting of OR neurons. Their role is to summarise the AND aggregates developed by the association layer.

• response unit (R), transforming the results of all these logic operations into a single physical quantity (numerical output of the fuzzy controller). The response unit is described accordingly:

$$u = \frac{z_1 \bar{u}_1 + z_2 \bar{u}_2 + \ldots + z_m \bar{u}_m}{z_1 + z_2 + \ldots + z_m} \tag{9.2}$$

where $z_1, z_2, \ldots, z_m$ are the logical values of the signals produced by the nodes of the post-association layer, while $\bar{u}_1, \bar{u}_2, \ldots, \bar{u}_m$ are the real values (real numbers) associated with the nodes. They can be treated as a collection of scaling (calibrating) factors. Their role is to convert the signals from the logical level into the physical level of the control signals. The parametric learning in this version includes two basic modes:

• learning the connections of the logic neurons of the network (while the calibrating signals are provided in advance and are not modified within the learning phase),

• learning the connections of the logic neurons, as well as the values of the $\bar{u}_j$s.

Since we studied the learning of the logic components in the previous Chapter, we concentrate now on the remaining parameters of the network.

We assume that the input of the network is given as **a** and **b**, both viewed as elements of the corresponding unit hypercubes. The pointwise control produced is equal to u = N(**a**,**b**). For the purposes of training of the network, we assume that a series of triples $(\mathbf{a}_k, \mathbf{b}_k, u_k)$ is provided. We also concentrate on an on-line version of learning, so that index 'k' in the data set (k) can be dropped.

The performance index is given as a squared error:

$$Q = [N(\mathbf{a},\mathbf{b}) - u]^2$$

The update scheme for the connections of the logic nodes was studied in Chapter 8. We are interested in the updates of the scaling factors $\overline{u}_j$. In general:

$$\frac{\partial Q}{\partial \overline{u}_j} = -[N(\mathbf{a},\mathbf{b}) - u]\frac{\partial u}{\partial \overline{u}_j}$$

The corresponding derivatives with respect to u_j are computed as follows:

$$\frac{\partial u}{\partial \overline{u}_j} = \frac{z_j}{z_1 + z_2 + \ldots + z_m}$$

Note also that the computations for the logic nodes involve the derivatives of Q with respect to z_j, $\partial Q/\partial z_j$.

The complete scheme of learning in the CMAC is an interesting example of 'reinforcement learning' [12], in which a single scalar quantity is used to determine a group of connections in the network. In comparison to the previous topology of the neural network, in this case we require that the fuzzy sets for the input variables be provided and that some pointwise representatives of the fuzzy sets of control be also available. In reinforcement learning involving the representatives, only the number of those categories to be recognised in the control space has to be specified in advance.

As a byproduct of this learning, we create a certain optimal quantisation of the space of control. The amount of *a priori* information that has to be prepared for the controller designed in this manner is limited in comparison to the previous construction. It is also possible to determine the fuzzy sets of control by displaying the degrees of activation of the hidden nodes with respect to the corresponding values of the pointwise control values produced by the fuzzy controller (9.2).

9.5 Reasoning by Analogy in Fuzzy Controllers

The well known paradigm of Artificial Intelligence, reasoning by analogy, can be used as a basic inference scheme [2,8] in the fuzzy controller. The underlying idea is that the inference is carried out on the basis of the similarity between the condition part in the 'if-then' statements and the fact provided for which the conclusion has to be inferred. The assumption made in this inference scheme is that similar conditions in the 'if' parts of the control rules induce a certain degree of similarity in the corresponding actions (consequences of the rules). The way in which this similarity is achieved in the condition parts becomes modified within the process of inference, implying that the corresponding level of similarity of the control actions has to be determined.

We now concisely represent the basic architecture and then discuss the relevant mathematical formalism [4,10]. The main concepts realised by the reasoning scheme here are shown in Figs. 9.8(a) and (b). The

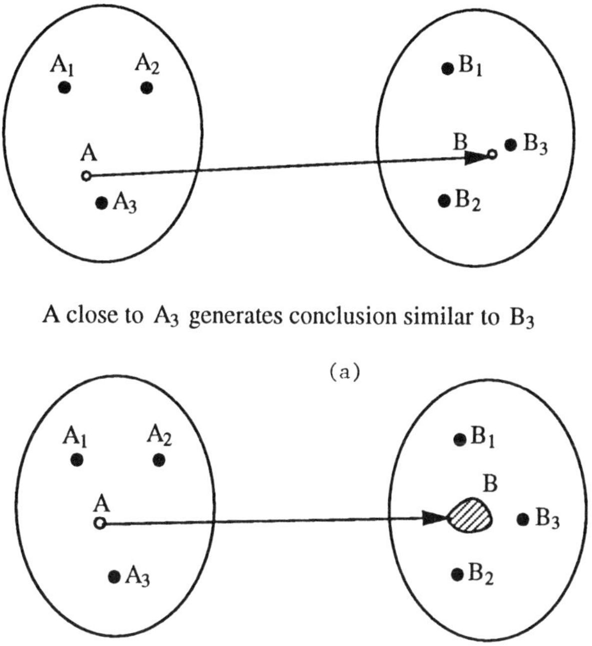

A close to A_3 generates conclusion similar to B_3

(a)

Condition A different from all A_k s leads to ambiguous conclusion B

(b)

Figs. 9.8(a) and (b) Basic structure of reasoning by analogy

three basic functional components correspond to the stages of information processing. We consider a single rule of the form 'if A_k, then B_k'; furthermore, all the fuzzy sets A_k and B_k are defined in the finite universes of discourse card(**A**) = n, card(**B**) = m. We can look at the reasoning by analogy by following a graphical visualisation in which fuzzy sets (or relations) standing in the control rules are treated as a relevant collection of 'anchor' points in the unit hypercube of the input variables of the controller. The inference scheme exposed to a new piece of information tries to reason on the basis of the information already available in the rules. The closer the new datum to one of the 'anchor' points, the higher the similarity of the conclusion to the corresponding point in the conclusion space. While the datum is distant from the information spanned over the antecedent space, the result of the analogical reasoning should pinpoint the resulting uncertainty and associate it with the inference result by converting it into a suitable form.

The exact idea of how the system functions is depicted in Fig. 9.9.

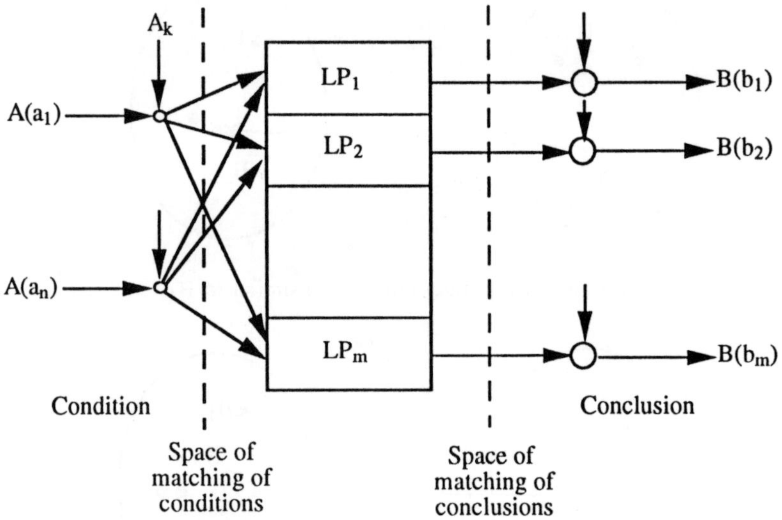

Fig. 9.9 Reasoning by analogy — functional blocks

The realisation of this concept is accomplished by:

(i) matching the input information and the existing antecedents,
(ii) transforming the result obtained into the space of matching of the conclusions, and finally
(iii) converting this derived level of matching into the fuzzy set of conclusion.

Below, we describe these phases in more detail:

1. Matching. At this stage, a new datum matches A_k and returns the

result of matching in a vector form where each of its entries takes on values between 0 and 1. As opposed to A and A_k, which are fuzzy sets in the universe of discourse with a certain physical meaning, this new vector, say **x**, becomes an element of an abstract input space of matching (called the antecedent matching space).

2. Transformation. During this transformation, we map elements from the antecedent matching space into the conclusion matching space. Its detailed form is determined by studying pairs of antecedents and the corresponding conclusions available in the rules.

3. Inverse matching. The idea here is to move from the conclusion matching space into the original space in which the conclusions of the rules (i.e. fuzzy sets of control) are defined. This operation is complementary to the matching procedure. Furthermore, we must ensure that the incorporation of a suitable level of confidence associated with the conclusion takes place within this step.

The matching phase requires 'n' matching neurons, each of which contributes to this operation in a coordinatewise manner. For each entry of A and the reference fuzzy set (one among A_ks), we determine the values of x_i, i = 1, 2, ..., n. This result (an element in the antecedent matching space) is processed by 'm' logic processors. The result obtained lies in the conclusion matching space and is then retransformed into the original space of conclusion. The fuzzy set of conclusion is computed with respect to another reference fuzzy set standing at this side of the scheme (B_k).

Where this matching is worked out with respect to the rules of the controller (of course, we can replace A_k by any other reference point that is of particular interest), the derived conclusion must be developed with respect to all the rules. The highest grade of matching is used to determine B as having the most evident impact on the results of inference. It may happen that the index k_0 pointing out the 'winning' rule is not the same for all the elements b_js of the space of the actions. The inference mechanism must be prudent enough to keep track of the values of the results of the matching obtained for the rules.

In a formal algorithmic setting, we write down the inference scheme thus:

initialisation

• set the initial values of the equality vector in the conclusion matching space to zero, i.e. $\gamma = \mathbf{0}$

repeat

• match A and A_k; transform the vector obtained lying in the condition matching space into the element of the conclusion matching space. Denote the result by γ-actual:

$$\gamma\text{-actual} = F(\mathbf{x})$$

where F denotes this mapping

• choose the maximal value among $\gamma(b_j)$ and γ-actual(b_j):

$$\max(\gamma\text{-actual}(b_j), \gamma(b_j)) \quad ;$$

if the current value γ-actual(b_j) is greater than or equal to $\gamma(b_j)$, update $\gamma(b_j)$ by accepting this maximal value and store the corresponding value of the membership function of the associated fuzzy set of control at b_j, $B_k(b_j)$; otherwise, do not perform any updates of the jth coordinate of γ

• process the next rule, increasing k, k := k+1

until the set of the rules has been completely examined.

The collected results are given in the form:

$$(\gamma, B')$$

(the fuzzy set B′ may constitute a combination of the grades of membership of different fuzzy sets of conclusion). By solving the inverse problem (Chapter 2):

$$B' = B > \gamma$$

we obtain an interval-valued fuzzy set $[B_-, B_+]$. This object reflects the level of confidence that we assign to the result of the completed reasoning.

To perform the above algorithm, a certain refinement must be made in order to cope with the cases where more than one rule matches the given factors to some (the highest) extent. If, for a certain b_j, several rules are invoked to the same maximal value $\gamma(b_j)$, the complete list of related $B(b_j)$s must be formed and the inverse problem be solved for the extremal membership values recorded there. This gives us the broadest range (pessimistic evaluation) of the values of the inferred conclusion.

Example 9.2
The control protocol consists of the three rules 'if A_k, then B_k' with fuzzy sets of conditions (A_k) and conclusions (B_k) defined below:

	a_1	a_2	a_3	a_4	a_5
A_1	1.0	0.6	0.5	0.2	0.0
A_2	0.2	0.4	1.0	0.4	0.2
A_3	0.1	0.3	0.5	0.7	1.0

	b_1	b_2	b_3
B_1	0.1	0.6	1.0
B_2	0.2	1.0	0.2
B_3	1.0	0.5	0.0

The matching process is completed with the aid of the equality index involving the Lukasiewicz implication. Then, the equality index becomes a piecewise linear function of both its arguments:

$$A_k(a_i) \equiv A(a_i) = \begin{cases} 1 - A_k(a_i) + A(a_i) & , \text{ if } A(a_i) \geq A_k(a_i) \\ 1 + A(a_i) - A_k(a_i) & , \text{ otherwise} \end{cases}$$

SOM-type logic processors use the operators of the product and the probabilistic sum. They perform the transformation between the spaces of matching. For learning purposes, we form a training set resulting from a pairwise comparison of the fuzzy sets standing in the control rules. The pairs used in the learning set are marked in the Table below:

Rule No.	Rule No.		
	1	2	3
1	*		
2	*	*	
3	*	*	*

The learning rate was set at 0.78, and the number of nodes in the hidden layer was equal to 3 (Fig. 9.10).

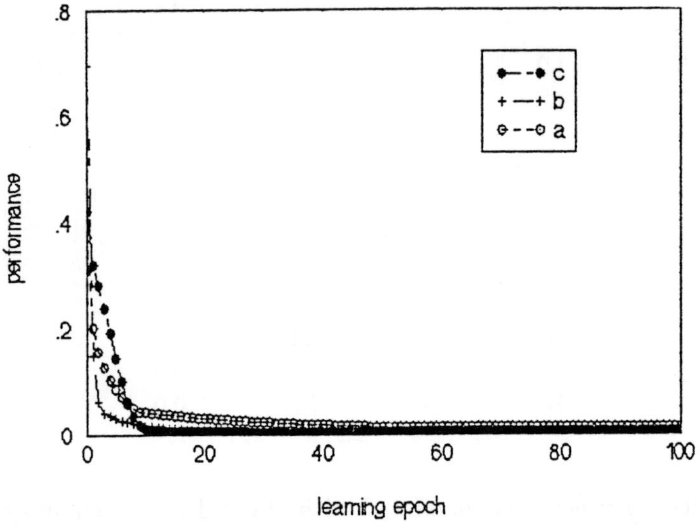

Fig. 9.10 Learning logic processors in the structure of analogical reasoning: $a-b_1$, $b-b_2$, $c-b_3$

The derived structure of the connections (Fig. 9.11) reveals that the degree of matching at the b_js is influenced by the input elements to very different levels. A qualitative assessment following the pruning of the network shows that the level of matching obtained in the conclusion space is mainly determined by a single element of the hidden layer.

A synthetic view on how the matching at the condition matching space affects the matching at the conclusion matching space can be obtained by accepting a certain uniform level of matching for all the inputs a_i, say $\alpha \in [0,1]$. The results produced by the logic processors invoked in this uniform way are given in Fig. 9.12.

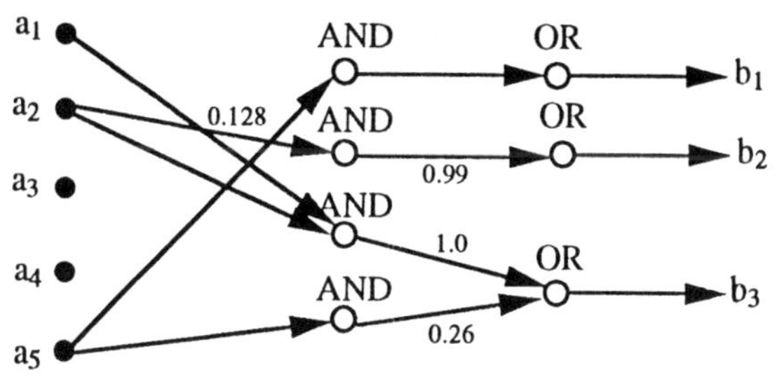

Fig. 9.11 Logic processors implementing analogical reasoning

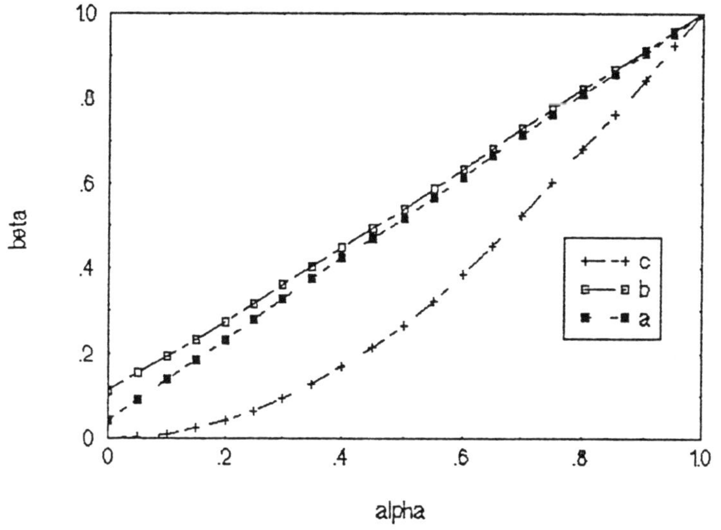

Fig. 9.12 Levels of matching in the conclusion space for a uniform activation of the input layer of the logic processors: $\beta = \beta(\alpha)$: $a-b_1$, $b-b_2$, $c-b_3$

Note that, while for b_1 and b_2 the resulting characteristics $\beta = \beta(\alpha)$ can be fairly well approximated to as two straight lines, at the third coordinate this relationship turns out to be rather nonlinear.

We analyse the resulting controller for some selected cases of the input information:

- First, let $A = A_1$, A_2, and A_3, respectively. The scheme returns B as equal to B_1, B_2, and B_3, respectively. Thus, the property of perfect mapping of the controller is achieved.

• Now, let A be a modified version of A_i. The original fuzzy sets are influenced by some linguistic hedges, namely A = *more or less* A_1 and A = *very* A_1. The influence of the hedges is modelled in a standard way:

$$\textit{very}\ A_1 = [1.0 \quad 0.36 \quad 0.25 \quad 0.04 \quad 0.0]$$

$$\textit{more or less}\ A_1 = [1.0 \quad 0.77 \quad 0.71 \quad 0.45 \quad 0.0]$$

Since A now differs from A_1, the result of the reasoning by analogy reflects this departure from the specific 'anchor' point, and the corresponding result reads:

$$\gamma = [0.99 \quad 0.84 \quad 0.85]$$

which, converted from the conclusion matching space, yields the interval-valued fuzzy set:

$$B = \begin{bmatrix} [0.09 & , & 0.11] \\ [0.44 & , & 0.76] \\ [0.85 & , & 1.00] \end{bmatrix}$$

This interval-valued fuzzy set is built around B_1 and, in fact, follows it quite well. For the second modified version of A_1, *very* A_1, we get:

$$\gamma = [0.99 \quad 0.81 \quad 0.80]$$

and

$$B = \begin{bmatrix} & 0.1 & \\ [0.41 & , & 1.00] \\ [0.80 & , & 1.00] \end{bmatrix}$$

which again includes B_1; however, the bounds are broader than in the first case.

• Let A be viewed as *unknown*. As an extremally nonspecific version of the input information, it perfectly matches all the A_ks (γ = 1.0). This, in turn, invokes B with bounds determined accordingly:

$$\left[\min_{k=1,2,3} B_k(b_i) \quad \max_{k=1,2,3} B_k(b_i)\right]$$

The result obtained shows that all the B_ks are activated. The fuzzy set of control covers all the B_ks:

$$B = \begin{bmatrix} [0.1 & , & 1.0] \\ [0.5 & , & 1.0] \\ [0.0 & , & 1.0] \end{bmatrix}$$

which highlights a significant level of ambiguity about the inference result.

Regarding some other existing approaches towards analogical reasoning applied either to general reasoning mechanisms, or to fuzzy controllers, in particular, we can refer to [7] and [13]. Both of them diametrically depart from the method discussed here.

First, the transformation between the condition and conclusion matching spaces is not available in the other approaches. Furthermore, the last transformation stage in the scheme discussed here is not included in those algorithms. This, in turn, makes it impossible to take advantage of the interval-valued extensions of the fuzzy sets in representing confidence about the outcome of the reasoning procedure.

9.6 Conclusions

We have studied combined approaches to the design of fuzzy controllers, including the techniques of fuzzy sets and neural networks. Fuzzy sets, with their properties of knowledge representation, are used in constructing the input interface of the controller. This causes all the input variables of the controller to be converted from the physical space into its logical counterpart. Neural networks are mainly applied to the computational (inference) part of the controller. Here, logic processors are of particular interest because their transparent structure is easily convertible into 'if-then' statements.

We have exploited the three architectures. First, we have translated the generic version of the fuzzy controller into a collection of LPs. The CMAC forms an interesting topology, with the reinforcement type of learning. Its peculiarity lies in the fact that one has to provide less initial design information than for the standard version of the controller (fuzzy sets of input variables and a number of categories of control action). Fuzzy controllers using the principles of analogical reasoning make it possible to cope with the uncertainty arising from the partial matching in the space of the antecedents. A unified approach, incorporating the components of fuzzy sets and neural networks into the same design platform, may result in a lot of other conceptual structures of the fuzzy controller.

9.7 References

[1] Albus, J.S. 1981. 'Brains, Behaviour and Robotics'. McGraw-Hill
[2] Davis, T.R., & Russel, T. 1987. 'A logical approach to reasoning by analogy'. Proc. IJCAI–87, Milan, Morgan Kaufmann, Los Altos, CA, pp. 264–269
[3] Hecht-Nielsen, R. 1990. 'Neurocomputing'. Addison-Wesley, Reading, Mass.
[4] Hirota, K., & Pedrycz, W. 1992. 'Prototype construction and evaluation as inverse problems in pattern classification'. *Pattern Recognition*, **23** (to be published)
[5] Jou, Ch. 1992. 'A fuzzy cerebellar model articulation controller'. IEEE Int. Conf. on Fuzzy Systems, San Diego, pp. 1171–1178
[6] Khanna, T. 1990. 'Foundations of Neural Networks'. Addison-Wesley, Reading, Mass.
[7] Koczy, L.T., Hirota, K., & Juhas, A. 1991. 'Interpolation of 2 and 2k rules in fuzzy reasoning'. Proc. Int. Fuzzy Engineering Symposium, Yokohama, Vol. 1, pp. 206–217
[8] Kodratoff, Y. 1988. 'Introduction to Machine Learning'. Pitman, London
[9] Ozawa, J., Hayashi, I., & Wakami, N. 1992. 'Formulation of CMAC-fuzzy model'. IEEE Int. Conf. on Fuzzy Systems, San Diego, pp. 1179–1186
[10] Pedrycz, W., Bortolan, G., & Degani, R. 1991. 'Classification of electrographic signals: a fuzzy pattern matching approach'. *Artif. Intell. in Med.*, **3**, pp. 211–226
[11] Pedrycz, W. 1991. 'Fuzzy logic in development of fundamentals of pattern recognition'. *Int. J. Approx. Reasoning*, **5**, pp. 251–264
[12] Sutton, R., & Barto, A. 1981. 'Toward a modern theory of adaptive networks: expectation and prediction'. *Psychol. Rev.*, **88**, pp. 135–170
[13] Turksen, I.B., & Zhong, Z. 1990. 'An approximate analogical reasoning schema based on similarity measures and interval-valued fuzzy sets'. *Fuzzy Sets & Syst.*, **34**, pp. 323–346

CHAPTER 10
Identification of Fuzzy Models

10.1 Introduction

In the analysis of complex systems, the principle of incompatibility established by Zadeh [17] helps us in modelling them:

'the closer one looks at a "real" world problem, the fuzzier becomes its solution. Stated informally, the essence of this principle is that, as the complexity of a system increases, our ability to make precise and yet significant statements about its behaviour diminishes until a threshold beyond which precision and significance (relevance) become almost mutually exclusive characteristics.'

This throws light on the place of fuzziness in the models we wish to construct. The form of uncertainty handled by fuzzy models stems from the overall perception process of the system, and is caused by the complexity and level of knowledge of the system. In [16], the notion of the fuzzy system (model) is introduced, its motivation being as above. As argued in [13] and [14], fuzzy models are appropriate where goals and constraints, as well as the physical mechanisms present, are not clearly defined. Constructing a fuzzy model where linguistics are significant does not require from the model-builder (who may not possess the strict mathematical formulas of the process) a deep formal insight. But he usually has good intuition and a mature experience of the system.

The two examples below use fuzzy-relational equations to express the relationship between the input, state, and output variables:

1. An activated-sludge-treatment process [15] lacks precise instrumentation, and has ill-defined control goals and poorly understood biological mechanisms. The river model has five input variables and two outputs. Input variables are:

(1) volumetric flow rate,
(2) river-water temperature,
(3) hours of sunlight incident on the river during each day,
(4) upstream biochemical oxygen demand,
(5) upstream dissolved oxygen.

The outputs are:

(1) downstream biochemical oxygen demand,
(2) downstream dissolved oxygen.

The fuzzy model working in discrete time intervals expresses the relationship between the linguistic values of the downstream biochemical oxygen demand (BOD) and the downstream dissolved oxygen (DO), in the fuzzy-relational equation:

$$DO_k = BOD_{k-3} \bullet DO_{k-1} \bullet R \tag{10.1}$$

with fuzzy relation R combining them together.

2. The second, classical, example is used in testing new identification or estimating algorithms. It comes from the Box and Jenkins book [1] dealing with the analysis of time series and concerns the data set of a gas furnace where the input (control) variable is the gas flow into the furnace and the state is the concentration of CO_2 in the outlet gases. Usually, models of this process are given in terms of a time series. To get a linguistic description of the system, linguistic labels are needed for the input and state variables. The structure of the model is specified by a fuzzy-relational equation. In [9], this model is proposed:

$$X_{k+1} = U_{k-\tau} \bullet X_k \bullet R \tag{10.2}$$

where a fuzzy set of control $U_{k-\tau}$ is delayed for a dead time τ. In general, we use sup-t composition here, since, as reported in [10], a product (instead of a minimum) leads to better model performance. For related studies on the same data set, see, for example, [14].

We now present some detailed model equations and identification procedures.

10.2 Fuzzy Models

For fuzzy models, we use fuzzy-relational equations and a state-space methodology. Their generality is sufficient to allow us to analyse a variety of systems.

Let **X**, **U** and **Y** be the spaces of state, control and output, respectively. All of them are assumed to be finite, say $\mathbf{X} = \{x_1, x_2, \ldots, x_n\}$, $\mathbf{U} = \{u_1, u_2, \ldots, u_m\}$, $\mathbf{Y} = \{y_1, y_2, \ldots, y_v\}$. Obviously, the results obtained can be extended to models with spaces consisting of an infinite number of elements, but then we have to restrict ourselves to those t-norms for which infinite distributivity is preserved.

These relationships govern the behaviour of the system modelled using a set of fuzzy-relational equations:

$$X_{k+p} = U_k \blacksquare X_k \blacksquare X_{k+1} \blacksquare \ldots \blacksquare X_{k+p-1} \blacksquare R$$
$$Y_{k+p} = X_{k+p} \blacksquare S \qquad (10.3)$$

where U_k denotes a fuzzy set of control (fuzzy control), X_{k+p}, X_{k+p-1}, ..., X_k refers to fuzzy sets of state, and Y_{k+p} stands for a fuzzy set of output. The first equation above is a state equation that relates the state at the (k+p)th time instant to those occurring at previous moments, namely k+p−1, ..., k, and the fuzzy control defined at the kth time moment. The second equation transforms X_{k+p} into the fuzzy set of output. The fuzzy relations R and S deal with the dynamics and output of the fuzzy model. They are defined in the Cartesian product of these spaces:

$$R : \mathbf{U} \times \mathbf{X} \times \underset{-p-\text{times}}{\mathbf{X}} \times \mathbf{X} - [0,1] \quad , \quad S : \mathbf{X} \times \mathbf{Y} - [0,1]$$

If **X** = **Y** and S is an identity relation, S(x,y) = 1 if x = y and 0 otherwise, then the second equation of the model is irrelevant and only the state equation is of interest. The model already given is called a fuzzy model of the pth order and can be viewed as a generalised version of a difference equation, not necessarily a linear one, of the pth order. For p = 1, we arrive at a model of the first order:

$$X_{k+1} = U_k \blacksquare X_k \blacksquare R \quad , \quad Y_{k+1} = X_{k+1} \blacksquare S \qquad (10.4)$$

With an extra assumption with respect to the output equation (S being the identity relation), this reduces to the single equation $X_{k+1} = U_k \blacksquare X_k \blacksquare R$.

By propositions, we show how a model of the pth order can be transformed into a model of the first order. This enables us to concentrate on the last form of the model, i.e. that of the first order.

Proposition 10.1
The fuzzy model given by (10.3) can be reduced to a model of the first order:

$$\tilde{X}_{k+1}\, U_k \blacksquare X_k \blacksquare \tilde{R} \qquad , \qquad \tilde{Y}_{k+1} = \tilde{X}_{k+1} \blacksquare S \tag{10.5}$$

where R is defined in $\mathbf{U} \times \mathbf{X} \times \mathbf{X}$, S in $\mathbf{X} \times \mathbf{Y}$, and X_{k+1} in $\mathbf{X}$. To prove this, we define a fuzzy relation R resulting from a composition of the fuzzy sets of state and R:

$$\tilde{R} = X_{k+1} \blacksquare X_{k+2} \blacksquare \ldots \blacksquare X_{k+p-1} \blacksquare R$$

We write:

$$\tilde{X}_{k+1} = X_{k+p} \qquad , \qquad \tilde{Y}_{k+1} = Y_{k+p} \qquad \text{and} \qquad \tilde{S} = S$$

Then, inserting all these into the original equations of the model, we obtain the reduced form of (10.5).

The use of this proposition is clear. In every case, we can omit reference to the method of reduction of fuzzy models of higher order. This also simplifies further consideration; without lack of generality, we can restrict ourselves to a model of the first order.

Now that we understand the structure of fuzzy models described by means of fuzzy-relational equations, it is worth while examining how the fuzziness processed in these models is handled. There are two sources of the fuzziness:

(i) First, in an intrinsic form of fuzziness, the ties between state and control variables are fuzzily known; i.e. one has only a fuzzy relation instead of a Boolean relation or a function dependence. This models a part of the linguistic description of the system analysed pertaining to a statement of the form:

'the state variables at consecutive discrete time intervals, say k and (k+1), are *more or less equal* if past control was *quite high*'

or

'at discrete time intervals, the state variable at time instant (k) is *greater* than in the past'.

In both statements, there are elements of the linguistic description of the process. More precisely, there is a relation that is known only in an approximate way.

(ii) Secondly, an external form of fuzziness stems from the fact that we are dealing with linguistic values attached to the state and control variables, whereas the ties existing between them are known exactly. An example is the behaviour of the pressure of an ideal gas in a container when its temperature and volume are specified only linguistically. Since we know the physical laws involved, the extension principle resolves the problem of system description.

To cope with these two sources of fuzziness, we specify these formulas:

If only the intrinsic form of fuzziness is present, U_k as well as X_k are degenerated fuzzy sets, with the membership functions defined as singletons, thus:

$$U_k(u) = \delta(u, u_k) \qquad , \qquad X_k = \delta(x, x_k)$$

Then X_{k+1} is equal to:

$$X_{k+1}(y) = \sup_{x,u} \left[U_k(u) \text{ t } X_k(x) \text{ t } R(u,x,y) \right] =$$

$$= \sup_{x,u} \left[\delta(u,u_k) \text{ t } \delta(x,x_k) \text{ t } R(u,x,y) \right] = R(u_k, x_k, y)$$

Thus the fuzziness conveyed by the fuzzy relation of the model made the state of the model X_{k+1} fuzzy. Its membership function is equal to the membership function of the fuzzy relation, with their arguments $u = u_k$ and $x = x_k$ already determined by the nonfuzzy variables.

In the second case, where U_k and X_k are fuzzy sets and R is a function, the membership function is:

$$R(u,x,y) = \begin{cases} 1 \ , & \text{if } y = f(u,x) \\ 0 \ , & \text{otherwise} \end{cases}$$

In other words, $R(u,x,y) = \delta(f(u,x), y)$. Performing formally the sup-t composition, we get:

$$X_{k+1}(y) = \sup\left[U_k(u)\ t\ X_k(x)\ t\ R(u,x,y)\right] = \sup_{u,x:y=f(u,x)}\left[U_k(u)\ t\ X_k(x)\right]$$

So we have the same result as we derived applying the extension principle to the fuzzy sets U_k and X_k transformed via the model equation $y = f(u,x)$. If U_k and X_k are nonfuzzy, we have the well known deterministic model of the system.

To allow fuzzy models formulated in the language of fuzzy-relational equations to be used effectively, we need to know how they are constructed and evaluated. To make fuzzy models applicable, we must solve the problem of system identification.

Nowadays, identification in deterministic and probabilistic (or statistical) models is a separate stream of engineering, with its own terminology and well developed methodology. When studying fuzzy models, we are in a different position. Since they are quite novel concepts and too specific in comparison to the models studied (e.g. statistical models), we need to develop some new tools of analysis while keeping to the previous methodology. We develop this idea on system identification in the presence of fuzziness in the following Sections.

10.3 System Identification in the Presence of Fuzziness

The methodology of system identification in this environment is not well established. Many fuzzy models are proposed, but they are given only *ad hoc* consideration against an intuitive, vague background, without details of estimation procedures. What is worse, the parameters of the fuzzy models, which are treated as fuzzy sets (or fuzzy numbers), are accepted from the model-builder without verification. It is assumed that the membership functions of the parameters in the model do not need to be estimated. This is unreal: intuition can help in creating the structure of a model, but one cannot expect to get the relevant membership functions of the model without performing any identification. As a loose analogy, consider regression analysis in constructing a model with one input and one output. Perhaps an assumption of linear dependence between the output and input variables may make sense in some cases where the model-builder has an intuitive view. Nevertheless, he cannot expect to get, even approximately, the parameters of the model. Despite a broad variety of models, their identification procedures consist of these three clearly defined steps [1,2]:

1. determination of the structure of the model,

2. estimation of the parameters of the model from the data set provided for identification,
3. validation of the model by testing its consistency with the data set.

These steps are treated broadly: thus the choice of the data set, which is not a trivial task, should be part of the first step. We need to know not only the estimated parameters of the model but also their precision, fundamental to the relevance of the fuzzy model. For a deeper look at the identification problem, we examine the first step in more detail:

Determination of the structure of the fuzzy model

This step is crucial to further investigation. As the fuzzy models refer to a linguistic description of our process, we have conditional statements of this format:

- if, in the past k, k–1, ..., k–ℓ time instants, the states of the system are given by fuzzy labels X_k, X_{k-1}, ..., $X_{k-\ell}$ and the control is equal to $U_{k-\ell}$, the state of the process is given by X_{k+1}.

This is directly implied by the form of the fuzzy model, with its order specified by the 'length' of these conditional statements. In fact, this length is not so large, since a conceptual model of the process realised by a human being consists of statements without a long condition part. Usually, a fuzzy model of low order is sufficient. In contrast, statistical models use sophisticated tools to estimate the order of the model equations and their type (i.e. linear or nonlinear). Working with fuzzy-relational equations, we are mostly concerned with the order of an equation rather than its type. The first notion is clear, while the second requires further optimisation. The order of the model equation has a straightforward impact on the form of the data set collected for the estimation stage. Consider, for instance, a model governed by:

$$X_{k+2} = U_k \blacksquare X_k \blacksquare X_{k+1} \blacksquare R$$

with U_k a fuzzy set of control, and X_k, X_{k+1}, X_{k+2} the fuzzy sets of state. The data set for determining the fuzzy relation R of the model consists of ordered triples. Starting with k = 1, we enumerate them thus:

$$\begin{array}{cccc} U_1, & X_1, & X_2, & X_3 \\ \vdots & & & \\ U_N, & X_N, & X_{N+1}, & X_{N+2} \end{array}$$

From Proposition 10.1, a model of higher order can be reduced to a model of the first order; therefore, we examine this equation:

$$X_{k+1} = U_k \blacksquare X_k \blacksquare R$$

Now the relevant data set consists of triples:

$$\begin{array}{ccc} U_1 , & X_1 , & X_2 \\ \vdots & & \\ U_N , & X_N , & X_{N+1} \end{array}$$

While considering estimation procedures, for notation simplicity we rewrite the model equation:

$$Y_k = X_k \blacksquare R \tag{10.6}$$

which is equivalent to the previous one, since X_k now replaces the fuzzy relation $U_k \blacksquare X_k$:

$$(U_k \blacksquare X_k)(u,x) = U_k(u) \text{ t } X_k(x)$$

This simplifies the format of the data set, which now converts into pairs (X_1, Y_1), (X_2, Y_2), ..., (X_N, Y_N). Thus the estimation problem is the task of solving a set of fuzzy relational equations $Y_k = X_k \blacksquare R$ for R unknown.

Estimation of the fuzzy relation of the fuzzy model

In our first approach, we calculate the fuzzy relation of the model as an intersection of the partial results of the φ-compositions of X_k and Y_k:

$$R = \bigcap_{k=1}^{N} (X_k \,\varphi\, Y_k) \tag{10.7}$$

This procedure consists of a sequence of steps:

1. put R = 1 (i.e. all the elements of R are set to 1.0),
2. put k = 1,
3. calculate $X_k \,\varphi\, Y_k$ and intersect R, $R = R \cap (X_k \,\varphi\, Y_k)$; increase k, $k = k + 1$,
4. if $k \leq N$, repeat (3); otherwise, stop.

Its output is the fuzzy relation of the fuzzy model. One remark should be made here. The fuzzy relation of the model computed using (10.6) assumes that the set of solutions of this family of equations is nonempty. The situation is made more difficult by the fact that this assumption is impossible to verify before the fuzzy relation of the model is calculated. This does not mean that this method of determining the relation should be discarded as theoretical but not practical. It can be used circumspectly, not overestimating the significance of its results and checking their validity carefully. Even if they are not satisfactory, they may give a good starting-point for applying more refined methods, to be clarified later. We call the method that utilises the findings for solving fuzzy-relational equations, without checking whether a solution exists, a brute-force one.

The model may also be constructed using fuzzy-relational equations of different types, as discussed in previous Chapters, e.g. equations with inf-s composition, adjoint equations or polynomial ones. Unlike the order of the equation, which is closely tied to the length of the statement, indicating linguistic labels, this method gives no general clues. In comparison to nonfuzzy models, the situation is more delicate, since no hypothesis (analogous to supporting linearity of the model) having any physical or at least numerical background can be formulated. It is simply a way of taking all the types of equation and verifying the extent to which each of them fits the data set provided.

To complete this presentation, we recall the model equation, using the remaining types of fuzzy-relational equation:

- for inf-s composition:

$$Y_k = X_k \bullet R \tag{10.8}$$

- for adjoint fuzzy-relational equations:

$$Y_k = X_k \varphi R \tag{10.9}$$

From the data set (X_k, Y_k), $k = 1, 2, \ldots, N$, the unknown fuzzy relation of the model is calculated:

$$R = \bigcup_{k=1}^{N} (X_k \beta Y_k) \tag{10.10}$$

and

$$R = \bigcup_{k=1}^{N} (X_k \times Y_k) \tag{10.11}$$

It will be recalled that such fuzzy relations fulfil the corresponding equations using some assumptions. Thus, the direct use of these formulas is in line with the brute-force method.

Two algorithms show how a fuzzy relation can be estimated. The first refers to the probabilistic set already mentioned in the design of fuzzy controllers. Here we focus on the role of higher monitors in defining the precision of the derived model. The second uses a probabilistic layer to determine the fuzzy relation of the model.

10.4 Estimating the Fuzzy Relation of the Model by Probabilistic Sets

In many cases, the fuzzy sets creating the data set (X_k, Y_k), k = 1, 2, ..., N, can be combined in (e.g. 'c') groups. Each cluster is an empirical representation of the probabilistic set, which thus reflects the structure of the data set. Their first two significant monitors are estimated from the partition matrix:

$$EX_i(x) = \frac{\sum_{c_i} X_k(x)}{n_i}$$

and

$$EY_i(y) = \frac{\sum_{c_i} Y_k(y)}{n_i}$$

The vagueness function is:

$$VX_i(x) = \frac{\sum_{c_i} [X_k(x) - EX_i(x)]^2}{n_i}$$

$$VY_i(y) = \frac{\sum_{c_i} [Y_k(y) - EY_i(y)]^2}{n_i}$$

$$x \in \mathbf{X} \quad , \quad y \in \mathbf{Y}$$

where the sum is taken over all the fuzzy sets X_k, Y_k belonging to the ith cluster; n_i is the number of elements in this cluster.

Instead of requiring equation (10.6) to be solvable, we seek to solve:

$$EX_i \blacksquare \tilde{R} = EY_i \qquad (10.12)$$

i = 1, 2, ..., c. Fuzzy relation R is estimated by the brute-force method:

$$\tilde{R} = \bigcap_{i=1}^{N} (EX_i \, \varphi \, EY_i) \qquad (10.13)$$

The number of clusters 'c' can be chosen by calculating the fuzzy relation R for the partition matrix, for c varying between 2 and N−1. The number of clusters is adjusted to minimise the index of quality of fit of the fuzzy model and the data set.

In a more general situation, the input-output data are probabilistic sets, given with their membership and vagueness functions. Thus:

$$(EX_i, EY_i) \quad , \quad (VX_i, VY_i) \quad , \quad i = 1, 2, \ldots, c$$

We may [5,8] regard any element of this data set, say a pair EX_i, VX_i, EY_i, VY_i (for brevity, we omit a subscript), as being distributed over intervals [a(x),b(x)] and [a(y),b(y)], so that their bounds are defined:

$$a(x) = EX(x)\,[1 - VX(x)]$$

$$b(x) = EX(x)\,[1 - EX(x)]\, VX(x)$$

$$a(y) = EY(y)\,[1 - VY(y)]$$

$$b(y) = EY(y)\,[1 - EY(y)]\, VY(y)$$

To obtain the bounds of the estimated fuzzy relation of the model, we restrict ourselves to sup-min composition in its equation. The upper bound of R is an α-composition:

$$R'(x,y) = a(x)\, \alpha\, b(y) \qquad (10.14)$$

while the lower bound is:

$$R''(x,y) = b(x)\ \alpha\ a(y) = \begin{cases} a(y) & , \text{ if } b(x) \geq a(y) \geq \max\limits_{x=x_0} b(x_0) \\ 0 & , \text{ otherwise} \end{cases} \tag{10.15}$$

This containment is satisfied:

$$R''(x,y) \leq R(x,y) \leq R'(x,y) \tag{10.16}$$

where R is the fuzzy relation of the model we are seeking. The closer R″ and R′ are, the better the estimation. Thus, we define the vagueness of the estimation of the fuzzy relation R, VR, as:

$$VR(x,y) = R'(x,y) - R''(x,y)$$

For any subset of the data set, we can estimate the fuzzy relation of the model and express the vagueness of the estimate. This implies a procedure called the dynamic estimation algorithm, which is realised by (i) calculating the fuzzy relations R′ and R″, and (ii) obtaining the relevant measures of vagueness and testing whether they are small enough to stop the algorithm. If the stopping criterion is not reached, the data set is modified or extended.

From the estimation procedure, we obtain two bounds of the fuzzy relation searched instead of just one as before, giving us much more information about the precision of the model derived. In further use of the model, we can take both bounds, leading us to an interval-valued fuzzy set.

Summarising, in the probabilistic approach, the uncertainty is conveyed by the data set or stems from the discrepancy of the model itself, and is seen in the bounds of the computed estimates. One example where fuzzy sets are applied and we deal with nonfuzzy data is examined below [7].

The fuzzy-relational equation can be used effectively where we do not know which class of model (e.g. linear, polynomial etc.) is suitable. The data set consists of input- output pairs of nonfuzzy data, (1,1), (2,2), (3,1), (2,4), (4,1), (4,3), (4,4), (5,5). Taking the spaces **X** and **Y** as discrete, **X** = **Y** = (1,2,3,4,5), the optimal fuzzy relation is obtained by clustering the data (FUZZY ISODATA is used), yielding a relation with these entries:

1.00	0.22	0.00	0.08	0.00
0.04	0.09	0.00	0.08	0.00
1.00	0.09	0.00	0.08	0.00
0.04	0.09	0.00	0.08	0.00
0.05	0.11	0.00	0.08	1.00

The performance index optimised by the clustering method is the sum of the squared errors, namely

$$\sum_{i=1}^{8} (y_i - \tilde{y}_i)^2$$

with y_i obtained by the centre-of-gravity method. We study the performance of this model. When an input has a nonfuzzy value, the output is given by the appropriate row of the fuzzy relation R (the relation is normalised for each row). Its row is called a fuzzy profile of the model, denoted for a certain x_i by $Fp(x_i)$. It is simply a fuzzy set defined in **Y**. Thus, for each i, i = 1, 2, ..., 5, one has a corresponding fuzzy profile. The closer the fuzzy profile to any fuzzy set of output (or nonfuzzy value) in the data set, the better the fuzzy model obtained. The model is perfect if, for each input, the fuzzy profile is a singleton with a nonzero value of membership assigned to the output recorded in the data set. The fuzzy profiles for the model already constructed are displayed in Fig. 10.1. Good agreement with the data set is achieved. A linear model does not fit this data set.

We now show how knowledge of a probabilistic form of uncertainty can be handled in the framework of fuzzy models. We call this class of models, structured fuzzy models, since the original fuzzy models are constrained (structured) by the probabilistic ties.

10.5 Structured Fuzzy Models

We focus on the static fuzzy model with max-min composition:

$$Y = X \bullet R \tag{10.17}$$

For clarity, we assume that **X** and **Y** are finite spaces. Each pair of elements in the Cartesian product $\mathbf{X} \times \mathbf{Y}$ is characterised by the joint probability function $p(y_j, x_i)$, $y_j \in \mathbf{Y}$, $x_i \in \mathbf{X}$. Evidently,

$$\sum_{i,j} p(y_j, x_i) = 1$$

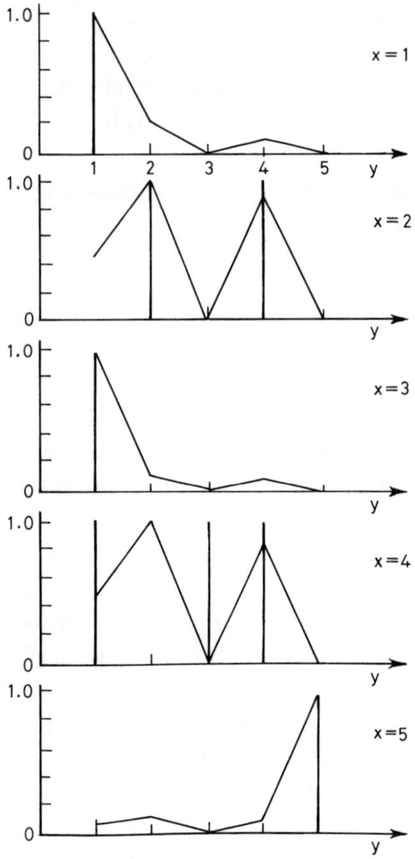

Fig. 10.1 Fuzzy profiles of the model

From this joint probability function, we obtain conditional probability functions $p(y_j \mid x_i)$:

$$p(y_j \mid x_i) = \frac{p(y_j, x_i)}{\sum_j p(y_j, x_i)} \tag{10.18}$$

expressing the strength of the transition (in the probabilistic sense) between the elements in spaces **X** and **Y**. All the transitions from a fixed x_i to u_j, j = 1, 2, ..., card(**Y**), give a sum of 1.0. This probabilistic layer is inserted into the model equation (10.18), in such a way that the indicated max-min composition is realised with respect only to those elements in **X** that have a sufficiently 'strong' probabilistic transition. The remaining elements are irrelevant, since their probabilistic transitions are negligible. In this light, we modify the model equation [12]:

$$Y(y_j) = \max_{x_i : p(y_j \mid x_i) \geq a} \left[X(x_i) \wedge R(x_i, y_j)\right] \tag{10.19}$$

'a' being a threshold level, $a \in [0,1]$. Its value allows us to eliminate the elements of **X** that have insignificant ties with the elements of **Y**. More compactly, we write the previous formula:

$$Y = X \underset{p}{\bullet} Y \tag{10.20}$$

where the index 'p' indicates the probabilistic structure of the model equation. We can derive another formulation equivalent to (10.19). We introduce a Boolean relation $B_a : \mathbf{X} \times \mathbf{Y} \to \{0,1\}$ with these entries:

$$B_a(x_i, y_j) = \begin{cases} 1 \ , & \text{if } p(y_j) \geq a \\ 0 \ , & \text{otherwise} \end{cases} \tag{10.21}$$

Then we get:

$$Y(y_j) = \max_{x_i} \left[X(x_i) \wedge B_a(x_i, y_j) \wedge R(x_i, y_j)\right] = \left(X \bullet B_a \bullet R\right)(y_j)$$

To show the value of 'a', we use the notation Y^a. A family of fuzzy sets $\{Y^a\}_{a \in [0,1]}$ is taken as a joint fuzzy-probabilistic representation of the output of the model. Two facts are evident:

- for $a = 0$, the influence of the probabilistic structure (layer) is disregarded,
- $a < b$ implies $Y^a \subseteq Y^b$.

For some measures of uncertainty ascribed to Y^a, refer to [12]. We estimate the fuzzy relation by using the fuzzy data (X_i, Y_i), $i = 1, 2, \ldots, N$. The data set is used in two ways: first to estimate the conditional-probability function; and second, in combination with the probabilistic characteristics, to calculate the fuzzy relation.

This computation uses a probabilistic strategy to eliminate some pairs in the data set that are biased and inconsistent with the probabilistic layer.

For estimating, we assume the above data set. Calculation of fuzzy relation R proceeds thus:

1. Start with fuzzy relation R having all entries equal to 1.0, $R(x_i, y_j) = 1.0$.
2. Put $k = 1$.
3. Calculate the α-composition of the kth pair from the data set, e.g. X_k and Y_k, and check simultaneously whether inequality

$$X_k \underset{p}{\bullet} (X_k \alpha Y_k) = Y_k$$

is satisfied. If so, modify R, intersecting it with $X_k \alpha Y_k$ thus:

$$R = R \cap (X_k \alpha Y_k)$$

Otherwise, leave fuzzy relation R unchanged.
4. Increase k, $k = k + 1$; if k is not greater than N, repeat 3; otherwise stop.

The output of the algorithm is simply the fuzzy relation of the model.

Let us explain the choice of threshold level 'a'. It is adjusted to meet a certain compromise. Too high an 'a' may cause only a few pairs of the data set to contribute to the computation; hence the fuzzy relation may be approximate. On the other hand, too low an 'a', e.g. about zero, may cause all the elements of the data set to contribute; thus the benefits of the probabilistic layer may be lost. It is reasonable to choose 'a' so as to minimise a performance index of the distance function between the fuzzy data and those fuzzy sets from the fuzzy model. Or we might use measures of the representative power of the identifying fuzzy sets (see Section 10.6).

We illustrate the above algorithm by an example of a Boolean data set with these input-output pairs:

X_k		Y_k		Frequency of Occurrence
x_1	x_2	y_1	y_2	n_k
1	0	0	1	1
1	0	1	0	n
0	1	0	1	n
0	1	1	0	1

The data set implies a matrix of joint probabilities:

$$[p(y_j, x_i)] = \begin{array}{c} \\ x_1 \\ x_2 \end{array} \begin{array}{c} \begin{array}{cc} y_1 & y_2 \end{array} \\ \begin{bmatrix} \dfrac{n}{2(n+1)} & \dfrac{1}{2(n+1)} \\ \dfrac{1}{2(n+1)} & \dfrac{n}{2(n+1)} \end{bmatrix} \end{array}$$

while the conditional probabilities are n/(n+1) for diagonal elements and 1/(n+1) for off-diagonal ones. Since this matrix contains only two values, say n/(n+1) and 1/(n+1), we can choose only a few values of the threshold level that may influence the fuzzy relation. Two are of special interest, say 1/(n+1) and n/(n+1). If this threshold is fixed below 1/(n+1), the probabilistic layer is omitted in further analysis. The relation obtained has all entries equal to zero, $R = \mathbf{0}$. Thus, it does not reveal any links between input and output and is of no value. Therefore, for all input fuzzy sets, the output is zero (null set $\mathbf{0}$). This gives a high performance index: e.g. if it is a sum of the Hamming distances between the fuzzy sets, we get:

$$\{|0-1|+|1-0|\}+n\{|1-0|+|0-0|\}+\{\|1-0\|+|0-0|\}+n\{|0-0|+|1-0|\}=2(n+1)$$

As n increases, this sum increases. So we cannot expect that additional elements in the data set will improve the model.

With a threshold level greater than 1/(n+1), some elements strongly inconsistent with the overall probabilistic layer are omitted. The remaining ones contributing to the relation of the model yield a diagonal matrix **1**. The sum of the relevant distances is 4, whatever the elements in the data set. For n = 40, it is only 4.5% of the performance index previously determined.

10.6 Detecting the Structure of the Data Set

This Section is devoted to algorithms for detecting the structure of the data set compiled for identification. The set is very useful, since only a few elements corrupted by noise strongly influence the estimation of the fuzzy relation. Our method compares the 'similarity' (or degree of equality) of the input and output fuzzy sets that are the elements of the data set. The degree of equality X_i and X_j is:

$$\gamma_{ij} = X_i = X_j = (X_i \subseteq X_j) \,\&\, (X_j \subseteq X_i) \tag{10.22}$$

where $X_i \subseteq X_j$ is the degree of containment of the fuzzy sets X_i and X_j. So, if $\gamma_{ij} = X_i = X_j$, $\xi_{ij} = Y_i = Y_j$, so that $\gamma_{ij} < \xi_{ij}$, we expect the data set consisting of two pairs (X_i, Y_i) and (X_j, Y_j) to be consistent. On the other hand, if $\gamma_{ij} > \xi_{ij}$, we can hardly solve the system of equations. We look at two situations:

First, $\gamma_{ij} = 1.0$ and $\xi_{ij} = 0.0$, which is the worst situation to solve, and, secondly, $\gamma_{ij} = 0.0$ and $\xi_{ij} = 1.0$, where the solvability is high.

Thus, we introduce this index:

$$\chi_{ij} = \gamma_{ij} \, \varphi \, \xi_{ij} \tag{10.23}$$

expressing the 'feasibility' of solving the equations. χ_{ij} is not equivalent in value to the solvability index, but is intermediately related to it. A modified version of the above equality index is of interest. We denote it by $(X_i = X_j)_{max}$ and define it:

$$\left(X_i = X_j\right)_{max} = \max_{x \in \mathbf{X}} \left\{ \left[X_i(x) \, \varphi \, X_j(x)\right] t \left[X_j(x) \, \varphi \, X_i(x)\right] \right\} \tag{10.24}$$

The above index gives an optimistic measure of the equality of the two fuzzy sets.

So far, the feasibility χ_{ij} has involved only a pair of data, X_i and X_j. For a global view of the data set *en bloc*, we refer to some common hierarchical clustering algorithms [2]. The objects grouped, i.e. the fuzzy sets, use the similarity of the pairs (X_i, Y_i) and (X_j, Y_j). The similarity matrix is $\chi = [\chi_{ij}]$, where:

$$\chi_{ij} = \chi_{ji} \quad , \quad \chi_{ii} = 1.0$$

To N clusters formed by single pairs from the data set, we apply an agglomerative procedure, so that the most similar clusters are merged. The number of clusters is reduced by 1, and the procedure repeated until one cluster is left. The distance between clusters **X** and **Y** is:

$$d(\mathbf{X}, \mathbf{Y}) = \max(\rho_{ij}) \tag{10.25}$$

where a maximum is taken over all pairs X_i, Y_i and X_j, Y_j belonging to **X** and **Y**. ρ_{ij} stands for the distance function between two objects, say the ith and jth ones, $\rho_{ij} = 1 - \chi_{ij}$. The hierarchy generated by the distance specified above is known in clustering techniques as the complete linkage method [2]. The distance ρ_{ij} has a straightforward interpretation. When

χ_{ij} specifies a degree of 'feasibility', the pair of fuzzy-relational equations allows us to solve the equations. ρ_{ij} is the 'difficulty' in solving this pair of equations. Hierarchical clustering methods help to represent the data set. The above example allows us to study the data set, which is the same as in [7], enabling us to compare these results with those achieved in setting probabilistic sets.

Example 10.1

The data set consists of N = 6 pairs of fuzzy sets with membership functions:

i	X_i	Y_i
1	[1.0 0.6 0.8 0.5]	[0.0 0.0 0.3 0.6]
2	[0.7 1.0 0.5 0.2]	[1.0 0.5 0.4 0.3]
3	[0.8 0.9 1.0 0.6]	[0.2 0.3 0.5 1.0]
4	[1.0 0.5 0.2 0.0]	[0.6 1.0 0.3 0.0]
5	[0.0 0.0 0.0 1.0]	[0.3 0.6 1.0 1.0]
6	[0.4 0.8 1.0 0.7]	[1.0 1.0 0.6 0.3]

The matrix of the model has these entries:

1.0	1.0	0.0	0.0	0.0	0.0
	1.0	0.2	0.0	1.0	1.0
		1.0	0.0	1.0	0.2
			1.0	0.0	0.0
				1.0	1.0
					1.0

From hierarchical clustering, we obtain the dendrogram in Fig. 10.2. At χ (calculated as a minimum of χ_{ij}, with i, j being the same objects in the same cluster) equal to 1.0, four clusters are seen: one containing three pairs {2,5,6} and the other three consisting of single pairs {3}, {1} and {4}.

As the number of clusters decreases, χ tends to zero; i.e. solvability of the system of equations (or its identification) is more difficult to achieve. The fuzzy relation, being an intersection of partial results (relations):

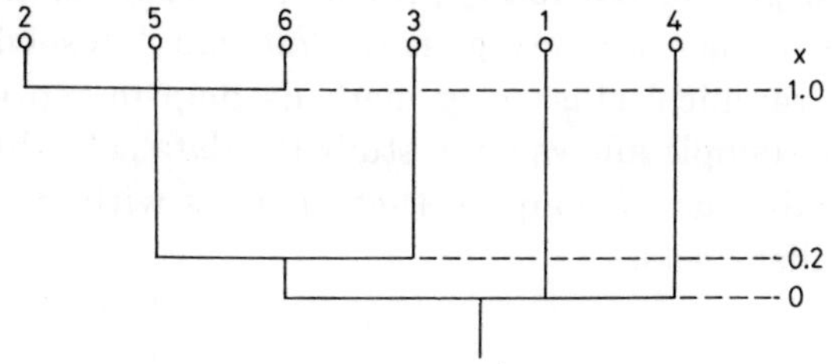

Fig. 10.2 Dendrogram from hierarchical clustering

$$\hat{R} = \bigcap_{i=1}^{6} \hat{R}_i$$

yields a poor performance, visible even without a formal specification of the index of solvability. We get this fuzzy relation:

$$\begin{matrix} 0.0 & 0.0 & 0.3 & 0.3 \\ 0.0 & 0.0 & 0.3 & 0.3 \\ 0.0 & 0.0 & 0.3 & 0.3 \\ 0.0 & 0.0 & 0.3 & 0.3 \end{matrix}$$

Thus, the fuzzy set $X_i \bullet \hat{R}$ differs significantly from the original Y_i: no one $X_i \bullet \hat{R}$ has a membership function greater than 0.3. The results of max-min composition are shown in Fig. 10.3. Taking the fuzzy relation equal to the intersection of $\hat{R}_2$, $\hat{R}_5$ and $\hat{R}_6$, say $\hat{R}_{256}$, based on the data forming one cluster in the previous dendrogram, we get:

$$\begin{matrix} 1.0 & 0.5 & 0.4 & 0.3 \\ 1.0 & 0.5 & 0.4 & 0.3 \\ 1.0 & 1.0 & 0.4 & 0.3 \\ 0.3 & 0.6 & 0.6 & 0.3 \end{matrix}$$

For comparison, the resulting fuzzy sets $X_i \bullet \hat{R}_{256}$ are also summarised in Fig. 10.3. The solvability index of the system of equations is higher now, but does not exceed the specified value of the index χ. Recalling the results of the probabilistic sets obtained by iterative clustering (in [7], ISODATA was used), we get this fuzzy relation:

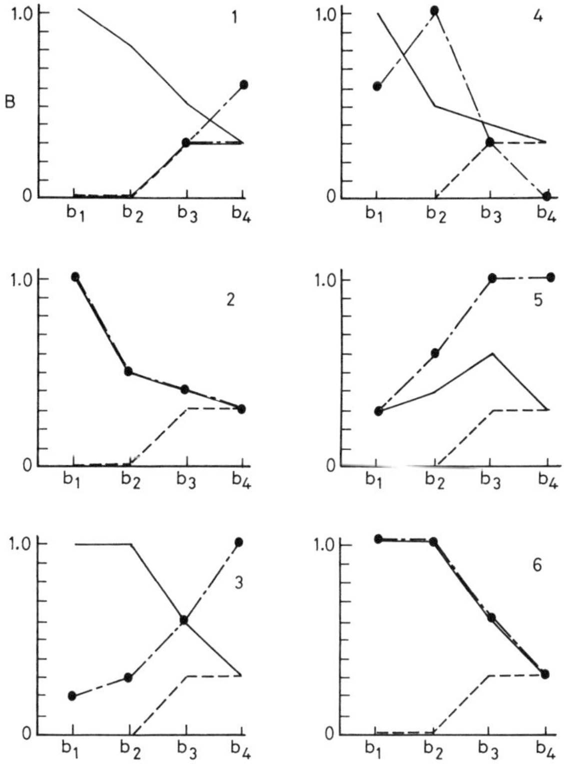

Fig. 10.3 Membership functions of the output of the model produced by various fuzzy relations

0.53	0.50	0.33	0.30
0.53	0.50	0.33	0.30
0.60	0.65	0.33	0.30
0.30	0.60	0.55	1.00

In this case, the data set contains three clusters, namely {1,2,4}, {5} and {3,6}. There are no significant differences between the entries of this fuzzy relation and that computed elsewhere, namely $\hat{R}_{256}$. Taking now χ_{ij}, defined by means of $(X_i = X_j)_{max}\ \varphi\,(Y_i = Y_j)$, we obtain:

1.0	0.0	0.0	0.0	0.0	0.0
	1.0	0.2	0.0	1.0	0.4
		1.0	0.0	0.2	0.2
			1.0	0.0	0.0
				1.0	0.3
					1.0

Using hierarchical clustering for this similarity matrix, we see a difference from the first dendrogram. Only two pairs, (X_2, Y_2) and (X_5, Y_5), give a system of equations that is completely solvable, $\chi = 1.0$. We get:

1.0	0.5	0.4	0.3
1.0	0.5	0.4	0.3
1.0	1.0	0.4	0.3
0.3	0.6	1.0	1.0

$X_2 \bullet \hat{R} = Y_2$, and $X_5 \bullet \hat{R} = Y_5$.

We focus on the previous relation, $\hat{R}_{256}$. The fuzzy sets $\hat{Y}_i$, i = 1, 3, 4, are greater than the original ones. This is not surprising, because fuzzy sets X_2, X_5, X_6 do not 'cover' the entire space **X**. Thus, fuzzy relation $\hat{R}_{256}$ does not map the data set very well. However, we add the next pair, e.g. (X_3, Y_3) (see the dendrogram), and fuzzy relation $\hat{R}_{2563}$ gives a low χ ($\chi = 0.2$); the fuzzy relation is:

0.2	0.3	0.4	0.3
0.2	0.3	0.4	0.3
0.2	0.3	0.4	0.3
0.2	0.3	0.4	0.3

In this analysis, we ask for the number of pairs of fuzzy data that come into the computation of the fuzzy relation. In other words, what threshold level in the dendrogram indicates the fuzzy sets that give the most reliable and consistent subset of the entire data set? This choice is not a trivial task. On the one hand, too many inconsistent fuzzy data in computing the fuzzy relation can give meaningless results, such as a fuzzy relation with nearly all its entries set to zero. On the other hand, too few fuzzy data prevent discovery of the entire relationship. It is vital to evaluate the 'representative' power of the fuzzy sets X_1, X_2, ..., X_N

contributing to the fuzzy relation. This problem is studied below.

10.7 Measuring the Representative Power of the Fuzzy Data

A concept enabling us to measure the representative power of a data set is based on a simple statement. Consider the fuzzy-relational equation:

$$X \bullet R = Y$$

with X, Y, R specified in the same spaces as before. From max-min composition, it is evident that only those elements of the fuzzy set X for which $X(x_j) \geq Y(y_k)$ holds contribute to the determination of the fuzzy relation. In other cases, the corresponding element of the fuzzy relation, namely (j,k), is equal to 1.0. We introduce an auxiliary vector defined in space **X**:

$$v = [v_1 \quad v_2 \quad \dots \quad v_n] \tag{10.26}$$

where each entry is:

$$v_j = \text{card}\{y_k \mid X(x_j) \geq Y(y_k)\} \quad , \quad j = 1, 2, \dots, n$$

This gives the number of events where the membership function X at point x_j exceeds or is equal to the membership function Y at point y_k. Replacing v by a probability vector that results from simple normalisation of v gives:

$$p = [p_1 \quad p_2 \quad \dots \quad p_n]$$

with

$$p_j = \frac{v_j}{\sum_{l=1}^{n} v_l} \tag{10.27}$$

If the sum of v_js is zero, p_j is zero as well. Vector p shows how well fuzzy relation R is determined and its best-estimated rows. The higher p_j is, the better the jth row of R is determined. If X is given as follows: $X(x_j) = \delta_{j,j_0}$, the corresponding probability vector is:

$$[0.0 \quad \dots \quad 0.0 \quad 1.0 \quad 0.0 \quad \dots \quad 0.0]$$

with 1.0 set at the j_0th position. The j_0th row of R is well estimated. But, taking X so that all v_j s are zero (vector of probabilities p = 0), we find that the fuzzy relation cannot be estimated. In an intermediate situation where X is 'unknown' (i.e. its membership function is 1.0 over the whole universe of discourse), for all x_j the value $v_j = m$ and, in consequence, $p = [1/n \quad 1/n \quad \ldots \quad 1/n]$. So all the rows of the fuzzy relation equal 1/n.

For a system of fuzzy-relational equations, we can assign to each pair of data a vector of probability p_i and perform a global evaluation:

$$p = \bigvee_{i=1}^{N} p_i \tag{10.28}$$

Notice that p_i allows us to eliminate some data in the preliminary analysis: all the fuzzy data whose vector of probability is **0** can be excluded. The highest p is attained when the elements of the data set are carefully chosen. Here are some general hints:

- if fuzzy sets X_i, i = 1, 2, ..., n (i.e. N = n), satisfy the conditions:

(i) $\max X_i(x_j) = X_i(x_i) \quad , \quad i = 1, 2, \ldots, n$

and

(ii) $\underset{i=1,2,\ldots,n}{\forall} \; \underset{k=1,2,\ldots,m}{\forall} \; \underset{j=i}{\forall} \left\{ y \,|\, X(x) \geq Y(y) \right\} = \varnothing$

then $p = [1.0 \quad 1.0 \quad \ldots \quad 1.0] = \mathbf{1}$.

Loosely speaking, fuzzy sets X_i should be maximum at just one point of the universe of discourse (x_i), and should be 'sharp' enough (condition (ii)). The value of p for our data set tells us how suitable are the fuzzy sets used to determine the fuzzy relation. The closer p is to **1**, the better the data set is for relation estimation; but only potentially, since all the pairs are taken separately. The whole picture is seen when we combine this approach with the overall analysis given by the dendrogram. This illuminates the data set from two different, competitive, points of view, namely:

- ability to solve the system of equations,
- ability to represent the fuzzy relation.

These are contradictory. Moving from the top of the dendrogram to the bottom, it is more and more difficult to satisfy the first requirement. In the opposite direction, the same holds for the second need. We return to our numerical example:

$$p_1 = [0.27 \quad 0.27 \quad 0.26 \quad 0.20]$$

$$p_2 = [0.33 \quad 0.44 \quad 0.23 \quad 0.00]$$

$$p_3 = [0.23 \quad 0.23 \quad 0.31 \quad 0.23]$$

$$p_4 = [0.67 \quad 0.33 \quad 0.00 \quad 0.00]$$

$$p_5 = [0.00 \quad 0.00 \quad 0.00 \quad 1.00]$$

$$p_6 = [0.11 \quad 0.22 \quad 0.44 \quad 0.23]$$

If we restrict ourselves to a compact cluster of the elements in the data set numbered 2, 5 and 6, for which $\chi = 0.1$, the overall vector p has these entries:

$$p = p_2 \vee p_5 \vee p_6 = [0.33 \quad 0.44 \quad 0.44 \quad 1.00]$$

Adding the pair (X_3, Y_3) resulting from the dendrogram does not change vector p. At the same time, χ decreases drastically. The ability of the whole data set to estimate the fuzzy relation is:

$$p' = [0.67 \quad 0.44 \quad 0.44 \quad 1.00]$$

This indicates that restricting the first subset of the data set to only three pairs does not significantly decrease its ability to determine the fuzzy relation. A change in p′ compared to p is observed for only one element of space **X**. The method presented is suitable for analysing the structure of the fuzzy data collected to determine the fuzzy relation. An index of the ability of a subset of data to estimate the fuzzy relation has been introduced. Nevertheless, the choice of threshold level in the dendrogram is open, and is up to the user. In every situation, some compromise is needed.

We discuss now a method of solving the system equations by introducing an additional space.

10.8 Fuzzy Models with Additional Variables

The essence of this estimation algorithm is an additional space in which the 'decoupling' fuzzy sets are defined. A loose analogy is found in model-building in a random environment. An output variable, say y, is tied to the input variable x by a function dependence, $y = f(x)$. If f is a linear function (it is usually assumed as a preliminary form of a model, if no clear indications are available), two parameters must be estimated, 'a' and 'b', so that $y = ax+b$. This is realised, for example, by minimising the sum of the squared errors,

$$\sum_{i=1}^{N} (a x_i + b - y_i)$$

with pairs (x_i, y_i) forming a data set (corresponding to the pairs of fuzzy data X_i and Y_i discussed here). Afterwards, the model is tested against the data set, using some powerful statistical-inference methods (e.g. the F-test). If the model is statistically inconsistent, it may be inadequate. This is detected by a statistical test. In this situation, it is reasonable to try a more complex form of model, not simply linear, but incorporating some terms of higher order. Thus, we are forced to introduce, for instance, a model of the form $y = cx^2 + ax + b$. Now, a new variable $w = x^2$ has to be added. The model is extended to one having two input variables, in which one is created in an artificial manner. Then the model is tested against the data set consisting of triples $(w_i, x_i, y_i) = (x_i^2, x_i, y_i)$. If necessary, new auxiliary explanatory variables can be introduced, so that an extended model is derived.

Along these lines, we reformulate the estimation of the fuzzy relation. To do this, we write:

$$(X_i \times C_i) \bullet G = Y_i \tag{10.29}$$

where C_1, C_2, ..., C_N are fuzzy sets defined in the auxiliary space $\mathbf{C}$, and fuzzy relation G is expressed in $\mathbf{X} \times \mathbf{C} \times \mathbf{Y}$. Fuzzy sets C_i are chosen so that the particular equations are decoupled (to be clarified later). The most 'difficult' situation (in the sense of satisfying the solvability of the system of equations) is:

$$\left(X_i = X_j\right) > \left(Y_i = Y_j\right)$$

We can improve this by introducing normal fuzzy sets C_i and C_j and calling them decoupling fuzzy sets, for which the inequality:

$$\left(X_i \times C_i = X_j \times C_j\right) < \left(Y_i = Y_j\right) \tag{10.30}$$

is satisfied. More formally, we state:

Proposition 10.2
For all fuzzy sets X_i, X_j, C_i, C_j, such that $C_i \cap C_j = \emptyset$, we have:

$$\left(X_i \times C_i = X_j \times C_j\right) \le \left(X_i = X_j\right)$$

To make the system of equations solvable (assuming that each equation of this set of equations does have a solution), it is sufficient to add a new space. Hence, the purpose of the extra fuzzy sets is obvious. They simply decouple our system of equations. Decoupling of fuzzy sets is permanent; thus, they can be found for any system of equations. With this modification, any system of equations (assuming each equation treated separately has a solution) is solvable. The only question is how to choose the decoupling fuzzy sets for the given system of equations. The simplest and almost self- evident way is to treat the space $\mathbf{C} = (c_1, c_2, \ldots, c_N)$ and the fuzzy sets $C_1, C_2, \ldots, C_N$ as singletons:

$$C_i(c_j) = \delta_{ij} \quad , \quad i, j = 1, 2, \ldots, N \tag{10.31}$$

But this choice is not unique. Now we offer a useful method of finding the biggest decoupling fuzzy sets for a system of equations. They have an interesting property. They express how strongly the membership functions in the extra space may overlap to ensure that the system of equations is still solvable. If an original system of equations is solvable, the biggest decoupling fuzzy sets have membership functions equal to 1.0 over the entire space $\mathbf{C}$. Consider some clusters of data detected by hierarchical clustering. Each cluster contains several pairs (X_i, Y_i) from the data set. We seek the decoupling fuzzy sets $C_1, C_2, \ldots, C_P$: $\mathbf{C} \to [0,1]$ with $\mathbf{C} = \{c_1, c_2, \ldots, c_p\}$. 'P' is the number of clusters. Thus, the data of a format $(X_{i1}, Y_{i1}, \ldots, X_{ip}, Y_{ip})$, $p = \text{card}(\mathbf{X}_r)$, $r = 1, 2, \ldots, P$, coming from the rth cluster, are extended into triples, $(X_{i1}, C_p, Y_{i1}), \ldots, (X_{ip}, C_p, Y_{ip})$. The membership functions of the fuzzy set C_p are computed:

$$\begin{aligned} &C_p(c_p) = 1.0 \\ &C_p(c_r) = \wedge\left(Y_i = Y_j\right) \end{aligned} \tag{10.32}$$

where a minimum is taken for all the pairs so that:

$$\left(X_i, Y_i\right) \in \mathbf{X}_r \quad , \quad \left(X_j, Y_j\right) \in \mathbf{X}_p \quad \text{and} \quad \left(X_i = X_j\right)_{max} \ge \left(Y_i = Y_j\right)$$

We adopt here the notation

$$\underset{0}{\wedge}\left(Y_i = Y_j\right) = 1.0$$

The results of this computation are shown in Fig. 10.4. This is an

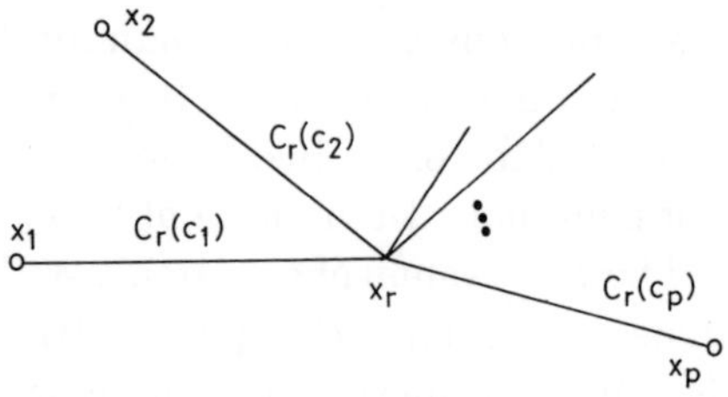

Fig. 10.4 Decoupling fuzzy sets

undirected graph with nodes formed by the clusters detected, and edges showing the strength of the relations between the clusters. Returning to our example and splitting the data set into four clusters, $\mathbf{X}_1 = \{(X_1, Y_1)\}$, $\mathbf{X}_2 = \{(X_2, Y_2), (X_5, Y_5), (X_6, Y_6)\}$, $\mathbf{X}_3 = \{(X_3, Y_3)\}$, $\mathbf{X}_4 = \{(X_4, Y_4)\}$, we have membership functions of the decoupling fuzzy sets equal to:

$$C_1 = [1.0 \quad 0.0 \quad 0.0 \quad 0.0]$$

$$C_2 = [0.0 \quad 1.0 \quad 0.2 \quad 0.0]$$

$$C_3 = [0.0 \quad 0.2 \quad 1.0 \quad 0.0]$$

$$C_4 = [0.0 \quad 0.0 \quad 0.0 \quad 1.0]$$

When every cluster consists of exactly one pair, $\mathbf{X}_i = \{(X_i, Y_i)\}$, the biggest decoupling fuzzy sets are:

$$C_1 = [1.0 \quad 0.0 \quad 0.0 \quad 0.0 \quad 0.0 \quad 0.0]$$

$$C_2 = [0.0 \quad 1.0 \quad 0.2 \quad 0.0 \quad 1.0 \quad 0.4]$$

$$C_3 = [0.0 \quad 0.2 \quad 1.0 \quad 0.0 \quad 0.2 \quad 0.2]$$

$$C_4 = [0.0 \quad 0.0 \quad 0.0 \quad 1.0 \quad 1.0 \quad 0.0]$$

$$C_5 = [0.0 \quad 1.0 \quad 0.2 \quad 1.0 \quad 1.0 \quad 0.3]$$

$$C_6 = [0.0 \quad 0.4 \quad 0.2 \quad 0.0 \quad 0.3 \quad 1.0]$$

10.9 Evaluation of the Fuzzy Model

Is our fuzzy model suitable? We distinguish between two situations, the first referring to a general mode of model testing and the second concerning evaluation methods for particular applications, such as prediction and control. Generally, evaluation is more synthetic, giving a yes-no response for the global model. If the special task of the model is solved, it is preferable for its evaluation to be pointwise. Thus, we can judge whether the model fits the data set in each element of the universe of discourse **Y**.

One of the representatives of the first group is a sum of distance functions between fuzzy sets Y_i and Y_i', where the second fuzzy set comes from the fuzzy model, $Y_i' = X_i \blacksquare R$:

$$D = \sum_{i=1}^{N} d(Y_i, Y_i') \tag{10.33}$$

where d(...) is the distance function, e.g. Hamming, Euclidean or, more generally, Minkowski. The acceptance or rejection of the model is now straightforward:

accept the model if $D < D_{crit}$

where D_{crit} is the critical value of the sum of the distances. Thus, if D is smaller, the model is acceptable. Shortcomings referring to this approach also are evident:

- we cannot express how the model fits the data set,
- the choice of critical value (D_{crit}) is open, with no reasonable underlying selection criterion.

The second situation is illustrated below.

10.9.1 Model evaluation by the fuzzy-measure approach

First, all the output fuzzy sets and the corresponding fuzzy sets are pointwise compared. Thus, for each Y_i and Y_i', we get a fuzzy set of their equality Γ_i: **Y** – [0,1], where

$$\Gamma_i = Y_i = Y_i' \tag{10.34}$$

i.e.

$$\Gamma_i(y_j) = \left[Y_i(y_j)\,\varphi\,Y_i'(y_j)\right] t \left[Y_i'(y_j) = Y_i(y_j)\right]$$

For a global evaluation of the equality set describing the model, an aggregation is needed. It is not unique, however, and these forms of aggregation are of interest:

- pessimistic:

$$\Gamma = \bigcap_{i=1}^{N} \Gamma_i$$

- optimistic:

$$\Gamma = \bigcup_{i=1}^{N} \Gamma_i$$

- average:

$$\Gamma = Av(\Gamma_i) \qquad , \qquad \Gamma(y_j) = \frac{\sum_{i=1}^{N} \Gamma_i(y_j)}{N}$$

If $\Gamma = \mathbf{1}$ (all elements of this vector are equal to 1), we assume that our model completely fits the data set. Conversely, if Γ tends to $\mathbf{0}$, we say that the model is weak.

The quality of the model using fuzzy set Γ is expressed by a fuzzy measure. For clarity, we consider the output space **Y** and say that the fuzzy model evaluated in terms of satisfaction of the property [4] is:

'the space **Y** *well mapped (well represented) by the fuzzy model under evaluation'* (10.35)

Thus, for any input X, the output of the model, and the fuzzy set describing the output of the system, are indistinguishable, or at least close to each other, especially for the fuzzy data used in constructing the fuzzy model (X_i, Y_i). For two fuzzy models with respective fuzzy sets Γ and Γ', we have two situations:

(i) $\Gamma \subset \Gamma'$; i.e. all the coordinates of Γ' are greater than or equal to the corresponding coordinates of Γ. Here, the model with Γ works worse than that with Γ' (it is less precise).

(ii) Γ and Γ' are not comparable.

Situation (ii) is more frequent than situation (i). This fact, as well as the pointwise construction of Γ, suggest a partial evaluation of the model for each coordinate, i.e. the fulfilment of a 'local' property:

'the ith coordinate of **Y** *is well mapped by the fuzzy model'* (10.36)

On the Gestalt principle, the global property (10.35) cannot be deduced from a simple, perhaps linear, aggregation of partial evaluations of the model using (10.36). This leads us to a fuzzy measure as a plausible tool for global evaluation of the fuzzy model. As we shall see, the structure of the fuzzy integral enables us to link the quality of the model and the controllability and predictability of the system.

From Chapter 2, fuzzy measure g_λ is a set function defined in the Borel field **B** as monotonic with boundary conditions $g_\lambda(\emptyset) = 0$ and $g_\lambda(\mathbf{Y}) = 1$. Fuzzy measure $g_\lambda(.)$ 'measures' the quality of the model for any subset of **Y**. If the known number of elements of **Y** with concrete levels of property (10.36) increases, our ability to judge the model as a whole increases. This is reflected by the monotonicity of the fuzzy measure. Without a point of **Y**, we can make no judgment. If we know our model quality in only one point of **Y**, say y_j, we know it from $g_\lambda(\{y_j\})$, which is identical with our previous evaluation through Γ_j. Here, we can make a partial judgment about the model, but it may be marginal. Considering only a single element y_j, significant overestimation of the quality of the model may occur if $\Gamma(y_j) = 1$. Conversely, if $\Gamma(y_j) = 0$, we underestimate its performance.

From fuzzy set Γ, the λ-fuzzy measure is determined numerically. From the λ-rule, the fuzzy measure is:

$$g_\lambda(\mathbf{Y}') = \frac{\left\{ \prod_{y_i \in \mathbf{Y}'} (1+\lambda\,\Gamma(y_i)) - 1 \right\}}{\lambda} \quad , \quad \lambda \in (-1,\infty)$$

10.9.2 Evaluation of the fuzzy model by using the induced confidence levels

Now we give some analogies to compare the methods used to evaluate the statistical (probabilistic) models. We discuss a model of first order:

$$X \blacksquare R = Y$$

assuming that the identifying fuzzy data (X_i, Y_i) are available. The fuzzy relation of the model (omitting its origin) yields a fuzzy set Y_i' for X_i treated as the input of the model. To summarise the membership functions of Y_i and Y_i' for all i = 1, 2, ..., N in a fixed element of the universe of discourse, say y_j, we construct an empirical distribution function of the equality index:

$$F_j(w) = \frac{\text{card}\left\{ i \,|\, \left(Y_i(y_j) = Y_i'(y_j) \right) < w \right\}}{N}$$

$$w \in [0,1]$$

The interpretation of $F_j(w)$ is straightforward: it articulates the probability that the equality index is not larger than 'w'. In the extreme:

- if the model fits perfectly the data set at the specified element of the universe of discourse y_j, all the values of the equality index $Y_i(y_j) = Y_i'(y_j)$ are set to 1.0, and therefore:

$$F_j(w) = \begin{cases} 0 \ , & \text{if } w < 1 \\ 1 \ , & \text{otherwise} \end{cases}$$

so the distribution function has one jump of height at 1.0, located at w = 1.0,

- if the model gives results different from the data in the data set, the distribution function is 1.0 for all arguments, with precisely one step at w = 0.0.

We can usually deal with intermediate situations, where the distribution function has a step-like function, by steps of different heights scattered over the whole unit interval. The better model has a distribution function with nonzero values moved towards higher values of the argument.

The equality index γ has this probability:

$$\alpha = P\left\{\omega \middle| \left(Y_i(y_j) = Y_i'(y_j)\right) \geq \gamma\right\}$$

(we skip index 'j' in the equality index γ, as well as α, assuming that the entire investigation is performed for each j, j = 1, 2, ..., card(**Y**)).

Probability α expresses the fraction of the grades of membership function for which the equality index exceeds γ. From the obvious relationship:

$$P\left\{\omega \middle| \left(Y_i(y_j) = Y_i'(y_j)\right) \geq \gamma\right\} = 1 - p\left\{\omega \middle| \left(Y_i(y_j) = Y_i'(y_j)\right) < \gamma\right\} = 1 - F_j(\gamma)$$

we get:

$$\alpha = 1 - F_j(\gamma)$$

For each fixed α, we can determine the corresponding γ. More precisely, we take a maximal γ, denoting by $\Gamma = \{\gamma \mid 1 - \alpha = F_j(\gamma)\}$ the solution γ_*, taken as

$$\gamma_* = \max_{\gamma \in \Gamma} \arg \gamma$$

Conversely, from the equality index γ, we get the probability. Thus, we see a function dependence between γ and α, $\gamma = f(\alpha)$, giving an interesting insight into the probability of occurrence of the specified values of the equality index. Further analysis of this expression allows us to discover some properties of the function $\gamma = f(\alpha)$. We rewrite it as $F_j(w) = 1 - \alpha$. If α increases, $1 - \alpha$ decreases; since the distribution function is a non-decreasing function, we get lower values of the equality index. This phenomenon can be expected: if we want to achieve high probability, it occurs at lower γ. Retransforming the grade of membership with the given γ, we get broader and broader equality intervals. The equality index has its analogue in a confidence interval found in statistics, especially in parameter estimation. For a higher value of confidence, say 0.01, as compared to a particular standard of 0.05, the intervals are quite wide.

Computing thus for all the elements of **Y**, we get the corresponding distribution functions of the equality indices. From the membership function of the output fuzzy set, we get the equality intervals for the prespecified γ or α.

We consider two extremes of behaviour of the above fuzzy model. In the first, the model fits completely the data set and the distribution function of the equality indices possesses one jump of 1.0 at w = 1.0. This implies that, for all $\alpha \in [0,1]$, $\gamma = \gamma(\alpha)$ = const = 1.0, and the corresponding equality interval decreases to a point. Thus, we get a genuine fuzzy set, not an interval-valued one, which shows that the model is extremely precise.

In the second case, the fuzzy model produces a completely different fuzzy set. Now, the distribution function has, as before, one jump of 1.0, but this jump is put at w = 0.0. By straightforward computation, we can verify $\gamma = \gamma(\alpha)$ = const = 0.0, which gives an equality interval of [0,1]. Thus, the model is irrelevant and the resulting fuzzy set conveys no useful information.

An example illustrates the use of the confidence intervals of the fuzzy model. The data set has already been discussed. The model has the form Y = X • R. Three ways of estimating the relation of the model have been used. The first (a) is a brute-force one; the second (b) is based on probabilistic sets; the third (c) considers a subset of the data set that has high consistency and is extracted by hierarchical clustering. The plots are shown in Figs. 10.5(a), (b) and (c). The brute-force method produces

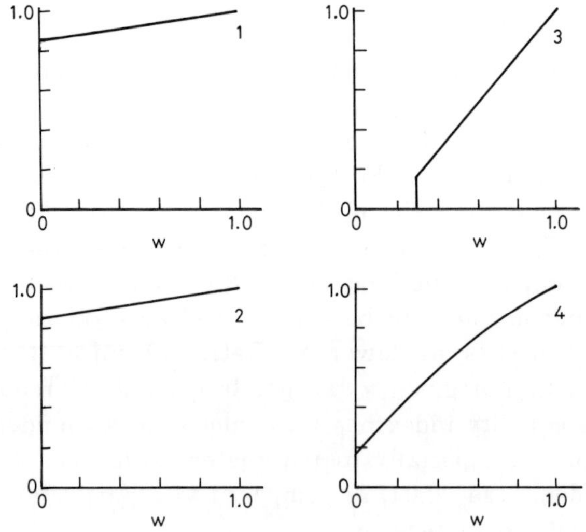

Fig. 10.5(a) Confidence intervals for the fuzzy models versus α

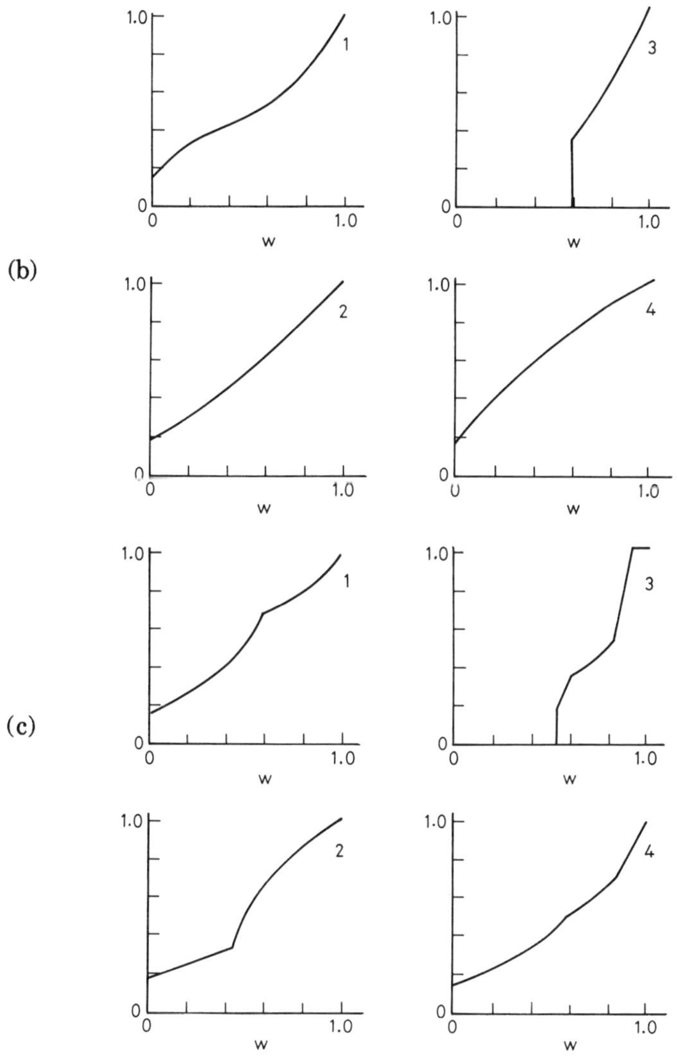

Figs. 10.5(b) and (c) Confidence intervals for the fuzzy models versus α

poor results, especially at the first and second elements of the universe of discourse **Y** (the distribution functions at w = 0.0 are near to 1.0). The other two methods give better results, allowing lower distribution functions at the origins of the coordinates. To show the quality of the model obtained by the second method, we compute its output when the input is a singleton X = [1 0 0 0] (Fig. 10.6).

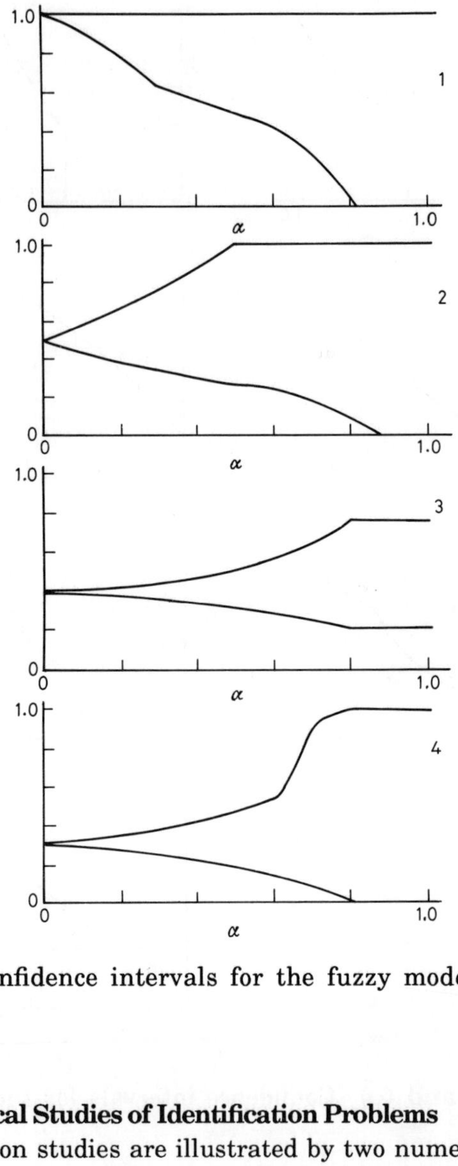

Fig. 10.6 Confidence intervals for the fuzzy model obtained by method (b)

10.10 Numerical Studies of Identification Problems

These identification studies are illustrated by two numerical examples:

Example 10.2

A gas-furnace data set (Box and Jenkins [1]) is a standard test for identification and estimation. In fuzzy sets, it has been used by Tong [14]. However, his approach, heuristic in nature, is completely different from that presented here. The data set consists of n = 296 parts of input-output observations, where the input is the rate of gas flow into the furnace and the output is the concentration of CO_2 in the outlet gases. The sampling interval is 9 s. Since the data are nonfuzzy, and we intend to build a fuzzy model describing the relations between the linguistic (fuzzy) labels, we start by looking for them. This can be done conveniently by inspecting the available data set. Without extra knowledge about the collected pairs of input and output, we are forced to use a clustering technique. We use FUZZY C-MEANS, a relative of ISODATA, which produces 'c' clusters and detects their centroids (fuzzy means). The choice of number of clusters depends on the user of the model. Formally, 'c' can vary between 2 and $N-1$. However, the number of clusters should accord with the number of linguistic labels, reflecting naturally the level of knowledge of the system under consideration, or the level of generality in the user's description of the system. We may need a concise and easy-to-use description without too many details, or we may need a detailed description for solving special types of task. If the number of clusters increases, a higher precision is obtained, but there is no evidence of a useful correspondence between the formal fuzzy description of the system and its linguistic representation. In the extreme, $c = N$, the establishment and assignment of linguistic labels are meaningless. But too low a number of categories (clusters) may cause low precision because the model obtained is too general. Clusters obtained for the state space are shown in Fig. 10.7 (putting five clusters). We shall call the clusters (fuzzy

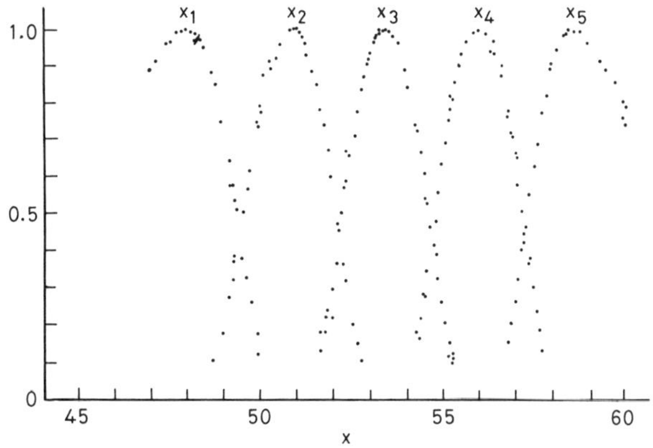

Fig. 10.7 Membership functions of the linguistic labels defined in the state space

sets) reference fuzzy sets. Thus, for space **X**, we distinguish X_1, X_2, ..., $X_{c\mathbf{X}}$, while the fuzzy sets in the control space **U** are denoted by U_1, U_2, ..., $U_{c\mathbf{U}}$. The form of the model is given by this fuzzy-relational equation:

$$X_{k+1} = U_{k-\tau} \blacksquare X_k \blacksquare R$$

$$\tau = 0, 1, 2, \ldots$$

The composition operator in the model equation is fixed as max-min or max-product. $U_{k-\tau}$, X_k, X_{k+1} are fuzzy sets defined in **U** and **X**, respectively. By **U** and **X** we mean a family of the reference fuzzy sets:

$$\mathbf{U} = \{U_1, U_2, \ldots, U_{c\mathbf{U}}\} \quad , \quad \mathbf{X} = \{X_1, X_2, \ldots, X_{c\mathbf{X}}\}$$

All the elements of the data set are transformed into fuzzy sets by means of our reference fuzzy sets. Based on the theory of fuzzy-relational equations, the fuzzy relation of the model is computed. Our performance index is the sum of the squared distances between the fuzzy sets obtained from the model, denoted by X'_{k+1}, and those from the data set X_{k+1}:

$$Q = \sum_{k=1+\tau}^{N-1} \sum_{i=1}^{c_{\mathbf{X}}} (X_{k+1}(i) - X'_{k+1})^2$$

The performance indices for different delay times τ and for the two composition operators are shown in Fig. 10.8. Hence, the max-product

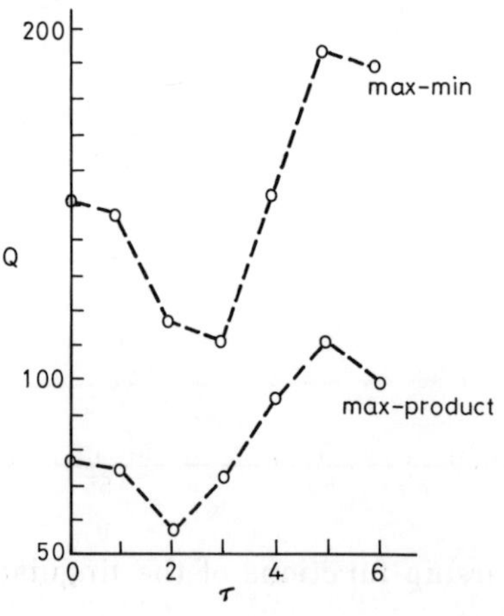

Fig. 10.8 Q versus τ for max-min and max-product compositions

composition operator is preferable with τ set to 2. Since there are five clusters in spaces **X** and **U**, the fuzzy relation consists of 5^3 elements. We can give it a more compact form. Each entry of R can be viewed as an implication statement with its own possibility measure. From this relation, we derive this linguistic description of the system:

if

the state of the system at the kth time instant is X_i *and the control applied is* U_j

then

the state at the (k+1)th time instant is X_l

The rules are given in decreasing order of the possibility measure, or they may be stored in this matrix form:

	X_1	X_2	X_3	X_4	X_5
U_1	--	--	X_5 0.92	X_4 0.76	X_5 0.98
U_2	--	X_3 0.70	X_3 0.82	X_4 0.92	X_5 0.95
U_3	X_4 0.49	X_2 0.89	X_3 0.99	X_4 0.98	X_5 0.99
U_4	X_1 0.79	X_2 0.96	X_3 0.81	X_4 0.69	X_5 0.10
U_5	X_1 0.97	X_2 0.7	X_2 0.27	--	--

The sign -- in the above Table indicates that the corresponding state and control produce no state with a possibility higher than 0.10. The fuzzy model can be used numerically. We replace the generated fuzzy set by a single numerical quantity by taking, for example, a weighted sum:

$$\tilde{x}_{k+1} = \sum_{i=1}^{c_X} X_{k+1}(i) *_i$$

where $*_i$ is given in the clustering procedure as the mean of the ith cluster. Measuring the quality of the fuzzy model by the sum of the squared errors:

$$\sum_{k=\tau_{opt}}^{N-1} (x_{k+1} - *_{k+1})^2$$

(where τ_{opt} is the optimised delay) or an average of this index, say $Q_{av} = Q/(N - \tau_{opt} - 2)$, we get, for 5, 7 and 9 clusters in **X** and **U**:

$c_X = c_U = 5$, $Q_{av} = 0.776$

$c_X = c_U = 7$, $Q_{av} = 0.478$

$c_X = c_U = 9$, $Q_{av} = 0.320$

The model becomes more precise as the number of reference fuzzy sets increases. Bearing in mind our remark above about the number of linguistic labels, we have a contradiction, stressed in Zadeh's principle of incompatibility (Section 10.1). The results of the model for five and nine reference fuzzy sets are shown in Fig. 10.9.

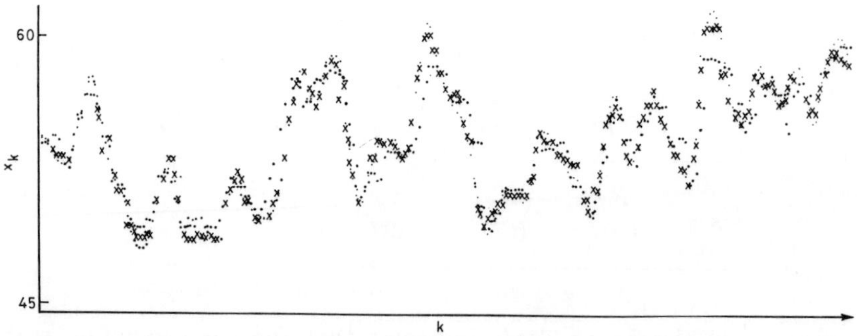

Fig. 10.9 Performance of the fuzzy model with nonfuzzy representation

Example 10.3

Our second example is the identification of a system described by a difference nonlinear equation:

$$x_{k+1} = 0.8\, x_{k-1} + 0.1\, x_{k-2} + 0.3\, x_{k-1}\, u_{k-1} + 0.8\, u_{k-1} + z_k$$

where z_k stands for the Gaussian disturbance with zero mean value and standard deviation $\sigma_z = 0.05$. The data set for identification is given in Fig. 10.10 and consists of 50 pairs of input and output (u_k, x_k). The fuzzy

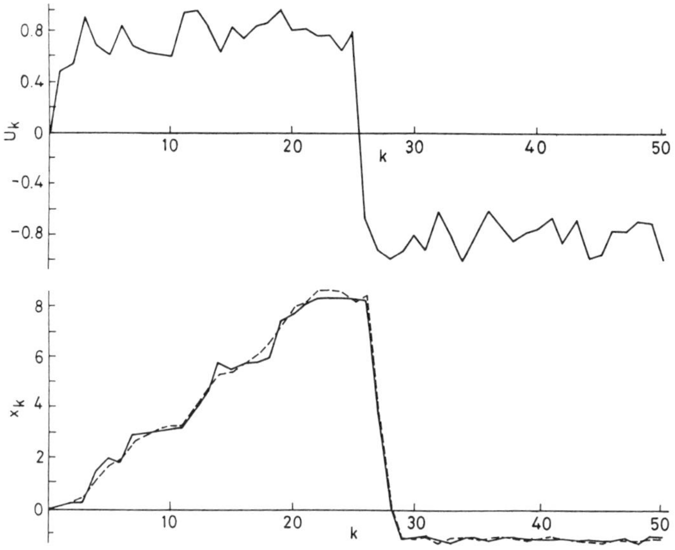

Fig. 10.10 Control and state variables of the identified process

model is:

$$X_{k+1} = U_k \blacksquare X_k \blacksquare R$$

with ■ treated as max-min and max-product compositions. Using the procedure of the first example, we get a model described by a set of implication statements (there are nine clusters):

if the control is U_1 and the state is X_1, then the state is X_1 with possibility 0.59,

if the control is U_1 and the state is X_2, then the state is X_2 with possibility 0.44,

if the control is U_1 and the state is X_3, then the state is X_3 with possibility 0.93, *if the control is U_9 and the state is X_1, then the state is X_2 with possibility 0.78,*

if the control is U_9 and the state is X_2, then the state is X_2 with possibility 0.97,

if the control is U_9 and the state is X_9, then the state is X_9 with possibility 0.76.

The nonfuzzy results of the model give us $Q_{av} = 0.103$ (for max-min composition) and $Q_{av} = 0.036$ (for max-product composition), shown in Fig. 10.10.

10.11 Distributed Modelling

The main thrust of distributed modelling is to develop a fuzzy model that is highly distributed and operates as an ensemble of logically coupled processing units. Each of them operates as an autonomous computing structure and is equipped with its own dynamics. These units can be realised in several ways; in our discussion we will concentrate on the use of logic processors (see Section 8.4).

A high level of autonomy in this class of modelling is achieved by associating each processor with an individual variable of the system. The dynamical behaviour of the system results as a sequence of interactions between its variables (processors) that are carried out in either a cooperative or a competitive manner. These interactions are reflected by the excitatory or inhibitory connections set up for the logic processors. The strength of the interactions is modelled by assigning different numerical values to the corresponding connections of the processors.

The dynamics of each of the processors can be effortlessly modelled by realising different feedback loops between the variables of the processor.

The underlying schematic structure within which the paradigm of distributed modelling is realised is portrayed in Fig. 10.11. Note that in addition to the internal links between the processors, some of them can also be exposed to various inputs from the environment.

The development of the distributed model is realised in a supervised mode. Each scenario in the collection of training events consists of the current status of the state variables (variables of the LPs), say $Q_1(k)$, $Q_2(k)$, ..., $Q_c(k)$ and inputs $I_1(k)$, $I_2(k)$, ..., $I_p(k)$ specified in the kth time instant, as well as the values of the state variables in the successive

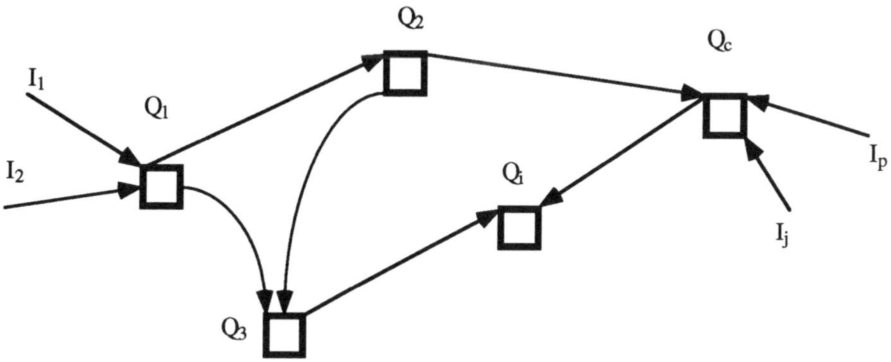

Fig. 10.11 Overall architecture of a distributed model (the nodes represent individual logic processors)

discrete time moment, say $Q_1(k+1)$, $Q_2(k+1)$, ..., $Q_c(k+1)$. We will assume that all I_is as well as Q_js take on numerical values situated in the unit interval. Using an abbreviated vector notation

$$\mathbf{Q}(k) = [Q_1(k) \quad Q_2(k) \quad \dots \quad Q_c(k)] \text{ and } \mathbf{I} = [I_1(k) \quad I_2(k) \quad \dots \quad I_p(k)]$$

to summarise the status of the model, the training set will then consist of triples

$$\{\mathbf{I}(k), \mathbf{Q}(k), \mathbf{Q}(k+1)\}$$

$$k = 1, 2, \dots, N$$

Based on that, learning can be realised separately for each of the logic processors. For this purpose the algorithms described in Section 8.4.2 can be fully exploited.

Within the same vein of supervised learning two essentially distinct learning situations can be distinguished, depending upon the character of the domain knowledge that is available about the system. In the first case, one has a global description of the system. This embraces all qualitative dependences between the variables and the inputs considered in the system. The knowledge provided a priori can be easily incorporated into the model by translating it directly into the form of relevant connections between the logic processors. An example architecture is shown in Fig. 10.12.

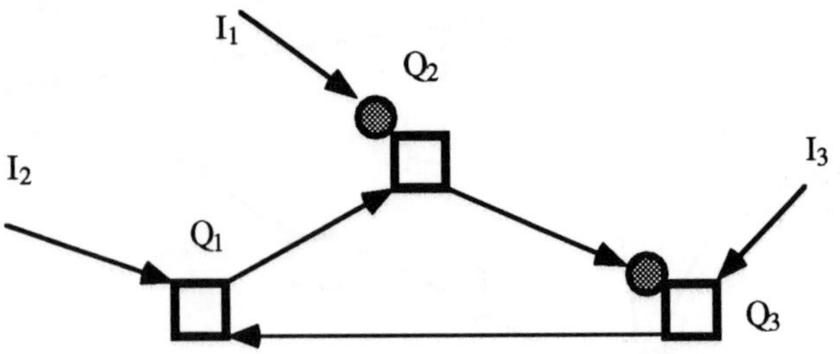

Fig. 10.12 An example of a distributed model with qualitative links formed based on the available initial qualitative domain knowledge (dots are used to summarise inhibitory connections)

This preliminary knowledge definitely eases and improves efficiency of learning, i.e. its speed, since only those connections indicated need to be quantified numerically.

On the other hand, if the available knowledge about the structure is not given in detail or becomes incomplete or unreliable, one should rather concentrate on a fully connected architecture as a primary topology and carry out learning in this context. Some of the weak connections can be eliminated afterwards through building the core or Boolean approximations of the processors. As the learning is performed individually for each Logic Processor, these need not be the same with regard to the details of their internal architecture.

Example 10.4

This example illustrates how the design of the distributed model proceeds. We will start with a collection of the training data summarised in tabular form on page 315.

We will assume no initial knowledge about the structural relationships between the variables so that the learning should be carried out in a fully connected topology. The performance of learning will be expressed as a sum of squared errors:

$$Q_j = \sum_{k=1} \left(t_k - LP_j(\mathbf{Q}, \mathbf{I}, \mathbf{connections})\right)^2 \quad , \quad j = 1, 2$$

I	$Q_1(k)$	$Q_2(k)$	$Q_1(k+1)$	$Q_2(k+1)$
0.70	0.32	0.87	0.45	0.21
0.56	0.78	0.12	0.34	0.18
0.21	0.98	0.05	0.65	0.23
1.00	0.03	0.86	0.45	0.78
0.60	0.45	0.12	0.76	0.23
0.89	0.45	0.32	0.64	0.11
0.43	0.34	0.65	0.84	0.05

The value of Q_j will be minimised separately for each of the processors. For both the processors the number of AND neurons situated in the hidden layer is 6. The results of learning are shown in Fig. 10.13.

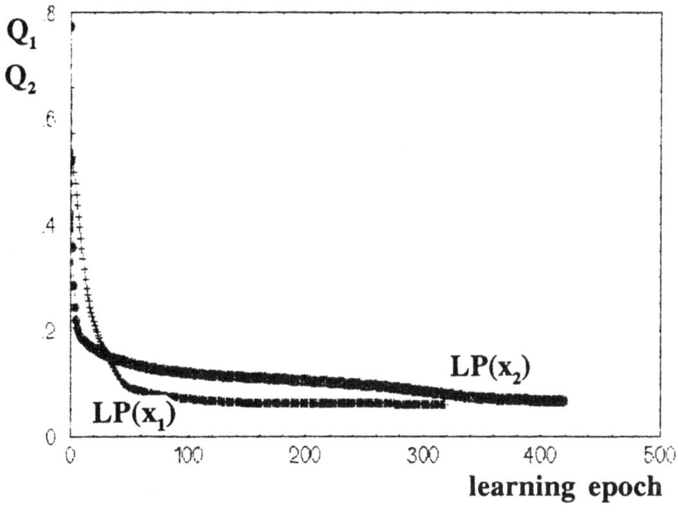

Fig. 10.13 The values of Q_1, Q_2 in the course of learning ($\alpha = 0.1$, t-norm: product, s-norm: probabilistic sum)

The results of learning expressed in terms of the connections (weights) of the processors are:

LP_1

output-hidden layer:	0.075	0.081	1.000	0.089	0.850	1.000
hidden-input layer:	0.328	0.167	0.368	0.426	0.057	0.467
	0.113	0.848	0.012	0.265	0.911	0.327
	1.000	1.000	1.000	0.939	0.010	0.260
	0.255	0.383	0.388	0.210	0.772	0.268
	1.000	0.598	0.969	0.008	0.933	1.000
	1.000	1.000	1.000	0.419	0.007	0.450

LP_2

output-hidden layer:	0.000	0.000	0.0852	0.000	0.000	0.0658
hidden-input layer:	0.318	0.160	0.371	0.424	0.081	0.474
	0.055	0.845	0.044	0.321	0.909	0.378
	0.000	0.994	0.001	0.995	0.000	0.995
	0.183	0.287	0.507	0.209	0.839	0.207
	0.513	0.000	0.945	0.157	0.840	0.856
	0.477	0.502	0.999	0.000	0.809	0.000

They clearly indicate that some of the connections can be easily eliminated. The approximation of the processors by their core versions produces small values of error for the optimal threshold levels; the plots of the surfaces of the approximation error displayed with respect to the threshold levels μ and λ support this finding.

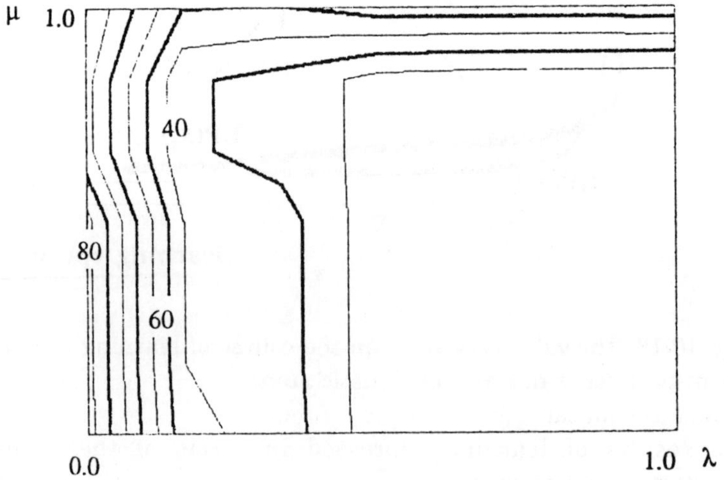

Fig. 10.14 Approximation error produced by the core structure of the logic processors for the inputs uniformly distributed over [0,1]

The optimal Boolean approximation gives rise to the higher values of the approximation error, see Fig. 10.15.

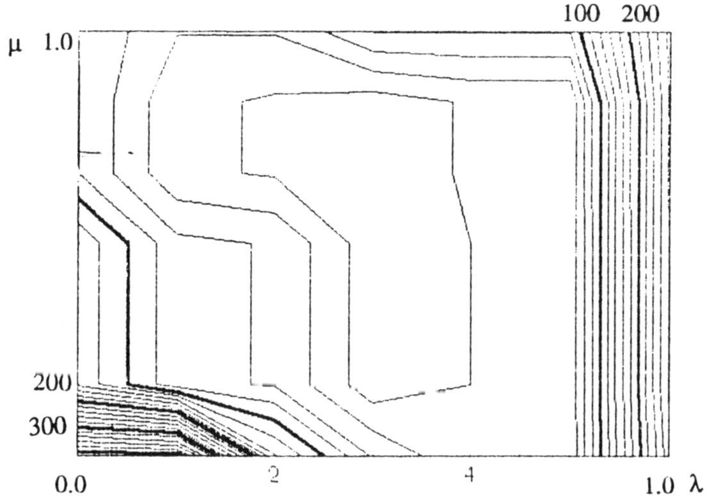

Fig. 10.15 Approximation error produced by the Boolean structure of the logic processors for the inputs uniformly distributed over [0,1]

The resulting formulas read as

$$Q_1(k+1) = \left(\overline{Q}_1 \text{ AND } \overline{Q}_2\right) \text{ OR } \left(\overline{I} \text{ AND } \overline{Q}_1 \text{ AND } \overline{Q}_2\right) = \overline{Q}_1 \text{ AND } \overline{Q}_2$$

$$Q_2(k+1) = I \text{ AND } Q_2 \text{ AND } \overline{Q}_1$$

As stated, the primary aim of distributed modelling is to capture relationships between the system variables formulated at the logical level. This implies that all of the processors when combined together describe interactions, and subsequently the dynamics, of the system as it has been manifested at the level of these labels. Obviously, linguistic labels with different levels of granularity could give rise to quite diversified dynamical patterns of the model.

Depending on applications, different interfaces between the model and the environment are worth considering. In particular, the one with the possibility-necessity transformation is interesting as being capable of expressing imprecision about the input numerical information. Let us recall, Section 2.3, that the possibility measure Poss(X | A) characterises an extent to which the input datum X and the linguistic label A overlap (coincide). The necessity measurement Nec(A | X) expresses the degree to which X is included in A.

For X being a simple numerical quantity $X = \{x_o\}$, $x_o \in \mathbf{R}$, these two measures coincide,

$$\mathrm{Poss}\left(\{x_o\} \mid A\right) = \mathrm{Nec}\left(A \mid \{x_o\}\right)$$

Let us treat x_i as the result of the possibility measure transformation realised for the input datum X (usually x_i constitutes one of the inputs of the logic processor). Furthermore, we will introduce another input defined as the complement of the necessity measure, $\bar{x}_i = 1 - \mathrm{Nec}(A \mid X)$. Bearing in mind this coincidence, for any numerical (pointwise) datum X the obvious relationship is preserved:

$$x_i + \bar{x}_i = 1$$

The equality, however, does not hold for X being less precise, see Fig. 10.16.

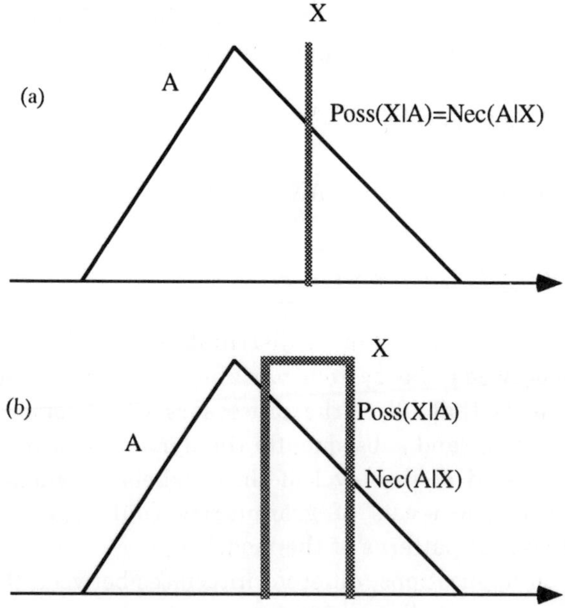

Fig. 10.16 Uncertainty in a numerical datum X conveyed by the possibility and necessity measures; (a) pointwise datum X, (b) interval-valued datum X

For the interval-valued information the above equality is violated, yielding $x_i + \bar{x}_i > 1$. This inequality results from a straightforward observation:

$$\mathrm{Poss}(X \mid A) + \left(1 - \mathrm{Nec}(A \mid X)\right) = 1 + \mathrm{Poss}(X \mid A) - \mathrm{Nec}(A \mid X) \geq 1$$

since $\mathrm{Poss}(X \mid A) \geq \mathrm{Nec}(A \mid X)$. The higher the uncertainty level (broader interval X), the more significant the departure from the original constraint. In the limit case, uncertain information produces $x_i + \bar{x}_i = 2$. This holds, for instance, when a normal fuzzy set A is contained in the interval-valued X, $A \subset X$. The value $\xi \in [0,1]$ resulting from the expression

$$x_i + \bar{x}_i = 1 + \xi$$

can be viewed as an indicator of imprecision (uncertainty) of X with respect to A. It should be stressed that ξ is a relative measure of uncertainty expressed in the context of A.

Bearing in mind that this factor of uncertainty could also occur in the training data set we can describe a way in which it is transformed by the logic processor by augmenting its basic one-output structure by an additional output node characterising $\bar{y}$. Then the uncertainty level produced by the processor can be expressed numerically through the values of $\eta \in [0,1]$, where

$$y_i + \bar{y}_i = 1 + \eta$$

10.12 Conclusions

We have studied in depth the problem of system identification in the presence of fuzziness. After a look at where fuzzy models are useful, we have concentrated on specific areas of fuzzy modelling. Techniques applying not only fuzzy sets but also probabilistic forms of information have been discussed. Identification is viewed as an iterative, three-phase procedure comprising a sequence of stages: (i) determination of the structure of the model, (ii) estimation of its parameters, and (iii) validation of the constructed model. The scheme is of a general nature: however, fuzzy models need novel techniques if they are to work within its framework. This is especially important in parameter estimation, where careful attention must be paid to the structure of the data set (input-output fuzzy sets). The notion of identifiability of the data set enables us to establish a rationale for finding the most readily estimated one. We have examined the evaluation of the fuzzy model by the fuzzy measure and the fuzzy integral and by confidence intervals. This strong tool for evaluation allows the user to decide whether a consecutive iteration loop is required. The methods proposed for model validation can be used to elicit its behaviour in a different application, such as control or prediction. We have studied the basic principles of identification before moving up to the relevant algorithms, thereby filling a gap in the literature.

10.13 References

[1] Box, G.E., & Jenkins, G.M. 1970. 'Time Series Analysis: Forecasting and Control'. Holden Day, San Francisco, USA

[2] Duda, R.O., & Hart, P.E. 1974. 'Pattern Recognition and Scene Analysis'. John Wiley, London, England

[3] Eyckhoff, P. 1974. 'System Identification: Parameter and State Estimation'. John Wiley, London, England

[4] Gottwald, S., & Pedrycz, W. 1986. 'On the suitability of fuzzy models: an evaluation through fuzzy integrals'. *Int. J. Man-Mach. Stud.*, **24**, pp. 141–151

[5] Hirota, K., & Pedrycz, W. 1980. 'On identification of fuzzy systems under the existence of vagueness'. *In* 'Summary of Papers on General Fuzzy Problems', **6**, pp. 37–40

[6] Hirota, K., & Pedrycz, W. 1982. 'Fuzzy system identification via probabilistic sets'. *Inf. Sci. (USA)*, **28**, pp. 21–43

[7] Hirota, K., & Pedrycz, W. 1983. 'Analysis and synthesis of fuzzy systems by the use of probabilistic sets'. *Fuzzy Sets & Syst.*, **10**, pp. 1–13

[8] Pedrycz, W. 1981. 'An approach to the analysis of fuzzy systems'. *Int. J. Contr.*, **3**, pp. 403–421

[9] Pedrycz, W. 1984. 'Construction of fuzzy relational models'. *In* 'Proc. Cybernetics & Systems Res.' (Ed. R. Trappl). North Holland, Amsterdam, pp. 545–549

[10] Pedrycz, W. 1984. 'On identification algorithm in fuzzy relational systems'. *Fuzzy Sets & Syst.*, **13**, pp. 153–167

[11] Pedrycz, W. 1985. 'Design of fuzzy control algorithms with the aid of fuzzy models'. *In* 'Industrial Applications of Fuzzy Control' (Ed. M. Sugeno). North Holland, Amsterdam, pp. 153–173

[12] Pedrycz, W. 1986. 'Structured fuzzy models'. *Cybern. & Syst.*, **16**, pp. 103–117

[13] Tong, R.M. 1977. 'A control engineering review of fuzzy systems'. *Automatica*, **13**, pp. 559–569

[14] Tong, R.M. 1980. 'The evaluation of fuzzy models derived from experimental data'. *Fuzzy Sets & Syst.*, **4**, pp. 1–12

[15] Tong, R.M., Beck, M.B., & Latten, A. 1980. 'Fuzzy control of the activated sludge wastewater treatment process'. *Automatica*, **16**, pp. 695–701

[16] Zadeh, L.A. 1971. 'Toward a theory of fuzzy systems'. *In* 'Aspects on Network and System Theory' (Eds. R.E. Kalman & N. De Claris). Holt, Rinehart, Winston, New York, USA, pp. 209–245

[17] Zadeh, L.A. 1973. 'Outline of a new approach to the analysis of complex systems, and decision processes'. *IEEE Trans. Syst., Man & Cybern.*, **1**, pp. 28–44

CHAPTER 11

System Analysis in Fuzzy-Relational Models

11.1 Introduction

In this Chapter, we cover different issues of system modelling, such as prediction, stability and control. For a clear exposition of the material, we focus primarily on fuzzy-relational models; however, the concepts developed here apply also to some other types (classes) of fuzzy model. The discussion starts with a group of forecasting problems. Here, we identify three basic prediction modes, including 'forward' and 'backward' forecasting as well as an interpolation task viewed as a sort of reconstruction algorithm. In this context, we investigate the notion of the horizon of prediction as an essential component in the evaluation of the results of prediction.

We show that both the imprecision of the input information and the limited accuracy of the fuzzy model imply building some confidence intervals associated with the results. Their length (and so the uncertainty of prediction) increases as the horizon of prediction increases. Regarding stability, we investigate several definitions. Since in fuzzy models we are dealing with set-like objects (a collection of objects), the classical definitions available in stability theory do not apply directly and must be greatly extended. The basic terms of stability are then carefully redefined.

11.2 Prediction Problems

Forecasting (prediction) constitutes an important class of problems in any modelling activity, whatever the nature of the model (deterministic, stochastic, set-theoretic etc.). The aim of this task is to make projections about the state of the system, either in successive discrete time moments or in a series of time moments in the past. Another interesting task is to reconstruct the states occurring between two discrete kth and (k+1)th time moments (interpolation).

Schematically, we can summarise these three classes of prediction problem ($t = 0$ stands for the current time instant) as follows:

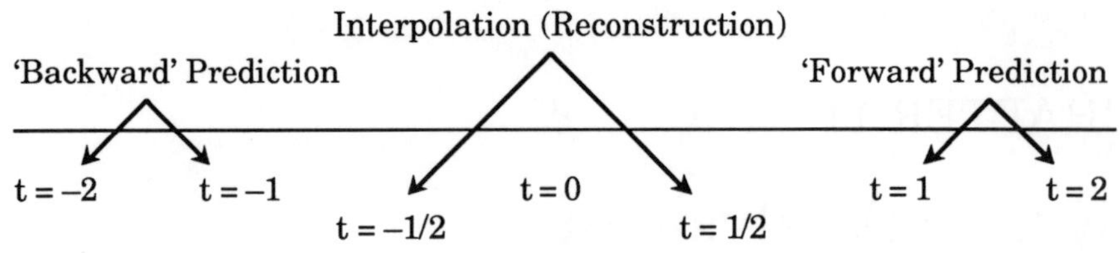

Two factors of fundamental importance in these problems should be highlighted:

• first, the notion of a horizon of prediction is central to the prediction problem. It relates to the interval between the actual time instant and the future (past) time instant at which we want to determine the state of the system, at least approximately. The longer the horizon of prediction, the more approximate the knowledge of the state of the system we can obtain. When a certain horizon of prediction is exceeded, prediction based on the model may be irrelevant. This phenomenon, in a highly qualitative form, reflects the real-world problem that long-term planning is less precise than short-term planning. Generally speaking, it is associated with the several iterations that are necessary to derive the state directly.

• Secondly, a model has finite precision; so a factor of uncertainty is assigned to it that cannot be eliminated, and this also applies to prediction. Uncertainty is carried from one step to another and becomes accumulated, leading subsequently to less-precise estimates of the state. For a graphical illustration of this phenomenon, see Fig. 11.1. Here, a point in the unit hypercube is converted into some regions that 'explode' from step to step.

For a quantitative picture of the simple prediction scenario, we discuss a fuzzy model of the first order:

$$X_{k+1} = X_k \bullet R \tag{11.1}$$

and assume that the universe of discourse in which X_k and X_{k+1} are defined is finite (i.e. the number of linguistic labels used in the construction of the fuzzy model is finite). Thus, X_k denotes a collection of activations of 'n' linguistic labels. We apply the notation:

$$X_k = [X_k(x_1) \;\; X_k(x_2) \;\; \ldots \;\; X_k(x_n)]$$

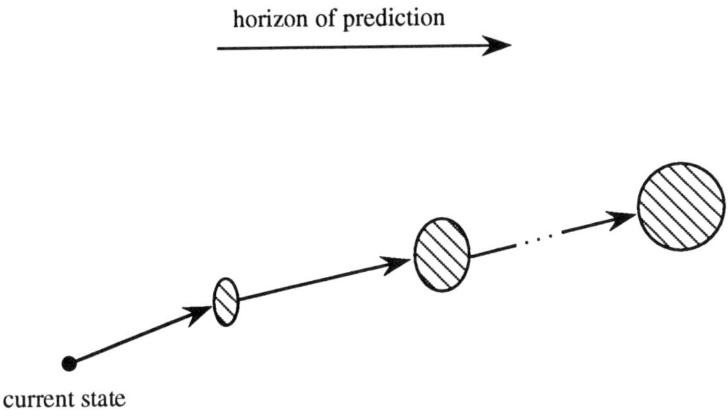

Fig. 11.1 Uncertainty associated with the prediction problem

which explicitly highlights the time instants.

Starting with the actual state X_1, the fuzzy set of state in successive time instants X_2, X_3, ..., X_n, X_{n+1}, ... is derived by iteration:

$$X_2 = X_1 \bullet R$$

$$X_3 = X_2 \bullet R = X_1 \bullet R \bullet R = X_1 \bullet R^2 \tag{11.2}$$

$$\vdots$$

and by induction:

$$X_{n+1} = X_1 \bullet R^n \tag{11.3}$$

The asymptotic behaviour of state X_{n+1} is governed by the nth power of the relation of the model when 'n' tends to infinity. The nth power of R is the fuzzy relation composed n-times: $R^1 = R$, $R^2 = R \bullet R$. As proved in [13], the powers of R either converge to a limit or cycle at a finite period. In situations of prediction, the first case complies with our intuition. The relevant condition leading to its emergence can be formulated accordingly.

Proposition 11.1
If, for any $x_i \in \mathbf{X}$, there exist x_j for which

$$X_k(x_i) \le X_k(x_j) \wedge R(x_j, x_i) \tag{11.4}$$

then

$$X_k \subseteq X_{k+1}$$

The proof is straightforward. Compute the membership function of X_{k+1}:

$$X_{k+1}(x_i) = \bigvee_{x_l \in \mathbf{X}} [X_k(x_l) \wedge R(x_l, x_i)] =$$

$$= \bigvee_{x_l \neq x_j} [X_k(x_l) \wedge R(x_l, x_i)] \vee [X_k(x_j) \wedge R(x_j, x_i)]$$

From assumption (11.4), we get inequality $X_{k+1}(x_i) \geq X_k(x_i)$, which holds for each x_i, and this completes the proof.

Proposition 7.1 leads to:

If

$R \supset I$, where I is the identity relation,

then

$$X_{k+1} \supset X_k \quad , \quad k = 1, 2, \ldots$$

Thus, when the diagonal elements of the relation of the model equal 1, the predicted states become more and more fuzzy:

$$X_1 \subseteq X_2 \subseteq \ldots \subseteq X_k \subseteq X_{k+1} \ldots \tag{11.5}$$

(i.e. more labels become activated to higher degrees).

We express this in terms of the energy of the fuzzy sets. Hence, a sequence of inequalities:

$$E(x_1) \leq E(X_2) \leq \ldots \leq E(X_k) \leq E(X_{k+1}) \leq \ldots \tag{11.6}$$

indicates that the model becomes more 'fuzzy' and conveys less useful information. We see this phenomenon in some communications areas: a message transmitted down a chain of people becomes fuzzier and less reliable, and in the limit is totally meaningless.

We have an upper bound of the energy of the predicted states when the horizon of prediction tends to infinity, $k \to \infty$. Now the transitive closure of R, cl(R), is of great help. The energy of the fuzziness of X_∞ (i.e. the state at an infinite horizon of prediction) is bounded by:

$$E(X_\infty) \le E(X_1 \bullet cl(R))$$

A similar analysis investigates prediction in structured fuzzy models [11]. With the model:

$$X_{k+1} = X_k \underset{p}{\bullet} G$$

we iterate to get a sequence:

$$X_2 = X_1 \underset{p}{\bullet} G$$

$$X_3 = X_2 \underset{p}{\bullet} G = X_2 \underset{p}{\bullet} G^2$$

For the tth time instant:

$$X_t = X_{t-1} \underset{p}{\bullet} G = X_1 \underset{p}{\bullet} G^{t-1}$$

For clarity, we expand this:

$$X_t(x_j) = \bigvee_{x_i \in \mathbf{X}} \left[X_{t-1}(x_i) \wedge B_a(x_i, x_j) \wedge G(x_i, x_j) \right] =$$

$$= \bigvee_{x_i \in \mathbf{X}} \left[X_1(x_i) \wedge B_a^{t-1}(x_i, x_j) \wedge G^{t-1}(x_i, x_j) \right]$$

Now it is obvious that:

$$X_t = X_1 \bullet \left(B^{t-1} \cap G^{t-1} \right)$$

Hence, fuzzy state X_t has two parts: the first one is just a contribution of the probabilistic layer. The second component results from the fuzzy relations used in the model. For an infinite horizon of prediction ($k \to \infty$), we derive this upper bound:

$$X_\infty \subseteq X_1 \bullet cl(B) \cap X_1 \bullet cl(G)$$

We have tackled the above prediction in a special model involving max-min composition. In no way is it general, and it should rather be seen as the result of a peculiarity of the model itself. In a general setting:

- The quality of prediction is a function of the mapping accomplished by the fuzzy model.

This implies that, instead of the X_{k+1} produced by the model, we get an interval-valued fuzzy set derived from the solution of the inverse problem:

$$X_{k+1} = X \geq \gamma$$

where X is sought, while vector γ is a byproduct of the identification procedure (we recall that γ includes all the grades of equality obtained for the individual coordinates of the universe). Iterating, we get X_{k+1} starting with X_1. From the relevance of the fuzzy model, fuzzy set X_2 may lie between the upper and lower bounds, denoted by X_+ and X_-, resulting from inequality $X_k = X \geq \gamma$. In other words, instead of X_{k+1}, we have an interval-valued fuzzy set. Now we can iterate over a finite horizon. Consider any fuzzy model of the first order producing X_{k+1} as a response to the previous state X_k. Let the quality of the model be evaluated with the aid of the equality index. More precisely:

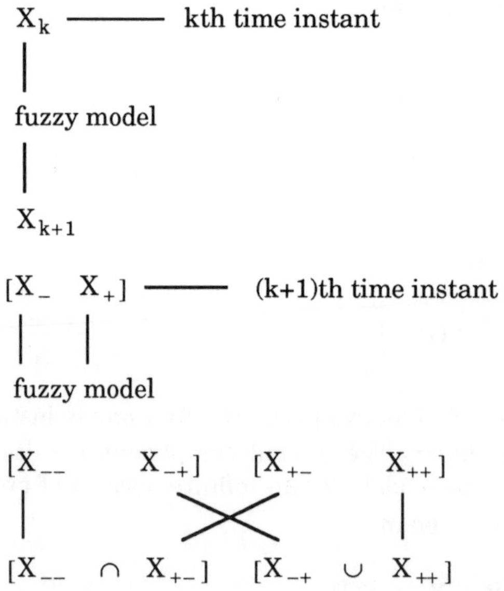

Note that the above aggregation includes the worst-case situation.

From further steps, our prediction is an interval-valued fuzzy set. Not only do we know the state; we observe how the equality intervals at each prediction horizon evolve from step to step. The length of the interval is a nondecreasing function of the horizon. This tells us the precision of the forecast and where it is uninformative; i.e. the attached equality interval covers almost the whole unit interval. Here, the fuzzy integral can greatly help to measure the quality of the prediction. The approach in [5] relies on a pointwise calculation of the equality index for the interval-valued fuzzy set, taking into account upper and lower bounds. We denote this fuzzy set by h:

$$h = [X_-(x_1) = X_+(x_1) \quad X_-(x_2) = X_+(x_2) \quad \ldots \quad X_-(x_n) = X_+(x_n)]$$

Thus h is aggregated by the fuzzy measure, giving a fuzzy integral:

$$E(h) = \int_{\mathbf{X}} h \bullet g_\gamma(.)$$

This integral is the grade of satisfaction of the predictability while the horizon of prediction increases: h decreases and so E(h) decreases. This reflects a common situation: long-term prediction is less reliable than short-term prediction. It allows us to fix a 'reasonable' horizon of prediction: i.e. the greatest number of iteration steps at which E(h) is larger than the imposed threshold level.

The above 'forward' prediction is usually discussed as a standard forecasting task ($t > 0$). We can look at the problem of extrapolation for $t < 0$, embracing situations where X_k is given. This sort of extrapolation ('backward' prediction) calls for expressing X_{k-1} for X_k specified and, in fact, converts into an inverse-type problem. As such, this mode may vary from model to model. Hence, the general methodological point of view is of a fairly limited use. Rather than follow this path, we concentrate on a specific form of the fuzzy model. For the relational model:

$$X_{k+1} = X_k \blacksquare R \tag{11.7}$$

with the general composition operator (s-t or max-t), we have to determine a minimal and maximal element (if they exist) in this collection of fuzzy sets satisfying the equation of the model:

$$X = X_{k-1} \blacksquare R$$

for any X fulfilling the conditions of the inverse problem:

$$X = X_{k+1} \geq \gamma$$

The above problem is significantly reduced with max-t composition. We recall (Chapter 5) that the theory applied to this equation:

$$X_{k+1} = X \blacksquare R$$

which has to be solved with respect to X, provides a maximal solution constructed as $R \varphi X_k$. Owing to some discrepancies in the fuzzy model, γ is not identically equal to 1. This produces an interval $[X', X'']$ of possible grades of membership instead of a single element X_{k+1}. Its bounds are used in solving the inverse problem. This in turn implies that the 'backward' prediction is:

$$[R \varphi X', R \varphi X'']$$

For instance, if $X' = \mathbf{0}$ and $X'' = \mathbf{1}$, the prediction implies a fuzzy set with the [0,1] range of membership values.

The propagation of uncertainty down the discrete time moments (k–1, k–2, ... etc.) is completed in the pessimistic manner described above for the 'forward' prediction.

The interpolation (reconstruction) problem tackles a situation in which we determine the state of the model for some intermediate time moments. The original model iterates $X_2, X_3, \ldots, X_k, X_{k+1}$. If we want to determine the state at an intermediate time instant k+1/2 (situated symmetrically between k and k+1), a reconstruction problem arises. In other words, X_k and X_{k+1} are given, and the $X_{k+1/2}$ we are seeking must satisfy these conditions:

$$\begin{cases} X_{k+1/2} = X_k \blacksquare G \\ X_{k+1} = X_{k+1/2} \blacksquare G \end{cases}$$

and has to be determined. Note that the fuzzy relation standing in the above relationships is different from R. The problem includes two unknown elements, namely $X_{k+1/2}$ and the fuzzy relation G.

By straightforward substitution, we derive:

$$X_{k+1} = (X_k \blacksquare G) \blacksquare G = X_k \blacksquare G \blacksquare G$$

which has to be solved with respect to G. Once G has been determined, computation of the interpolated state $X_{k+1/2}$ is straightforward. A similar procedure can be used to determine the state of the model at successive time subintervals, such as k+1/4, k+1/8 etc. The reconstruction problem is then solved with respect to G (for k+1/4):

$$V \blacksquare V = G$$

and then for V, P ■ P = V (k+1/8), and so forth.

In more detail, we formulate the reconstruction as a particular optimisation task:

- given an (n × n) fuzzy relation $\mathbf{r} = [r_{ij}]$, determine $\mathbf{g} = [g_{ij}]$, i, j = 1, 2, ..., n, so that the s-t composition implies that this collection of conditions is satisficd:

$$r_{ij} = \underset{k=1}{\overset{n}{S}} \left[g_{ik} \; t \; g_{kj} \right]$$

Usually, we cannot achieve equality. The feasible way of handling this problem is to minimise a particular performance index reflecting the 'similarity' of the entries of **r** and the corresponding results of the s-t composition. The standard method will be to introduce this performance index:

$$Q = \sum_{i}^{n} \sum_{j}^{n} \left[r_{ij} - \underset{k=1}{\overset{n}{S}} \left(g_{ik} \; t \; g_{kj} \right) \right]^2 = \sum_{i}^{n} \sum_{j}^{n} Q_{ij}$$

where Q_{ij} is used to denote the ijth component in the above sum. The modifications of the elements of **g** are standard:

$$g_{i_1 j_1} = g_{i_1 j_1} - \alpha \frac{\partial Q}{\partial g_{i_1 j_1}}$$

To calculate this derivation, note that the sum can be partitioned into several groups to facilitate further steps in the computation:

$$Q = \sum_{i}^{n} \sum_{j}^{n} Q_{ij} = \sum_{i \neq i_1}^{n} Q_{i j_1} + \sum_{j \neq j_1}^{n} Q_{i_1 j} + Q_{i_1 j_1} + \sum_{i \neq i_1}^{n} \sum_{j \neq j_1}^{n} Q_{ij}$$

The derivation is nonzero for all the terms but the last one. The detailed calculations require further detailed specifications of the triangular norms standing in the above expression.

11.3 Stability of Fuzzy Models

Now our concern is the stability of the above fuzzy models. We explain this by a first-order discrete time model:

$$X_{k+1} = U_k \bullet X_k \bullet R \tag{11.8}$$

By free motion of the model (trajectory induced by fixed control), we mean a sequence of fuzzy states generated by (11.8) when fuzzy control U_k is fixed over time, say $U_k = U$. We plug this fuzzy set U into the model:

$$X_{k+1} = X_k \bullet (U_k \bullet R)$$

and denote by G the fuzzy relation computed as:

$$X_{k+1} = X_k \bullet G \tag{11.9}$$

The behaviour of the sequence X_1, X_2, ..., X_k in (11.9) strongly relies on a G equivalent to a power series of Gs, say G, G^2, G^3, ... From [13], the sequence converges to a limit G_∞ or cycles at a fixed period. Bearing this in mind, we define [8, 9,14]:

A free-motion fuzzy model is stable in the limit if a relation G_∞ exists such that:

$$\lim_{k \to \infty} G = G_\infty$$

otherwise we call the model oscillatory with period τ, i.e.:

$$\forall_{k=1,2,\ldots} G^{k+\tau} = G^k$$

Obviously, the stability refers to a fuzzy set of control; if it changes, the corresponding free-motion fuzzy model changes, e.g. from stable to oscillatory.

These notions concern infinite stability. In practice, this means a long-enough stability from the initial instant. For finite universes of discourse, the transient response of the model directly depends on the fuzzy relation G. After several iterations, its value does not change if there is no oscillation.

Another definition of stability is proposed in [7] that is based on the energy of fuzziness of the fuzzy set. For a fuzzy set of state X_k, a functional $\varepsilon(X_k)$ is assigned so that:

$$\varepsilon(X_k) = \frac{\sum_{\mathbf{X}} f_1(x_i) f_2(X_k(x_i))}{\mathrm{card}(\mathbf{X})}$$

where $f_2 : [0,1] \to [0,1]$ is a nondecreasing function (energy-type function) and $f_1 : \mathbf{X} \to \mathbf{R}$ represents the contribution of every element of $\mathbf{X}$ to its global value. This function is of special interest when $\mathbf{X} \subseteq \mathbf{R}$.

Concerning the changes of $\varepsilon(.)$ with time, the free-motion fuzzy model is:

- stable if:

$$\varepsilon(X_k) - \varepsilon(X_{k-1}) \le 0 \qquad \text{for } k \to \infty$$

- unstable if:

$$\varepsilon(X_k) - \varepsilon(X_{k-1}) \ge 0 \qquad \text{for } k \to \infty$$

- oscillatory with an oscillation of τ if:

$$|\varepsilon(X_k) - \varepsilon(X_{k-1})| = |\varepsilon(X_{k+\tau}) - \varepsilon(X_{k-1+\tau})| \qquad \text{for } k \to \infty$$

Our definitions are global because they cover a limited (in time) behaviour of the model.

We must closely study its transient behaviour, i.e. its performance at time instants k = 1, 2, 3, ... Two situations are observed:

- the states generated by the fuzzy model are different, say $X_1 = X_2 = X_3 = \ldots$

- all the states from the fuzzy model for an initial X_1 are equal, $X_1 = X_2 = \ldots$. This occurs when X_1 forms an eigen fuzzy set of the fuzzy relation G.

The first situation is common. In the evolution of the states, to visualise how stable the initial state is, or to express how far a sequence of fuzzy states moves from X_1, we introduce a definition of α-stability, $\alpha \in [0,1]$. It borrows some elements from previous definitions of stability [9,14]. For two time instants, k and (k+1), we denote by X_k^α and X_{k+1}^α the α-cuts of fuzzy sets X_k and X_{k+1}. We take the intersection of their supports, $\mathrm{supp}(X_k^\alpha) \cap \mathrm{supp}(X_{k+1}^\alpha)$. We call X_k α-stable if the intersection of the supports of the respective α-cuts is nonempty. The value of α in this definition is the highest ensuring that:

$$\mathrm{supp}\left(X_k^\alpha\right) \cap \mathrm{supp}\left(X_{k+1}^\alpha\right) \neq \varnothing \tag{11.10}$$

The α-stability refers to two consecutive time instants, and the threshold level α may vary for different time instants in a given time horizon. We take a sequence $X_1, X_2, \ldots, X_L$ and express α in a definition of α-stability for successive time instants, $\alpha_1, \alpha_2, \ldots, \alpha_{L-1}$. The minimal value:

$$\alpha = \min(\alpha_1, \alpha_2, \ldots, \alpha_{L-1})$$

determines the stability of the model in an L-step horizon of prediction.

11.4 Control Problems

In control-system studies, we look first at the controllability for which the fuzzy model is built, together with the associated fuzzy set of equality h, which expresses the relevance of the fuzzy model.

In controllability, our goal is expressed by the fuzzy set G defined in the output space **Y**. Having the fuzzy model, we specify a fuzzy control (input) X. From the model equation:

$$G = X \blacksquare R \tag{11.11}$$

This equation is solved with respect to X when G and R are given. If the set of solutions:

$$\aleph = \left\{X : \mathbf{X} \to [0,1] \mid X \blacksquare R = Y\right\}$$

is nonempty, then $X = R \, \varphi \, G$ is the greatest element of $\aleph$. Otherwise, if this equation has no solution with respect to X, we look for an approximate solution or simply take the one resulting from the above φ-composition as a preliminary approximation. The quality of the control is verified by:

$$\overline{G} = \hat{X} \blacksquare R$$

To assess how well the goal G is reached by using our fuzzy control, we compute:

$$h = \left(G = \overline{G}\right)$$

in a pointwise manner; so the ith coordinate of h is:

$$h_i = \left(G(x_i) = \overline{G}(x_i)\right)$$

The higher the elements of the fuzzy set h, the better the controllability of the model with respect to the goal G. We must remember that h indicates how easily the model may be controlled. But we are interested in the system, not the model. h expresses the controllability of the model only. The controllability of the system depends on both the controllability of the model and its overall quality (characterised with the aid of the fuzzy measure $g_\lambda(\,.\,)$). This relationship is expressed by a fuzzy integral of h with respect to $g_\lambda(\,.\,)$:

$$I(h) = \int_X h \bullet g_\lambda(.)$$

For a deeper insight into the evaluation, we consider several situations:

(i) Let $h = \mathbf{1}$; i.e. the model can reach the goal G. Whether the satisfaction of this property is high also for the system depends on the quality of the model ($g_\lambda(\,.\,)$): the better the model, the higher the values of I(h) reported. Without this evaluation of the model, we may be too optimistic about controlling the system itself. This may arise from a misleading interpretation of the result, which should be interpreted as pertaining to the controllability of the model (not the system).

(ii) Let $h \cong \mathbf{1}$. This indicates that the model can be controlled only approximately, since G is reached inexactly. Again, I(h) allows us to judge whether the system is controllable to state G (the fuzzy set G is reachable).

Until now, we have proposed the simplest control problems, in which our aim was to reach the goal G without restricting ourselves to a fuzzy set of constraint imposed on the fuzzy control; so any fuzzy control satisfying this condition is feasible. Our formulation of the control problem does not take into account the imprecision of the fuzzy model. Instead of the fuzzy goal G, it is more convenient to refer to a set of available goals, implying for a fixed γ a lower and an upper bound of the goal G (solving $G = G' \geq \gamma$ for given G and γ). Then, instead of solving:

$$X \blacksquare R = G$$

for known R and G, we can discuss a set of two inequalities:

$$G_{-} \subseteq X \blacksquare R \subset G_{+}$$

In fact, the limited precision of the model makes any other type of precise formulation of the control targets totally unrealistic. Thus our control task becomes:

- find the fuzzy set of control that preserves the above inequality. For a model of specified accuracy, it is better to achieve a set of goals, i.e. a family of fuzzy sets between G_{-} and G_{+}, rather than a single fuzzy set G. If the fuzzy model is perfect, this is valuable; but if not, we cannot satisfy G by any control applied. For an imperfect model, the goal is defined with similar precision (modelled by the length of the grades of membership in the interval-valued fuzzy set). The better the model, the more precisely may the goal be defined.

In (11.11), fuzzy sets G_{+} and G_{-} depend on γ. With decreasing γ, in the limit ($\gamma = 0$), the bounds formed by G_{-} and G_{+} are broad enough to ensure that (11.11) has a nonempty set of solutions. But too small a value of γ leads to meaningless results.

Our next step is to consider a fuzzy set of the goal G and a fuzzy set of the constraint C, $C : \mathbf{X} \rightarrow [0,1]$ [1] [2] [3] [4] [6] [12]. Our problem is to determine the fuzzy control X fulfilling both requirements, which may be contradictory, as in this formulation:

'Achieve high accuracy and speed of response
in the system with minimal cost of control'

We may expect the problem to have only an approximate solution. So at the outset it is worth estimating how contradictory are the constraints. When X = C, the constraint is completely satisfied. Numerically, $(X = C) = \mathbf{1}$, and using the model equation we calculate its output, $Y = C \blacksquare R$. Then

(Y = G) can tell us how closely the goal G is attained by the model. Conversely, put Y = G and solve G = X ■ R with respect to X, assuming that the set of solutions $\aleph = \{X | X \blacksquare R = C\}$ is nonempty. Then compare the solution of the above set with C, (X = C), with X belonging to $\aleph$. So now we get a case where (Y = G) = **1** and usually (X = C) = **1**. This second case, with higher equality, is preferred. Nevertheless, in a general formulation, we must look for an X that maximises the performance index:

$$\max\{(X = C)\,\text{AND}\,(Y = G)\}$$

subject to

$$Y = X \blacksquare R$$

i.e.:

$$\max_{X:\mathbf{X}\to[0,1]}\{(X = C)\,\text{AND}\,(X \blacksquare R = G)\}$$

Another construction assigns a grade of acceptability to each pair of control and state. It expresses how far the control and the state considered together fit the concepts of the constraint and the goal, respectively. It has a strong link with the construction of the fuzzy controller.

The simplest case of a dynamic fuzzy model is the model of the first order:

$$X_{k+1} = U_k \blacksquare X_k \blacksquare R \tag{11.12}$$

in which card(**X**) = n, card(**U**) = m. We assume a set of linguistic values of state $X_1, X_2, \ldots, X_{N_1}$ and control $U_1, U_2, \ldots, U_{N_2}$ defined in appropriate spaces, so that for each pair we can assign a quantity ξ_{ij} lying in the [0,1] interval and expressing how far the situation described by X_i and U_j is acceptable from the point of view of the goal and the constraint. U_i is the control at time instant (k), while X_j is the fuzzy state at the (k+1)th instant. Neither the goal nor the constraint is fixed for the entire evaluation of the model but may vary from one pair (X_i, U_j) to another; moreover, they are not given explicitly. They are regarded as being known to the designer using them to evaluate ξ_{ij}. The values are:

	X_1	X_2	X_j	X_{N_1}
U_1	ξ_{11}	ξ_{12}	ξ_{1j}	ξ_{1N_1}
U_2	ξ_{21}	ξ_{22}	ξ_{2j}	ξ_{2N_1}
	$\vdots$	$\vdots$	$\vdots$	$\vdots$
U_i	ξ_{i1}	ξ_{i2}	ξ_{ij}	ξ_{iN_1}
	$\vdots$	$\vdots$	$\vdots$	$\vdots$
U_{N_2}	$\xi_{N_2 1}$	$\xi_{N_2 2}$	$\xi_{N_2 j}$	$\xi_{N_2 N_1}$

Usually, matrix $[\xi_{ij}]$ is not symmetrical. From the model given by (11.12) and the Table above, we can formulate the control strategy in the format of the control rule 'if state, then control'.

The algorithm described below leads us to search the fuzzy control for the specified fuzzy sets of state X_i. For a given X_i at a time instant, and imposing fuzzy control U_{j_0}, we calculate X_{k+1}, which predicts the effect of the control applied. Hence:

$$X_{k+1} = U_{j_0} \blacksquare X_i \blacksquare R$$

Next, X_{k+1} is matched against the previously listed linguistic categories. The result of the matching (we can use any of the matching procedures discussed in Chapter 2) is λ_k (from matching X_{k+1} against X_l). Performing calculations for all l, l = 1, 2, ..., N_1, we get a sequence of numbers $\lambda_1, \lambda_2, \ldots, \lambda_{N_1}$. The overall suitability of fuzzy control U_{j_0} to control the system is a weighted sum:

$$\bar{\xi}_{j_0} = \frac{\sum_{l=1}^{N_1} \lambda_l \, \xi_{j_0 l}}{\sum_{l=1}^{N_1} \lambda_l}$$

Finally, the fuzzy control is chosen by taking the index j_0 for which $\bar{\xi}_{j_0}$ is maximum. We denote it by $U_{j(i)}$:

$$\max \bar{\xi}_{j_0} = \bar{\xi}_{j(i)}$$

Thus, by changing X_i from i = 1, 2, ..., N_1 for each of them, the corresponding fuzzy control $U_{j(i)}$ is determined. Moreover, we have generated a list of control rules of the format:

if X_i, then $U_{j(1)}$ with $\bar{\xi}_{j(1)}$

if X_2, then $U_{j(2)}$ with $\bar{\xi}_{j(2)}$

$\vdots$

if X_{N_1}, then $U_{j(N_1)}$ with $\bar{\xi}_{j(N_1)}$

with values of suitability attached to each of them (equal to $\bar{\xi}_{j(i)}$). Thus, we have obtained the control protocol common in the construction of the fuzzy controller. Unlike the knowledge coming from the operator of the process, the control rules presented here are consistent, since they are generated on the basis of the fuzzy model and the state-control evaluation table. The suitability values allow us to identify the most significant and most reliable subset of the control rules. Consequently, we have a clear picture of how closely the requirements imposed on the control strategy can be fulfilled in a specific fuzzy model.

Further steps leading to the synthesis of the relation of the fuzzy controller are the same as analysed in this Chapter, with just the modification that $\bar{\xi}_{j(1)}$ contribute to the aggregation of the particular relation resulting from the individual rules. If a sup-t composition is used, the relation of the controller R_{cont} is:

$$R_{cont} = \bigcap_{i=1}^{N} \left(X_i \,\varphi\, U_{j(i)}\right)^{\bar{\xi}_{j(i)}}$$

the φ operator being appropriately adjusted; the power in the formula takes the respective membership function up to the power $\bar{\xi}_{j(i)}$.

Now we give some technical details, and indicate the characteristic features of the building and use of fuzzy models.

We discuss an example using the Box-Jenkins data set analysed for identification purposes in Chapter 10. The data set is of a combustion process in a furnace. Its format is (i) the gas-flow rate into the furnace; (ii) the percentage of CO_2 in the exhaust gases. From the point of view of the model constructed, the gas-flow rate is treated as the control variable and the concentration of CO_2 is viewed as the state variable. Numerical studies and optimisation procedures produced this fuzzy-relational model:

$$X_{k+1} = U_{k-2} \blacksquare X_k \blacksquare R$$

where the model works with reference fuzzy sets specified in the control space and the state space (the t-norm is specified as the product). The entire relation R describing the model is:

			X			
	0.03	0.01	0.01	0.01	0.01	
	0.02	0.04	0.01	0.05	0.02	
	0.01	0.03	**0.39**	**0.92**	0.05	
	0.00	0.02	0.05	**0.76**	**0.74**	
	0.00	0.02	0.04	**0.11**	**0.98**	U_1
	0.04	0.03	0.03	0.01	0.01	
	0.04	**0.64**	**0.73**	0.04	0.02	
	0.01	0.05	**0.83**	**0.72**	0.04	
	0.01	0.02	**0.41**	**0.92**	**0.35**	
	0.01	0.02	0.04	**0.50**	**0.95**	U_2
	0.48	**0.49**	0.05	0.02	0.01	
	0.13	**0.89**	**0.39**	0.03	0.01	
X	0.02	**0.29**	**1.00**	**0.13**	0.01	
	0.01	0.04	**0.49**	**0.98**	**0.18**	
	0.01	0.01	0.04	**0.69**	**0.99**	U_3
	0.79	**0.32**	0.04	0.02	0.01	
	0.23	**0.96**	**0.21**	0.02	0.00	
	0.04	**0.54**	**0.81**	**0.60**	0.04	
	0.02	0.05	**0.64**	**0.69**	0.05	
	0.00	0.01	0.05	0.03	**0.13**	U_4
	0.97	0.09	0.04	0.02	0.01	
	0.51	**0.75**	0.06	0.02	0.01	
	0.05	**0.27**	0.03	0.01	0.00	
	0.03	0.02	0.03	0.02	0.02	
	0.02	0.01	0.00	0.01	0.02	U_5

The most significant elements of the fuzzy relation of the model have been identified. Each entry in this matrix can be translated into a statement of this format:

if

the state at the kth instant is *X* and the control at the (k–2)th instant is *U*

then

the state at the (k+1)th instant is *X′*

Now we design a fuzzy-control algorithm by the method previously discussed. We are now dealing with a model with control delay, which requires a modification in the basic algorithm [10]. Performing iteration:

$$X_{k+1} = U_{k-2} \blacksquare X_k \blacksquare R$$

$$X_{k+2} = U_{k-1} \blacksquare X_{k+1} \blacksquare R$$

$$X_{k+3} = U_k \blacksquare X_{k+2} \blacksquare R$$

Inserting the first two expressions into the third one:

$$X_{k+3} = U_k \blacksquare U_{k-1} \blacksquare U_{k-2} \blacksquare X_k \blacksquare R \blacksquare R \blacksquare R$$

Then, if the goal is given by the fuzzy set G, we force the above equality, writing $X_{k+3} = G$. The fuzzy control of our interest depends on G and the control actions U_{k-2} and U_{k-1}. The greatest possible fuzzy set of control, $\hat{U}_k$, is:

$$\hat{U}_k = (U_{k-1} \blacksquare U_{k-2} \blacksquare X_k \blacksquare R \blacksquare R \blacksquare R)\,\varphi\, G$$

Now we derive the control rules for the fuzzy goal specified by the membership function:

$$G = [0.059 \quad 0.11 \quad 0.49 \quad 0.97 \quad 0.11]$$

defined in the space of the reference fuzzy sets of state (Fig. 11.2). These rules depend, of course, on past control actions, i.e. at the (k–2) and (k–1) instants. After computation, we get this set of control rules:

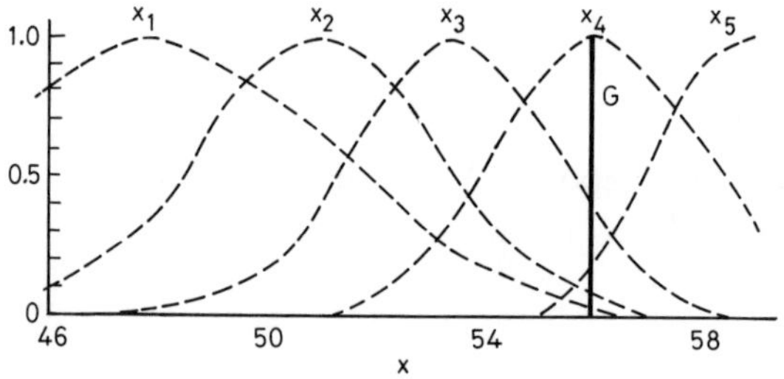

Fig. 11.2 The reference fuzzy sets X_k and the fuzzy set of the goal G

U_1 & U_1 &	X_1 then	[0.32	0.33	0.32	1.00	1.00],
	X_2 then	[0.20	0.21	0.20	1.00	1.00],
	X_3 then	[0.16	0.17	0.16	1.00	1.00],
	X_4 then	[0.16	0.16	0.13	1.00	1.00],
	X_5 then	[0.12	0.12	0.17	0.88	1.00].
U_2 & U_2 &	X_1 then	[0.35	0.51	0.37	0.34	0.34],
	X_2 then	[0.28	0.42	0.30	0.28	0.28],
	X_3 then	[0.22	0.46	0.38	0.29	0.35],
	X_4 then	[0.18	0.18	0.17	0.39	0.64],
	X_5 then	[0.12	0.13	0.12	0.60	0.97].
U_3 & U_3 &	X_1 then	[0.60	0.27	0.19	0.17	0.18],
	X_2 then	[0.35	0.22	0.16	0.15	0.15],
	X_3 then	[0.20	0.27	0.19	0.18	0.18],
	X_4 then	[0.16	0.17	0.16	0.27	0.33],
	X_5 then	[0.11	0.12	0.11	0.41	0.50].
U_4 & U_4 &	X_1 then	[0.60	0.22	0.16	0.20	0.10],
	X_2 then	[0.38	0.19	0.13	0.12	0.12],
	X_3 then	[0.31	0.23	0.17	0.15	0.16],
	X_4 then	[0.31	0.42	0.31	0.28	0.29],
	X_5 then	[0.39	0.62	0.45	0.42	0.42].
U_5 & U_5 &	X_1 then	[1.00	0.37	0.13	0.08	0.06],
	X_2 then	[1.00	0.31	0.16	0.09	0.07],
	X_3 then	[1.00	0.38	0.27	0.19	0.15],
	X_4 then	[1.00	0.69	0.50	0.34	0.28],
	X_5 then	[1.00	1.00	0.76	0.52	0.42].

The control rules derived are put into the equation of the controller. This gives us an explicit formula for calculating fuzzy control U for a given fuzzy state and defined past controls:

$$U = X \blacksquare U_{k-2} \blacksquare U_{k-1} \blacksquare G \qquad (11.13)$$

The fuzzy control depends on the state of the model. Eq. (11.13) is similar in shape to the formula for feedback control in the optimal control of deterministic or stochastic tasks. Determination of the relations of the controller takes account of the fuzzy sets specified above. The computations are collected concisely:

	0.25	0.26	0.25	1.00	1.00
	0.20	0.21	0.20	1.00	1.00
	0.17	0.17	0.17	1.00	1.00
	0.15	0.15	0.13	1.00	1.00
	0.12	0.12	0.12	0.88	1.00 (U_1, U_1)
	0.34	0.51	0.37	0.34	0.34
	0.28	0.41	0.30	0.28	0.28
	0.22	0.24	0.23	0.30	0.35
	0.16	0.16	0.15	0.39	0.46
	0.12	0.13	0.12	0.55	0.90 (U_2, U_2)
	0.43	0.26	0.19	0.17	0.18
	0.25	0.22	0.16	0.14	0.15
U	0.20	0.22	0.19	0.18	0.18
	0.14	0.15	0.14	0.16	0.23
	0.11	0.12	0.11	0.22	0.38 (U_3, U_3)
	0.46	0.22	0.16	0.20	0.10
	0.38	0.19	0.13	0.12	0.12
	0.31	0.23	0.17	0.15	0.15
	0.31	0.30	0.22	0.20	0.20
	0.39	0.59	0.43	0.39	0.40 (U_4, U_4)
	1.00	0.37	0.13	0.08	0.06
	1.00	0.31	0.16	0.09	0.07
	1.00	0.38	0.20	0.11	0.09
	1.00	0.50	0.36	0.20	0.16
	1.00	0.97	0.70	0.48	0.39 (U_5, U_5)

The calculated control is interpreted as follows. Let U_k be the fuzzy set of control from (11.13), suggesting which control among the linguistic labels U_1, U_2, ..., U_5 should be applied. Nevertheless, we do not specify a nonfuzzy value to be treated as the control action. There is no unique way of establishing such a value. When we face the same problem in probabilistic decision-making, we can take the median, mean or modal probability function expressed in the relevant space. We now adopt the mean method (centre of gravity). Fuzzy set $U = [a_1 \;\; a_2 \;\; \dots \;\; a_5]$ is replaced by its nonfuzzy representation, i.e. a number $\bar{a}$ obtained as:

$$\bar{a} = \frac{\sum_{i=1}^{5} a_i \bar{x}_i}{\sum_{i=1}^{5} a_i}$$

where $\bar{x}_i$ is the centre of the reference fuzzy set U_i obtained by the clustering method; for details, the reader is referred to the identification method in the previous Chapter. The numerical characteristics of the controller are shown in Fig. 11.3. We stress that the characteristics of the fuzzy controller are nonlinear. For middle values of the state variable, the control values do not depend heavily on x; the characteristics are

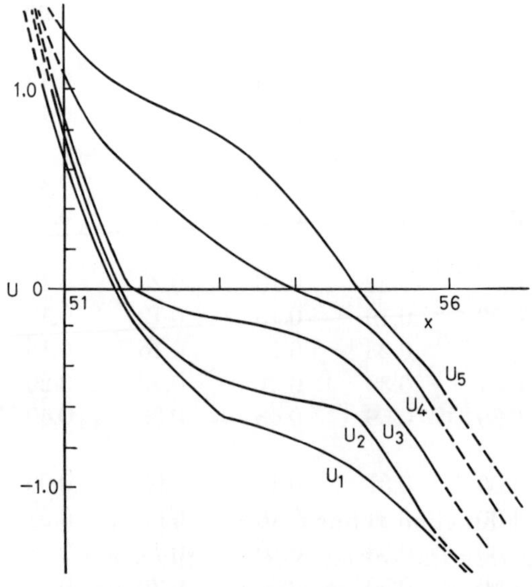

Fig. 11.3 Characteristics u = u(x) derived with the aid of the fuzzy model

rather flat. For smaller and larger values of the state variable, slight changes in x generate significant changes in the control variable. This type of nonlinearity is in accordance with the general control strategy of the fuzzy controller; changes in the control variable should be small near the goal and large when the state is far from the goal.

11.5 Fuzzy-Model-Based Control Algorithms

The detailed control algorithms can be developed on the basis of the fuzzy model. We introduce three of them. The first two approaches look at the appropriate fuzzy-relational structures for use as a control algorithm. The third one determines a fuzzy set of control in an iterative way, following the model of the system.

The model takes on the general form of the first-order fuzzy-relational equation:

$$X_{k+1} = FR(U_k, X_k) \tag{11.14}$$

where FR stands for an abbreviation of the equation of the model (the type of the equation should be known in advance; however, this is not necessary at this level of generality of the foregoing discussion).

The first control structure is characterised as a single-level feedback structure utilising a current fuzzy set of state:

$$U_k = X_k \blacksquare T$$

where T denotes the fuzzy relation of the controller. Furthermore, we assume that a certain fuzzy set of the goal G defined in the state space and the constraint C defined in the control space are also provided.

The ultimate target of the control policy is to keep the goal satisfied and to cope with the constraints over time of the control. In other words, we require that the fuzzy relation of control (T) be determined by minimising the performance index:

$$\sum_{X_k \in X} \left[\text{distance}(G, FR(X_k \blacksquare T, X_k)) + \text{distance}(X_k, C)\right]$$

with the distance function defined over the two fuzzy sets specified above and aggregated over the initial X_ks coming from a certain family X. The fuzzy controller determined in this manner is referred to as an optimal one in the class of initial conditions provided by X. The fuzzy control obtained in this way is an example of a single-step control. The structure of this controller is the simplest one, and in general it reminds us of the control law introduced in the theory of linear systems. Of course, the fuzzy models are not linear, and neither is the derived control scheme.

We can enhance this control rule by permitting logic processors to perform the mapping from the state to the control space, i.e.:

$$U_k = LP(X_k, W)$$

where W comprises all the connections of the network. The logic processor has to be constructed by optimising the performance index:

$$\sum_{X_k \in X} \left[\text{distance}(G, FR(LP(X_k, W, X_k)) + \text{distance}(X_k, C)\right]$$

The minimisation is carried out by changing all the connections of the logic processor.

In the third approach, we tend to determine the fuzzy control U_k so that the control objectives are satisfied to the highest extent. Thus, the aim of the control is to maximise the satisfaction of the objectives (C and G). Since these are defined in the two different spaces, we have to transform one of them (the constraint) into the space of state using the fuzzy model. In schematic form (Fig. 11.4), we can visualise the structure as follows (we assume that the U_ks are defined over an m-dimensional discrete space, while the space of state consists of 'n' elements).

Let the fuzzy model of the system be described by (11.14). The control U_k is evaluated with respect to the constraint C through the evaluation block. In general, we can talk about any relevant operation (i.e. relevant from the control-engineering point of view) in which U_k and C are involved. We denote this operation by 'f'. For instance, 'f' may serve as a matching operation, causing U_k and C to be matched coordinatewise, $U_k(u_i) = C(u_i)$. We may also be interested in considering an inclusion constraint that states that U_k should not exceed C. In general, the higher the value $f(U(u_i), C(u_i))$, the better the consistency between U and C at this element of the universe. The matching operation is also completed with respect to the result of the application of U in the fuzzy model, which gives rise to the values $m(x_1)$, $m(x_2)$, ..., $m(x_n)$. The results are aggregated with respect to the rows of the grid obtained (Fig. 11.4).

We denote the derived values by

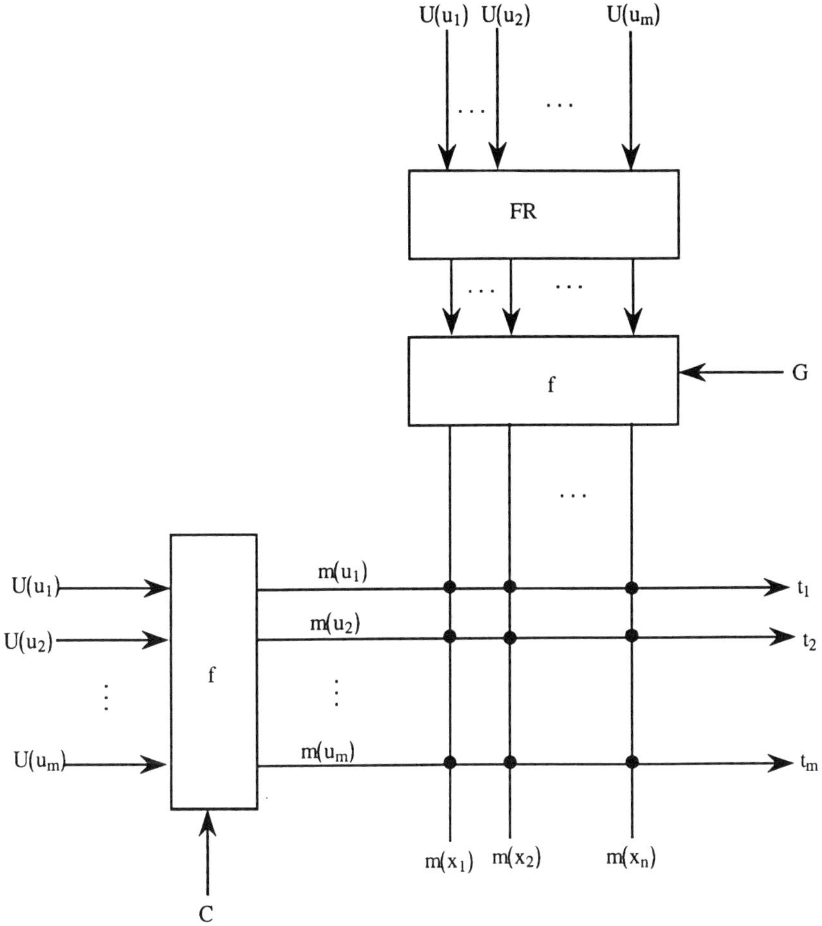

Fig. 11.4 Derivation of the fuzzy set of control U for the goal G and the constraint C

$$t_i = \sum_{j=1}^{n} m(x_j)\,\text{AND}\,m(u_i) \quad , \quad i = 1, 2, \ldots, m$$

where the AND operation highlights the fact that both the objectives have to be satisfied. The value of t_i states the degree to which the grade of membership $U(u_i)$ is preferred as satisfying the above objectives of the control. Assuming that all the $m(x_j)$s are maximal (equal to 1), t_i attains $m(u_i)n$ (this holds for AND operations implemented by the use of t-norms).

The value of $U(u_i)$ has to be modified so that t_i increases. The fuzzy set of control $U(u_i)$ can be derived by applying a standard Newton-like scheme. Sometimes, the derivative can also be approximated to by some auxiliary data points. The simplest option appears in this format: in addition to the point $U(u_i)$, we compute t_i for $U(u_i) - \varepsilon$, where ε stands for a small increment, $\varepsilon > 0$, $\varepsilon \to 0$. The corresponding value of t_i is denoted by t_i'. The update scheme for the grade of membership $U(u_i)$ reads:

$$(u_i) = U(u_i) + \alpha(t_i - t_i')$$

$$\alpha \in [0,1]$$

Example 11.1
Let n = m = 1 and the scalar fuzzy-relational model be:

$$X_{k+1}(x_1) = \min(0.75, 1 - U_k(u_1))$$

Furthermore, the goal and the constraint are equal to $G(x_1) = 0.725$ and $C(u_1) = 0.26$. The results of the learning, summarised in terms of t_1 with $\alpha = 0.15$ and $\varepsilon = 0.05$, are shown in Fig. 11.5.

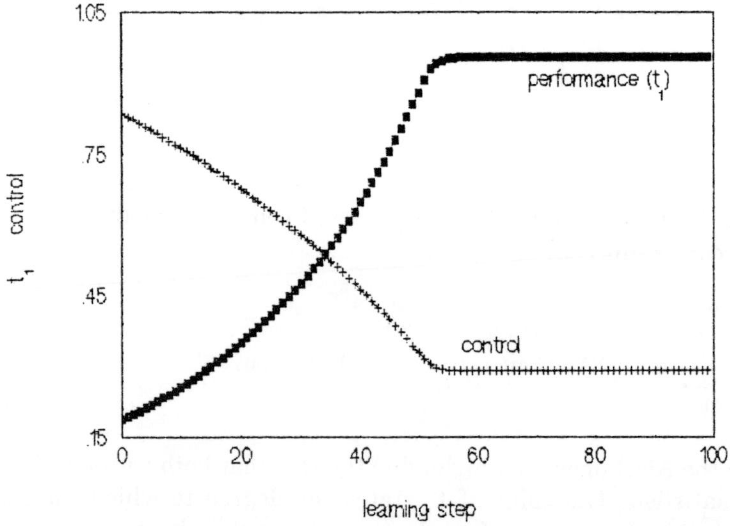

Fig. 11.5 Results of the learning (t_1) and the derived fuzzy control $U(u_1)$

The Figure contains the successive values of the t_1s as well as the derived values of the control $U(u_1)$.

11.6 Conclusions

We have discussed the basic tasks of system analysis. The first significant observation we can make is that our fuzzy models require new basic definitions of the well known terms of system analysis and call for new algorithms. The notion of stability is one of the evident examples where extensions of this type are necessary to cope with the collective characteristics of many objects (elements) encapsulated within fuzzy sets.

The prediction problem studied in its forward, backward and intermediate models demands new methods. The important aspect of evaluating prediction results is accomplished by studying the behaviour of the associated confidence intervals. Since they become increasing functions of the horizon of prediction, we can use them as suitable stopping criteria to avoid meaningless prediction outcomes. The confidence intervals may be found useful in formulating control problems.

Different control algorithms have been developed on the basis of fuzzy models of the system. Contrary to the classical approach (with pointwise models), both the control algorithm and the model cooperate at the same conceptual level of the fuzzy sets. The methodology discussed constitutes a viable alternative to the standard design method used in the construction of fuzzy controllers. Here we can benefit fully from the fuzzy model so that the different design options can be easily exploited. Furthermore, 'what-if' questions referring to the design alternatives discussed can be posed and analysed.

11.7 References

[1] Bellman, R.E., & Zadeh, L.A. 1970. 'Decision-making in a fuzzy environment'. *Manage. Sci.*, **17**, pp. 141–164

[2] Chang, S.S.L. 1972. 'On a fuzzy algorithm and its implementation'. *IEEE Trans. Syst., Man & Cybern.*, **8**, pp. 142–144

[3] Chang, S.S.L., & Zadeh, L.A. 1972. 'On fuzzy mapping and control'. *IEEE Trans. Syst., Man & Cybern.*, **2**, pp. 30–34

[4] Gluss, B. 1973. 'Fuzzy multi-stage decision-making, fuzzy state and terminal regulators and their relationship to non-fuzzy quadratic state and termination time'. *Int. J. Contr.*, **17**, pp. 177–192

[5] Gottwald, S., & Pedrycz, W. 1986. 'On the suitability of fuzzy models: an evaluation through fuzzy integrals'. *Int. J. Man-Mach. Stud.*, **24**, pp. 141–151

[6] Kacprzyk, J. 1982. 'Decision processes in a fuzzy environment: a survey'. *In* 'Fuzzy Information and Decision Processes' (Eds. M.M. Gupta & E. Sanchez). North Holland, Amsterdam, pp. 251–263

[7] Kiszka, J.B., Gupta, M.M., & Nikiforuk, P.N. 1985. 'Energistic stability of fuzzy systems'. *IEEE Trans. Syst., Man & Cybern.*, **6**, pp. 783–792

[8] Pedrycz, W. 1979. 'Stability in systems described by fuzzy relation'. *Zesz. Nauk. Politech. Gliwice (Poland)*, pp. 36–43

[9] Pedrycz, W. 1981. 'An approach to the analysis of fuzzy systems'. *Int. J. Contr.*, **34**, pp. 403–421

[10] Pedrycz, W. 1985. 'Design of fuzzy control algorithms with the aid of fuzzy models'. *In* 'Industrial Applications of Fuzzy Control' (Ed. M. Sugeno). North Holland, Amsterdam, pp. 153–173

[11] Pedrycz, W. 1985. 'Structured fuzzy models'. *Cybern. & Syst.*, **16**, pp. 103–117

[12] Stein, W.E. 1980. 'Optimal stopping in a fuzzy environment'. *Fuzzy Sets & Syst.*, **3**, pp. 177–259

[13] Thomason, M.G. 1977. 'Convergence of powers of a fuzzy matrix'. *J. Math. Anal. & Appl.*, **57**, pp. 476–480

[14] Tong, R.M. 1978. 'Analysis and control of fuzzy systems using finite discrete relations'. *Int. J. Contr.*, **27**, pp. 431–440

Index